Working Knowledge

Working Knowledge

MAKING THE HUMAN SCIENCES
FROM PARSONS TO KUHN

Joel Isaac

HARVARD UNIVERSITY PRESS
Cambridge, Massachusetts, and London, England • 2012

For Cara

sine qua non

Library of Congress Cataloging-in-Publication Data
Isaac, Joel, 1978–
    Working knowledge : making the human sciences from Parsons to Kuhn / Joel Isaac.
        p. cm.
    Includes bibliographical references and index.
    ISBN 978-0-674-06574-1 (alk. paper)
    1. Social sciences—Study and teaching—Massachusetts—Cambridge. 2. Social
sciences—Reseach—Massachusetts—Cambridge. 3. Harvard University—History.
4. Universities and colleges—Massachusetts—Cambridge—History. I. Title.
    H62.I77 2012
    300.71'17444—dc23   2011044604

Contents

We naturally believe ourselves far more capable of reaching the center of things than of embracing their circumference. The visible extent of the world visibly exceeds us; but as we exceed little things, we think ourselves more capable of knowing them.

—Pascal, *Pensées*

Prologue

How Paradigms Are Made

In the early summer of 1960, the historian of science Thomas Kuhn found himself in a quandary. His long-gestating book manuscript, which he had christened *The Structure of Scientific Revolutions,* was almost ready for publication. But he was vexed by an unresolved problem that lay at the heart of his theory of scientific development. It concerned the source of the stability of what Kuhn termed "normal science": the "mopping-up operations" surrounding a major discovery that "engaged most scientists throughout their careers." What was the social or intellectual glue that bound scientists together in a tradition of normal scientific inquiry? "For some years," Kuhn explained in a draft of his manuscript dated June 1960,

> I took it for granted that a set of rules sufficient to determine normal scientific practice was implicit in the textbooks and training procedures that initiate the student to the profession. During that time my discussions of normal science made no reference to paradigms. Instead they referred to the specific constellation of commitments— methodological, theoretical, and ontological—which make a particular problem-solving tradition possible. On that view periods of normal science were periods of consensus, during which the entire scientific community agreed about the rules of the game. And scientific revolutions were then episodes through which the rules of the game were changed.

Kuhn still found this viewpoint "extraordinarily tempting," and maintained that a scientific community's commitment to certain general rules kept the mopping-up work of normal science on track. Yet he also admitted that his suggestion that consensus was the bedrock of normal science was "very probably wrong."[1]

Kuhn had come to believe that the rules guiding normal science could be found only in the model problem-solutions or "paradigms" that a scientific community enshrined in its textbooks and training regimes. However, the "rules" for the conduct of inquiry that these paradigms implied were usually not strong enough to explain why a particular model problem-solution held sway in a tradition of normal science. As Kuhn ruefully observed, "the number of rules that can be educed by such study never seems sufficient to define the puzzles that scientists normally undertake or to restrict scientific attention to their pursuit." If consensus on rules implicit in paradigms was not the answer to the question of the stability of normal science, then what was? By 1960, Kuhn was inclined to conclude that the "sample problems and applications" provided in a scientific education were what held a tradition of normal science together, even in the absence of consensus on methodological or theoretical fundamentals. "Applications and problems need not imply rules in order to determine science," Kuhn claimed. "Rather than learn rules the scientist can, and in some part clearly does, learn by practicing on paradigm problems." It was the trainee scientist's working knowledge of how to perform an experiment or solve a model problem that gave coherence to the scientific enterprise, at least under "normal" circumstances. This was the way science operated, Kuhn argued, from high school to the PhD and into the professional enterprise of normal science itself. Paradigms were a means of education, a guide for research, and the key to understanding how scientists created knowledge—they defined pedagogy, research, and epistemology all at once. A paradigm was therefore "both more and less than a set of rules for the conduct of the scientific life. It is because they learn in this way [i.e., through paradigms] that scientists can so regularly agree in their evaluations of particular problems and particular solutions without manifesting any similar agreement about the full set of rules that appear to underlie their judgments. One can model work upon a paradigm or recognize work modeled on one without being entirely able to say what it is that gives the model its status."[2] So much for misleadingly mechanical notions of "rules" and "consensus"; science was now to be understood as a practical art embodied in concrete activities of learning and research.

Kuhn's Breakthrough: Two Readings

Reflecting on these unpublished pages of Kuhn's soon-to-be-famous manuscript, it is difficult to resist the conclusion that we are witnessing the first stirrings of a revolt against traditional, science-centered conceptions of knowledge in the human sciences—a revolt in which the idea of the paradigm played a key part. If we take this view of Kuhn's solution to the problem of normal science, the rest of the story almost writes itself. Kuhn recorded his qualms about consensus in a draft chapter entitled "Normal Science as Rule Determined." The chapter was replaced in the 1962 first edition of *The Structure of Scientific Revolutions* by two chapters of largely new material: "Normal Science as Puzzle-Solving" as Chapter 4, and, in the clearest evidence of Kuhn's Damascene conversion, a slender Chapter 5 entitled "The Priority of Paradigms." In the latter, Kuhn appealed to Wittgenstein's description, in the *Philosophical Investigations* (1953), of how so-called kind terms like "chair," "leaf," or "game" came to be applied "unequivocally and without provoking argument." Kuhn took Wittgenstein's point to be that the evident agreement among members of a speech community on how to apply these words did not rely on each member knowing the same rules or possessing identical criteria, but instead rested on an overlap between the model applications by means of which individual speakers learned how to use a given word. The overlap of these sets of model applications, from speaker to speaker, was sufficient to ensure mutual understanding between members of the speech community. Something similar happened, as Kuhn saw it, when scientists shared a paradigm. Just as speakers of a language did not learn how to use words according to strict, universal rules, so scientists "never learn concepts, laws, and theories in the abstract and by themselves. Instead, these intellectual tools are from the start encountered in a historically and pedagogically prior unit that displays them with and through their applications." Consequently, "the process of learning a theory depends upon the study of applications, including practice problem-solving both with a pencil and paper, and with instruments in the laboratory."[3]

On this seemingly natural reading of Kuhn's breakthrough from "consensus" to "paradigms," Wittgenstein's philosophy gave Kuhn the terms for his critique of the positivist treatment of scientific theories as logical systems of propositions while, at the same time, pointing to the "role for history" in the philosophy of science Kuhn so ardently sought. In the brave new Wittgensteinian world, epistemology was to be rooted in the actual practices of research and training: "How," Kuhn speculated rhetorically, "could history of science fail to be a source of phenomena to which theories about knowledge may legitimately be asked to apply?"[4] Here, according to a range of influential commentators, was the moment in which the grip of positivism in Western philosophy and social science was decisively weakened. It was the "apparently unquestionable, supposedly uncrossable line separating science as a form of intellectual activity, a way of knowing, from science as a social phenomenon, a way of acting," wrote the anthropologist Clifford Geertz, "that Kuhn in *Structure* first questioned and then crossed."[5] Richard Rorty, in his most sustained attack on the epistemological tradition of the modern West, credited Kuhn with showing how to replace the theory of knowledge with the sociology of scientific communities.[6] More generally, it is often asserted that Kuhn's paradigm encouraged the "restructuring of social and political theory" during the last third of the twentieth century.[7] In the post-*Structure* era, knowledge was a social, and not a purely cognitive, phenomenon.

Compelling though it is, this is not the right story to tell about Thomas Kuhn's late embrace of the paradigm. Interpreting Kuhn's breakthrough as an episode in the overcoming of "positivism" or "traditional epistemology" in the human sciences conflates the reception of Kuhn's book with the historical context of its composition. It is one of the purposes of *Working Knowledge* to show that Kuhn's arguments in *Structure* belong not only to the world of "postpositivism," but also to a largely forgotten conversation concerning the sources and validity of knowledge in the human sciences. In this language of epistemology, practices of scientific research, regimes of professional training, and the theory of knowledge were seen to be intricately related and mutually reinforcing enterprises. What is more, this language shaped the thought of a number of important twentieth-century philosophers and social scientists.

We can gain a first glimpse this style of thinking when we consider a little more carefully Kuhn's response to the failure of consensus-talk in his theory of science. His proposal was that careful attention to the way in which scientific education was carried out (namely, by working with "sample problems and applications") *and* of the way in which everyday research was conducted (again, by modeling work upon paradigmatic "puzzle-solutions") would lead the philosopher of science inexorably toward a historical theory of knowledge. Kuhn's conviction that the consideration of training regimes and research practices in the sciences could provide the basis for a theory of knowledge was shared by a community of scholars and scientists at Harvard University during the middle decades of the twentieth century. In the human sciences today, we are enthralled by the war of the methodological schools, by the divisions between "positivism" and "interpretivism," or between "naturalism" and "historicism." But these all-or-nothing contrasts have obscured the distinctive epistemological vision of Kuhn and his teachers and colleagues. These figures came to identify the creation of knowledge not with the abstract cogitation of "pure reason" or with the iron laws of induction, but with the working knowledge and craft-like skill that typified the education and practical investigations of professional scientists. Knowledge in the human sciences, these thinkers asserted, was to be "made" in a similarly artisanal fashion. In his appeal to the practical understanding embodied in paradigms, Kuhn was joining a chorus of earlier voices—albeit at a unique pitch. *Working Knowledge* seeks to recover this way of thinking about knowledge in the human sciences.

The key characteristics of this style of thinking can be summarily described. For nearly half a century after World War I, there existed in the United States, and especially within the institutional matrix of Harvard University, a culture of inquiry in which the pursuit of knowledge about human beings was connected both to the philosophical discourse of epistemology and to the practices of sciences. The thread that bound these ideas together was an intellectual tradition I shall term "scientific philosophy." Scientific philosophy was at once more and less than what has come to be recognized as professional philosophy of science. Although indebted to Immanuel Kant's theory of knowledge, and especially to the German sage's description of the structures of cognition in virtue of which subjects acquired knowledge of the world, the exponents of scientific philosophy searched for practical replacements for Kant's transcendental notions of concepts, intuitions, and a priori justifications. They looked for these replacements in the actual techniques—cognitive, social, and instrumental—through which the natural and mathematical sciences were learned and conducted. The influence of scientific philosophy was widely felt in American intellectual life during the later nineteenth and early twentieth centuries. While giving the extensive reach of scientific philosophy in the United States its due, I want to show that the tradition reached its apotheosis at mid-twentieth-century Harvard. Talcott Parsons's "grand theory" of social systems, B. F. Skinner's behaviorist psychology, W. V. Quine's "analytic" philosophy, and Kuhn's theory of science are some of the research programs that trace their origins to this Harvard milieu.

Placing the ideas of figures such as Parsons, Skinner, Quine, and Kuhn in context, this book offers an intellectual history of a family of philosophical, sociological, and psychological concepts. But it is in the very nature of the ethos of "working knowledge" examined in this study that we must constantly relate seemingly esoteric ideas about epistemology, the nature of theory, and the proper study of humanity to concrete techniques and scientific activities; in our investigation "ideas" will ultimately give way to practices—of thinking, calculating, experimenting, and teaching. This is because the explanation for the flourishing of scientific philosophy at midcentury is to be found in the peculiar importance for Harvard's human scientists of the research practices and pedagogical techniques in which scientific philosophers found their replacements for Kant's "transcendental" conditions of knowledge. Harvard had consigned many of its rising human scientists to an intramural grey zone of marginal professional schools, special seminars, interfaculty discussion groups, and nonprofessionalized societies and teaching programs. In these interstitial, underinstitutionalized academic spaces, scientific philosophy's identification of epistemology with pedagogy and research practice offered a means of defending the integrity of the human sciences. At mid-twentieth-century Harvard, then, relations between the human sciences, the theory of knowledge, and the natural sciences were mediated by research protocols, pedagogical reform, and institutional growth. In this "Harvard complex"—which consisted in a set of irregular institutional arrangements and the attitudes toward science and knowledge that those institutions fostered—epistemology was embodied by and sustained in practices of pedagogy and inquiry. It was this practice-oriented tradition of epistemology that fueled Kuhn's discussion of paradigms in *Structure* and many other research programs in the American human sciences.

In pursuit of an intellectual history of the mid-twentieth-century human sciences, then, *Working Knowledge* will present an institutional history of the Harvard complex. In its opening chapter, it traces the sources of the complex as far back as the nineteenth century. The major portion of the book examines the intellectual products of the Harvard complex. These include many concepts and motifs central to the contemporary human sciences: "system," "theory," "operational definition," "meaning," and "paradigm." *Working Knowledge* also marks the Harvard complex's dissolution, which occurred when champions of postpositivist and natural-scientific approaches to the human sciences began to lock horns over questions of rationality and objectivity—questions associated with works like Kuhn's *The Structure of Scientific Revolutions*. My overarching aim is to show that the options presented by these rival camps are not the only way of conceiving of the place that philosophy and the natural sciences have held in the self-conception and the historical development of the human sciences. In particular, the practical, "everyday" aspects of the theory of knowledge I seek to uncover in the Harvard complex present a salutary contrast to the inflated role often granted to epistemological rubrics like "positivism" and "interpretivism" in the formation of the human sciences. As practitioners of the human sciences, we fall too quickly into talk of ideologies of knowledge— into disputes over "objectivism," "relativism," "naturalism," "historicism," "positivism," and "scientism." In doing so, we forget the diverse practices and commitments these catchall labels fail to distinguish. My task in what follows is not to show how to steer between the extremes represented by these positions or to insist that the thinkers of the Harvard complex give us that kind of longed-for via media in the human sciences.[8] Rather, my goal is to add some historical weight and complexity to the "isms" listed above and especially to the science-leaning orientations with which many of my protagonists (although not, of course, Kuhn) are identified. Even those readers predisposed to consider *Working Knowledge* a chronicle of positivism in the modern human sciences will, I hope, come to recognize that, for members of the Harvard complex, the modeling of the human sciences on their physical counterparts involved a refined sense of the importance of research practice and pedagogy, and not merely the desire to reduce the content of the human sciences to the material world limned by the natural sciences.

The Human Sciences in Crisis

In the rest of this Prologue, I want to describe in more detail the contributions an investigation of the Harvard complex can make to the history of the human sciences. Those potential contributions will become clearer once we

understand why the theory of knowledge has become the object of such bitter ideological division in the human disciplines. Answers to this question
are sought, in the first instance, in the postwar development of the human
sciences, and in the way historians and philosophers have written about
those developments. While the postwar upheaval in the human sciences has
created very real divisions in methodology and practice, such differences
have unhelpfully been treated as reflections of deep-lying metaphysical positions in what is grandly called the intellectual culture of modernity. It is
this combination of fragmentation and the ideological significance attached
to that fragmentation which explains the bitter and seemingly intractable
character of debates across methodological divides. After exploring these
issues, I present an alternative strategy of inquiry, which is designed militate
against the internecine conflict of epistemic ideologies in discussions about
the constitution of the human sciences.

Since World War II the Anglophone human sciences, especially in the
United States, have developed through three distinct phases. The first phase
was one of millennial expectation. During the epoch of the high Cold War—
roughly speaking, the years between the announcement of the Truman
Doctrine in 1947 and the Cuban missile crisis in 1962—many commentators believed, or professed to believe, that the sciences of society, politics,
and mind were finally joining the experimental and mathematical sciences
as value-neutral, technical disciplines.[9] This was so despite the underlying
diversity of theory and method in the postwar human sciences.[10] Eclecticism
could be downplayed because the general attitude among practitioners was
that scientific maturity had come to the human disciplines. "The sociology,"
Talcott Parsons told the 1949 meeting of the American Sociological Society,
was not merely "about to begin," as his children had jokingly barked in hom-
age to their father's frequent prophecies; it had "been gathering force for
a generation and is now really underway."[11] These grand expectations for
social science were intertwined with the emergence of an assertive, ecumenical liberalism in the postwar United States.[12] The belief that the human sciences were becoming value-free disciplines rested on, and in turn reinforced,
notions of a liberal consensus in American history and a faith in the importance of professional elites in managing democratic regimes. These ideas
encouraged the conviction that the United States had entered a postideo-
logical age, one geared toward the ethos of science and social engineering.[13]

A neologism captured the mood: behavioral science.[14] World War II had been the crucible of the behavioral vision; the premium placed on technical problem solving and discipline-blind collaboration by those engaged in war work banished talk of metaphysical differences between the social and human sciences. This accounts for why those with different visions of the field could nonetheless adopt the same millenarian attitude.[15] No single research program held sway in the Cold War behavioral sciences, but a web of family resemblances between different methods and research orientations ensured a certain comity of spirit among such diverse enterprises as rational choice theory, structural-functional sociology, information theory, behavioralism, operations research, systems engineering, modernization theory, and cognitive science. Explanations in each of these fields dealt with causal chains, or at any rate with "systems" of variables whose interrelations could be formally stated. The fundamental unit of explanation, meanwhile, was the individual, conceived as a culturally programmed human agent.[16]

The second phase in the postwar history of the human sciences opened with an assault on the assumptions of behavioral science. These critiques emerged in the wake of the appearance of Wittgenstein's *Philosophical Investigations,* and they reached their apogee in the years following the publication of *The Structure of Scientific Revolutions* in 1962.[17] This period saw the revival of disciplines and methodologies consigned to the scrap heap of scientific history by the behavioral sciences. Important essays on the return of normative political theory by Isaiah Berlin and Alasdair MacIntryre, along with attacks on the pretensions of "scientific" sociology by C. Wright Mills, exposed—or at any rate were thought by many to have exposed—the ideological commitments of the behavioral scientists.[18] By the early 1970s, the computational theory of mind, which drove forward the related enterprises of cognitive psychology and Artificial Intelligence (A.I.), was subjected to phenomenological attack at the hands of the philosopher Hubert Dreyfus.[19] Meanwhile, the models of scientific explanation that underpinned the behavioral approach were thrown into question both by postpositivist philosophers of science and by students of narrative understanding in historiography such as Louis Mink and Hayden White.[20]

Behind these polemical interventions lay a constructive program. As the philosopher Charles Taylor sought to demonstrate in an agenda-setting essay of 1971, insofar as the human sciences were supposed to explain patterns of behavior—intentional actions like deciding, exclaiming, bargaining, and placating—then they had to deal with the meanings—self-understandings, presuppositions, intentions, or descriptions—without which actions could not be performed by agents.[21] Not only were the human sciences inextricably "hermeneutic" or "interpretive" sciences: their findings were unavoidably based on the subjective valuations and meanings held by the agents under analysis. *Pace* the behavioral scientists, there could be no value-free explanations or modes of inquiry that did not undertake the circular, messy, and decidedly unlawlike task of making explicit the meanings implicit in social practices.[22]

In each of the two principal postwar modes of human science, a specific vision of philosophy reigned. During the heyday of the behavioral sciences, many "analytic" or "ordinary language" philosophers assigned themselves a second-order role among the sciences. Philosophical naturalists cast philosophy as the intellectual dependent of vanguard disciplines such as physics and psychology. On this view, philosophy was tasked with the logical clarification of the ontology and epistemology implicit in the theory of the world propounded in current science.[23] At the same time, practitioners of the new philosophical subfield of philosophy of science sought to uncover the "structure" of scientific explanations, so as to demarcate science from other, less well-defined areas of inquiry. The explanatory ambitions of the sciences, natural or human, were codified in programmatic statements from philosophers on the "logic" of scientific theory.[24] By contrast, interpretivist critics of the behavioral project gave philosophy a primary role. No longer abandoned to the toil of clarifying the "languages" of science, philosophy was brought into the heart of the interpretive human sciences. Taking a lead from speech act theorists like J. L. Austin, many historians and social scientists conceived their work as part of the post-Wittgensteinian enterprise of showing how words and concepts allowed one to find one's way about in a particular social world.[25] In a related development, continental European exponents of hermeneutic philosophy, including Martin Heidegger, Paul Ricoeur, Hans-Georg Gadamer, and Michel Foucault, were increasingly cited as lodestones of "interpretive social science."[26] To some, these developments seemed to portend a fundamental "restructuring" or "refiguration" of the human sciences around the grand themes of meaning and interpretive understanding.[27] The revival of pragmatism during the early 1980s appeared to confirm this cultural shift in the human sciences.[28]

But the revolution did not come to pass. The third phase in the history of the Anglophone human sciences since 1945 is marked by a sense of bathos and stalemate. In the influential edited book *The Return of Grand Theory in the Human Sciences*, first published in 1985, Quentin Skinner sounded the classic themes of the hermeneutic insurgency. The "turmoil," "upheavals," and "transformations" of the past generation, Skinner suggested, made the restructuring of the human sciences an accomplished fact: "amidst all this turmoil the empiricist and positivist citadels of English-speaking social philosophy have been threatened and undermined by successive waves of hermeneuticists, structuralists, postempiricists, deconstructionists, and other invading hordes."[29] Belying such bold declarations of success, the interpretivist revolution had in fact run out of steam. By the late 1980s and early 1990s, three figureheads of the interpretivist revolution—Thomas Kuhn, Clifford Geertz, and Richard Rorty—had rejected the notion that the human sciences could be divorced from the natural sciences.[30] Meanwhile, the basic tenets of interpretivist philosophy were being domesticated as yet another methodological "paradigm"; interpretivists had won an important niche in the human sciences, but the revolution had stalled before hegemony was secure.[31]

The behavioral scientists were never decisively routed from their citadels. Variants of rational choice theory, for example, remain dominant in economics, political science, and sociology.[32] Modernization theory has enjoyed numerous revivals since the end of the Cold War.[33] A.I. and cognitive science live on, despite the attempts of Jerry Fodor, Hubert Dreyfus, and John Haugeland to explode their scientific pretensions. The second-order, naturalist vision of philosophy today dominates the discipline.[34] Indeed, if there has been a general trend in the human sciences over the past two or three decades, it has centered on the increasing influence of biological forms of explanation—most notably the alliance of evolutionary accounts of human psychology with experimental work in the neurosciences.[35] "Biologism" has come to claim more and more of the intellectual terrain previously controlled by the social sciences; many social scientists—as well graduate students who may in the past have trained in disciplines like economics, political science, and social psychology—have migrated toward fields on the border of the biological and social sciences.[36] Despite these shifts toward formal methods and the explanatory insights of biology, the armies of interpretation continue to lay siege to some key fortresses, with rational choice theory a favorite target.[37] Some battlements have been decisively overrun—in history, for example, or in anthropology—while inroads have been made by interpretivists in political theory, sociology, and philosophy. But the impression persists that human scientists today live in a fractured world. It comes as little surprise that recent surveys of the field have pointed to the existence of two cultures in the Anglophone human sciences, one dominated by ahistorical rational choice paradigms, the other by historicism and hermeneutics.[38]

Because these rival conceptions of how knowledge is to be made have become so deeply embedded, the discourse of the contemporary human sciences recurs to the trope of steering between extremes and finding common ground.[39] In political science, for example, a generation of austere rule by rational-choice modelers and statisticians has given way to a bout of academic perestroika. The American Political Science Association and its house journal, the *American Political Science Review,* have let down the methodological drawbridge (if not yet all the way) and allowed "heterodox" students of politics into the professional fortress.[40] Elsewhere, a growing band of scholars has sought to minimize the differences between the canonical figures of the two cultures of the Anglophone human sciences. This is the sort of thinking that makes the purported vanquisher of positivism Thomas Kuhn a "close ally" rather than a mortal intellectual enemy of the archlogical positivist Rudolf Carnap.[41] It converts Clifford Geertz, the advocate of "thick description" in cultural inquiry, into a disciple of the schematizing social theorist Talcott Parsons.[42] In the third phase of the postwar crisis—our own, post-revolutionary moment—the search for strategic alliances across the ideological divide is underway.

The Epic Vision of Epistemology

Despite these fitful efforts at reconciliation, there is as yet no common set of terms in which negotiations may be conducted. We lack a vision of the tradition in which the affinities between a Carnap and a Kuhn might appear natural rather than exceptional. What we have instead is a historical discourse in which the epistemological differences between the several camps have been detached from intramundane academic contexts of research and professional training and inflated into conflicting "visions" of the human condition. By attaching epistemology to rival conceptions of selfhood and social life, parties to the postwar *Methodenstreit* have translated everyday struggles over pedagogy and the techniques of inquiry into epochal clashes of ideology. With the stakes of the debate raised so high, it has become very easy to consider the human sciences locked into a permanent crisis over method and epistemology. This has made the recognition of shared ground especially difficult to bring about.

For an illustration of how the recent history of the human sciences has been unhelpfully rendered as a philosophical morality play, we may turn to Charles Taylor's attacks on the doctrine he terms "naturalism." In the name of a hermeneutic account of human subjectivity, Taylor has assailed what he sees as the destructive influence of naturalism across a spectrum of disciplines. Among the instances of naturalism he finds in the contemporary scene are behaviorism and cognitivism in psychology, rational choice theory in political science, and truth-conditional semantics in philosophy.[43] According to Taylor, these "serpentine heads" of naturalism belong to a single hydra: epistemology. In Taylor's reading, epistemology is elevated from a technical branch of philosophy to the general cultural dogma "that we can somehow come to grips with the problem of knowledge, and then later proceed to determine what we can legitimately say about other things."[44] Even in the more esoteric manifestations of epistemology in the human sciences, Taylor argues, "there is still a strong draw toward distinguishing and mapping the *formal* operations of our thinking" as the necessary prolegomenon for a science of economics, of mind, or of politics. This indictment is directed, inter alia, at the psychologist who invokes computers to define intelligent performance, and at the economist who insists on developing formal equilibrium models as the precursor to an account of market behavior. "In certain circles." Taylor asserts, "it would seem an almost boundless confidence is placed in the defining of formal relations as a way of achieving clarity and certainty about our thinking, be it in the (mis)application of rational choice theory to ethical problems or in the great popularity of computer models of mind."[45]

Fundamental to Taylor's philosophical position is the claim that naturalism is itself an expression of "the intellectual culture of modernity."[46] This culture, for Taylor, is predicated upon an epistemological understanding of the human condition. The economist's concern for the "formal operations" of cognition is for Taylor an echo of the general "reflexive turn" in Western

thought inaugurated by Descartes and developed by Locke and Hume. In
this philosophical history of modernity, Taylor describes a two-stage conceptual revolution. First, the rise of the mechanical world picture during the
Scientific Revolution spelled the end for the Aristotelian "participational"
theory of knowledge and prepared the way for a Cartesian "representation-
alist" view. In the seventeenth-century concept of mechanical causality
an agent's perception had to be seen "as another process in a mechanis-
tic universe." This meant it had to be construed "as involving as a crucial
component the passive reception of impressions from the external world.
Knowledge then hangs on what is 'out there' and certain inner states that
this external reality causes in us."[47] The second revolutionary step, taken by
Descartes in his *Meditations,* was to find a method for determining which
of our inner states or representations could count as reliable knowledge. The
method in question was that of rigorous analysis of the contents of the mind
so as to pick out the clear and distinct ideas that could be taken as the foundation of certain knowledge. This is the original sin of our "epistemological"
modernity: "Descartes is thus the originator of the modern notion that certainty is the child of reflexive clarity, or the examination of our own ideas in
abstraction from what they 'represent', which has exercised such a powerful
influence on Western culture, way beyond those who share his confidence in
the power of argument to prove strong theses about external reality."[48]

We find in Taylor's philosophical history a vivid example of how local
struggles over epistemic values in the human sciences can be projected into
an epic dispute about the human situation: in Taylor's hands, a compressed
foreground of postwar research programs—analytic philosophy, behaviorism, mathematical economics—is thrown into relief against an expansive
background of Cartesian epistemology and the rise of the modern ideal of
the "disengaged" self. Taylor's compulsion to broad-brush cultural diagnosis when reflecting on the recent history of the human sciences is by no
means an isolated case. In addition to Taylor's *Sources of the Self* (1989),
philosophical histories written in a similarly epic fashion include Jürgen
Habermas's *Knowledge and Human Interests* (1968) and *The Philosophical
Discourse of Modernity* (1985), Michel Foucault's *The Order of Things* (1966),
Alasdair MacIntyre's *After Virtue* (1981), and Richard Rorty's *Philosophy
and the Mirror of Nature* (1979).[49] Many of these histories recount the
anti-behaviorist counterrevolution of the 1960s and 1970s as an epochal

attempt to move Western culture away from invidious Cartesian dualisms about mind and world. Such histories are (sometimes avowedly) instances of *Geistesgeschichte:* their purpose is "to give plausibility to a certain image of philosophy" by constructing a canon in which the philosopher's favored program will come to seem compelling (or a rival program less so).[50] In philosophical histories of this sort, epistemology in the postwar human sciences is tied to epic accounts of "modernity" or the "modern self." The differences between alternative pictures of inquiry are thereby inflated into contrasts between competing visions of human nature, which have supposedly defined the intellectual culture of the modern West.[51] Consequently, the philosophical history of the human sciences has made the postwar fragmentation of the disciplines seem at once inevitable and—given the incommensurable theories of human nature that supposedly underpin differences between practitioners—intractable.

Although historians have usually avoided the epic narratives of their philosophical colleagues, there is, in much of the historiography of the human sciences, a concern for "epistemology" conceived as an ideology. This is particularly true of the literature on the United States. Studies of the foundation of the disciplines of sociology, political science, economics, psychology, history, and philosophy have tended to emphasize the connections between epistemological commitments to "value-freedom," "objectivity," "scientific naturalism," or "scientism," on one hand, and, on the other, the attempt by America's ruling classes to assert their cultural authority on the basis of secular, professional expertise. From this perspective, the promotion of the epistemic values of objectivity and technical expertise, vested in the professions, was an ideological enterprise designed to reinstate or redraw the boundaries of authority in the face of the political, social, and economic upheavals of the post–Civil War decades.[52] Investigations of "scientism" and "positivism" in the twentieth-century human sciences have found similar ideological motives at play in the appeal to the methodology of the natural sciences.[53]

These studies have helped us to see some of the ways in which the epistemic commitments of human scientists have been associated with broader waves of social and intellectual change. As I have tried to indicate, however, invocations of epistemology *qua* ideology have distinct limitations. These accounts are usually self-interested: to accede to the philosophers' story about a Cartesian epistemological modernity is to adopt, however tacitly, the critical position of the interpretivists, which is itself part of the war of methods in the human sciences we are seeking to throw into historical relief. Some historians have been explicit about their partisanship in this regard.[54] In addition, by reading the intellectual labors of analytic philosophers or mathematical economists as mere instances of the clash of epochal ideologies of knowledge, we rob such work of its concrete meaning for the actors involved. As the philosopher of science Joseph Rouse has observed, grand narratives about "modern knowledge" obscure the immediate, often local, stakes involved in the scientific practices being narrated.[55]

Embedded Epistemology

The emphasis recent scholarship has laid on epistemology as a form of ideology has unhelpfully ramped up the conflict between the two cultures of the postwar human sciences. More significantly, the philosophical historians' epic narratives about the reach of "Cartesianism" across the centuries bypasses what has been for the human sciences a consequential set of historical changes in prevailing conceptions of knowledge. These changes involved the breakdown of the Kantian program in epistemology and the construction in its place of a tradition of thinking about the sources of knowledge I have called scientific philosophy. This alternative, post-Kantian tradition of epistemology was shaped not only by grand ideologies and epic narratives about knowledge, but by the everyday practices and analytical toolkit of scientists and mathematicians. It construed epistemology as a technical, instrumental, even worldly enterprise. Many research programs that are today thought to be straightforwardly either "behavioralist" or "interpretivist" are more accurately seen as part of the intellectual culture, not of "the modern West," but of scientific philosophy.[56]

We can grasp the animating concerns of scientific philosophy in terms of the profound challenges modern science and mathematics posed to Immanuel Kant's theory of human knowledge. These challenges opened a conceptual space in which practice-oriented ideas about the sources of knowledge were articulated by the scientific philosophers. In the *Critique of Pure Reason* (1781/1787) Kant contended that, in the mathematical and physical sciences, there are things we know that we cannot possibly have gathered from experience. Vital for empirical knowledge yet independent of all experience: this was a seeming paradox about human knowledge of the natural world that Kant sought to resolve by examining the conditions of possibility of synthetic a priori truths. Kant's best examples of synthetic a priori knowledge were pure mathematics (specifically arithmetic), geometry, and Newtonian physics. In each case, the "transcendental" element upon which such knowledge relied was located among what Kant called the faculties of sensibility (roughly, sense perception) and of the understanding (the realm of basic or "pure" concepts). The truths of geometry and arithmetic, Kant claimed, were grounded in the "pure forms" of intuition—space and time—which filled out and gave shape to the deliverances of the sensibility. The laws of Newtonian mechanics, meanwhile, rested on spontaneous acts of intellectual synthesis in which pure concepts of the understanding, mediated by the pure forms of intuition, were brought to bear on raw sense data. Crucially, these examples left Kant's account of the synthetic a priori in a precarious position. If it were shown that intuition played no role in the validity of the propositions of arithmetic or geometry, there would be strong reason to doubt the transcendental status of Kant's synthetic a priori truths. Likewise, if the logical deductions that guided the synthesizing activities of the intellect in cognizing the physical world could be detached from the operations of the understanding, it would again appear that the a priori elements of knowledge were by no means necessary features of human cognition. This is exactly what happened during the course of the nineteenth and early twentieth centuries. In place of Kant's pristine, ahistorical synthetic a priori there emerged a notion of the nonempirical, constitutive element in knowledge as something not transcendental but man-made—a structure embedded in mathematical and scientific practices of one sort or another.[57]

The first rupture came in geometry. Like many of his contemporaries, Kant had assumed that the science of geometry had been definitively laid down in Euclid's *Elements*. Not only were Euclid's famous axioms grounded in an intuitive grasp of space, the "outer sense" as Kant referred to it; the deduction of theorems on the basis of those axioms was also "always guided by intuition."[58] These assertions of the saturation of geometry by intuition were undermined in two ways. The certainty and centrality of the Euclidean system was shaken by a wave of mathematical proofs in the first half of the nineteenth century. These included the proof that it was impossible to deduce Euclid's notoriously troublesome fifth postulate (which stated, in effect, that only one parallel line can be drawn through a point lying outside a given line) from the other postulates. What this showed was that Euclid's axioms were not self-evident but instead open to revision and were perhaps even optional. New work in physical geometry then demonstrated how alternative, non-Euclidean systems could be constructed in which the fifth postulate did not hold. This was an era of interconnected but often independent discoveries, in which the field of geometry was reinvigorated by such august figures as Gauss, Bolyai, Lobachevski, and Riemann.[59] At the turn of the twentieth century, Einstein's relativity theory showed that non-Euclidean geometries could be applied to representations of space-time. No longer were they "mere" formal possibilities; after Einstein, the choice among alternative systems was empirically consequential.[60]

Another pillar of Kant's system fell in the last year of the nineteenth century, when the German mathematician David Hilbert gave a full axiomatization of Euclidean geometry that did not rely on intuition, "pure" or otherwise. Hilbert's achievement in his *Foundations of Geometry* (1899) was to show that Euclidean geometry could be treated as a purely abstract and formal deductive system whose primitive terms, although undefined in any explicit sense, relied on no properties (such as the one Kant called "intuition") beyond those indicated in the axioms or foundational assumptions of the system.[61] Hilbert's key claim was that the terms "line," "point," "being on," and so on, meant nothing more than *how they were used* in the axioms; they rested not at all on any transcendental intuitions. Indeed, such terms could be interpreted to mean anything—it did not matter so long as the rules stipulating their use were adhered to.

The elimination of the intuitive vestiges from Euclidean geometry was part of a wider axiomatization movement in mathematics. Hilbert was appalled by the idea that many proofs of theorems in analysis, number theory, and geometry appealed to poorly explicated notions such as "spatial intuition" and "infinitesimal magnitude."[62] Like many mathematicians of the nineteenth century, Hilbert wanted to get away from the notion that mathematics was the science of magnitude or number. The purpose of axiomatization, for Hilbert, was to establish the purely abstract and deductive status of a given branch of mathematics. For if such an enterprise were carried through successfully, then the primitive terms of a given discipline (e.g., "point" and "line" in geometry) could be drained of their traditional meanings and defined as pure abstractions with respect to a given set of axioms. Hilbert's *Foundations* came late in the day of the axiomatization movement in mathematics. More than a quarter of a century earlier, the core of post-Newtonian pure mathematics, analysis, had been "arithmetized" by Weierstrass, Dedekind, and Cantor. Arithmetization aimed to obviate the appeal to intuitions and infinitesimals in the calculus through proof techniques rooted in number theory.

Meanwhile, arithmetic itself was being grounded in new forms of logic that were likewise independent of the claims of intuition. The key players in this development were the Jena-based logician Gottlob Frege, the Italian mathematician Giuseppe Peano, and, in due course, the British philosopher Bertrand Russell.[63] Such moves toward formal rigor and abstraction in mathematics dealt a further blow to Kant's faculty-based grounding of the synthetic a priori. The arithmetization of analysis, as Hermann Hankel had predicted in 1867, entailed the abandonment of the Kantian claim that arithmetic was synthetic a priori. The real numbers, for Hankel, were not magnitudes given to the intuition, but "intellectual structures";[64] Hilbert's axiomatization of geometry thus ratified the formalism that already underpinned the arithmetization movement.

By the turn of the twentieth century, both geometry and arithmetic had been decoupled from the pure forms of intuition. The emergence of new forms of logic also cast into severe doubt Kant's most intricate grounding of the synthetic a priori in the faculties: what was described in the *Critique of Pure Reason* as the transcendental schematism of the understanding. For Kant, the deductive steps whereby the pure concepts of the understanding were combined with the unconceptualized manifold of the sensibility were inseparable from the spontaneous, synthetic activity of the understanding. Deduction was thus pegged to a transcendental psychological faculty. Gottlob Frege undid this argument and thereby pulled yet another element of the synthetic a priori out of the realm of the transcendental subject and into the world of mathematical calculation. First, in his *Begriffsschrift* (1879), Frege argued that the contents of judgments were not captured in, and were in fact distorted by, the subject-copula-predicate form that logicians had long relied upon and which Kant had appropriated as his model of judgment. For two statements with entirely distinct subjects and predicates could nonetheless express one and the same content, as in the propositions "The

Greeks defeated the Persians at Plataea" and "The Persians were defeated by the Greeks at Plataea." The proper distinction to be made in a proposition, if one wanted to analyze its content, was between *functions* and *arguments,* as those terms were understood in mathematics.[65] On the basis of this distinction, Frege went on to show that a system of logical inference could be just as formal and abstract as Hilbert's axiomatized geometry. Deduction, in other words, did not need to be grounded in anything like the actions of the intellect for its validity. Another bulwark of the synthetic a priori had crumbled.

Now that logical deduction had been hived off from the faculties of the transcendental subject, the epistemological status of the truths of Newtonian mechanics—indeed of pure physical theory *tout court*—was placed in the same uncertainty as the propositions of geometry and arithmetic. On the one hand, it was clear that the statements of these disciplines were, to some extent, analytic rather than synthetic in character: they were true by virtue of the meaning of the terms involved, rather than by virtue of empirical data. It also seemed to be the case that the meaning of the primitive terms of these systems was, as Hilbert had shown in the case of Euclidean geometry, arbitrary, a matter of human convenience. On the other hand, even if geometric axioms or theorems of mathematical logic could be postulated without appeal to Kant's transcendental faculties, one could not make scientifically valuable empirical statements, or posit physical laws, without them. On what, then, did their warrant rest, and how was it that they could be *necessary* yet still *optional* conditions of empirical knowledge? These were similar questions to those that Kant had confronted in the 1780s, but now his answers had seemingly been ruled out of court. The reduction of geometry to algebra, of analysis to number theory, and of arithmetic to logic merely displaced the problem. What could take the place of Kant's synthetic a priori, now that all science had to go on was its own intellectual forms and practices?

The obvious answer to the question was already at hand. What Kant had considered transcendental features of human cognition were in fact man-made constructs, hand-crafted with the formal tools of logic and mathematics. More than that, they were conventional systems that could be redesigned or replaced by the skilled scientist or mathematician. This was the lesson taught by the nineteenth-century shift toward formalization in logic, geometry, and mathematical analysis: although constitutive of empirical knowledge, Kant's a priori had become thoroughly analytic and instrumental. If, for mathematicians like Frege, that still meant that the a priori existed in an abstract realm of pure forms, other scientists and philosophers wrestling with epistemology took the instrumental view to heart and attempted to define the conditions of knowledge using the tools of experimental science. In the mid- to late-nineteenth century, the German polymath Hermann von Helmholtz stood at the forefront of those who called for the radical reappropriation of the Kantian critical project as an "epistemology" based on the findings and research practices of the natural sciences. Although those involved in the back-to-Kant movement in Germany would eventually become sharply distinguished from one another, the basic conviction that philosophy had to be imbued with the scientific spirit was sufficiently compelling across the philosophical community in Germany to issue in a journal, founded in 1877, devoted to *"wissenschaftliche Philosophie."*[66] The attempt to produce a suitably "naturalized," non-Idealist Kant was a delicate hermeneutic task. Proponents of scientific naturalism like Ernst Haeckel, Wilhelm Ostwald, and Richard Avenarius admired the Kant who took Newtonian mechanics and pure mathematics as the acme of the experience of nature, the Kant who shrank metaphysics to the delimitation of the synthetic a priori. Conversely, they disavowed the dualistic Kant who spoke of limiting knowledge to make room for faith, and the Kant who divorced the "thinkable" from the "cognizable" worlds.[67]

Some felt able to abjure the adaptation of Kant's system altogether. For them, the psychophysiological turn in epistemology urged by Helmholtz made speculation on the nature of the a priori unnecessary. Ernst Mach espoused a scientific monism according to which the knowing subject or ego was itself one construction out of an underlying phenomenon common to all worldly things, which Mach called "elements."[68] Mach's doctrine of elements was intended to pull the rug out from under the entire project of epistemology, with its division between "subject," "object" and "sensation." "Perceptions, presentations, volitions, and emotions, in short the whole inner and outer world," Mach wrote in *The Analysis of Sensations,* "are put together, in combinations of varying evanescence and permanence, out of a small number of homogeneous elements. Usually, these elements are called sensations. But as vestiges of a one-sided theory inhere in that term, we prefer to speak simply of elements, as we have already done."[69] William James's "radical empiricism," which substituted a theory of experience for the doctrine of elements, owed much to Mach's monism. James, in fact, was only one conduit for the transmission of Mach's rigorous empiricism into

American philosophy and social science. Mach's idea that knowledge was constructed from purely natural phenomena would influence key members of the Harvard complex, notably L. J. Henderson, P. W. Bridgman, B. F. Skinner, and W. V. Quine.[70]

An additional position in scientific philosophy soon presented itself in the study of the foundations of geometry. At the turn of the twentieth century, Henri Poincaré, the celebrated French mathematician, captured the imaginations of a generation of philosophers and scientists with his argument that the use of geometric axioms, empirically consequential though they were once deployed, were matters of convention and thus of pragmatic convenience. Later logical empiricists, such as Moritz Schlick and Rudolf Carnap, took Poincaré to have stated what we now call the Duhem-Quine thesis: namely, that because experience alone could not dictate the adoption of a specific geometrical system, other hypotheses were needed to justify the deployment of a particular system. Those hypotheses, because they could not be dictated by intuition, had to be conventions, chosen with an eye for convenience and empirical fruitfulness. Poincaré's conventionalism, detailed in *Science and Hypothesis,* was in fact much more restricted than the "anything goes" holism advocated by early proponents of logical positivism.[71] The deployment of a given set of axioms (e.g., those of Euclid or hyperbolic geometry) was for Poincaré a matter of convention; the geometry thereby chosen was both a priori and constitutive, but not synthetic a priori in Kant's sense of a necessary and universal condition of knowledge. Hence analytic axiomatic systems in knowledge—in this case those of geometry— were to be taken up not as features of a transcendental subject to be incorporated into the discipline of pure reason, as in Kant's *Critique,* but were to be viewed as pragmatic instruments by means of which fruitful empirical knowledge could be constructed. The analytic a priori had been instrumentalized. These notions too would be included in the repertory of scientific philosophy as it was received at twentieth-century Harvard.

The Harvard Complex

Scientific philosophy offered a theory of knowledge in which mathematical conventions and scientific practices filled in the gaps left in epistemology by the erosion of Kant's system during the nineteenth century. Epic histories of the human sciences miss the salience of this chastened, practice-oriented, worldly conception of the constitution of knowledge. Intellectual historians have, to date, tended to follow the philosophers in neglecting this vision of epistemology. Although the history of scientific philosophy and its interaction with the human sciences does not present us with anything like a key to a new synthesis in the study of human affairs—once again, we cannot offer here a stable via media between behavioralism and hermeneutics—this history can teach us to think about the *stakes* of contemporary debates about the epistemic status of the human sciences in less "epic," less ideological, and less abstract terms.

It remains to be asked, however, why scientific philosophy should have mattered so much to practitioners of the human sciences, and why at Harvard in particular. It is the answer to this question, in fact, that makes clear the importance of thinking about the history of epistemology in a less elevated, more concretely historical manner. The reply put forward in this book centers on the uptake and transformation of scientific philosophy by thinkers operating within the "interstitial academy" that Harvard nurtured within its walls during a one-hundred-and-fifty-year process of reform and adaptation—a process I describe in Chapter 1. Baldly stated, it was the new human sciences—in particular sociology, anthropology, social psychology, management science, "analytic" philosophy, and the history, sociology, and philosophy of science—that were consigned to Harvard's interstitial academy, and it was in that institutional space that the technical and practical epistemology inherent in scientific philosophy became attractive and—from the perspective of discipline formation—effective to those who sought to develop those disciplines.

By "interstitial academy" I mean the realms of intellectual engagement that existed between more established or specialized discipline-based departments or schools. The principal vehicles for the Harvard faculty's and student body's repeated engagements with scientific philosophy were a cluster of clubs, societies, seminars, and inchoate disciplinary ventures in the human sciences. The Royce Club, the Pareto circle, the Society of Fellows, Alfred North Whitehead's Cambridge salon, the Science of Science Discussion Group, the Inter-Science Discussion Group, the Institute for the Unity of Science in Boston, the Department of Social Relations, the General Education program of James Bryant Conant—these were some of the extradisciplinary, preprofessional, or avocational academic structures that did

most to promote the culture of scientific philosophy. This network of associations cut across to the prevailing institutional model of the American
research university. Within these interstitial forums, there flourished a
number of research programs in the human sciences. What emerged from
this Harvard complex forms a key part of the heritage of the mid-twentieth-
century human sciences: the systems theory espoused in the Harvard Pareto
circle; the operationist theories of science advanced by Percy Williams
Bridgman, B. F. Skinner, and S. S. Stevens; the American embrace of logical empiricism and the rise of analytic philosophy; Talcott Parsons's general
theory of action; and the history and sociology of science promoted by figures such as Robert Merton and Thomas Kuhn. These intellectual stirrings
around Harvard Square at midcentury have usually been considered sin-
gly.[72] By providing a description of the interstitial academy from which they
arose, I shall demonstrate that they share a common institutional sociology.

Part of my task throughout this study will be to assess what was unique
or local about the Harvard complex and what belonged to wider national
and transnational trends in higher education and intellectual life. There is
no single answer to this question, no linear narrative in which Harvard is
straightforwardly the "cause" of a reorientation in the modern human sciences. Nor will I treat Harvard as a "case study" in the history of the human
sciences, if that is taken to imply that we find in the Harvard complex a
microcosm of national or international developments in philosophy, social
science, and the theory of scientific inquiry. Rather, Harvard merits our
attention because it was there that the tradition of scientific philosophy
and the maturation of the American university came together in a way that
produced a set of striking and influential ideas about knowledge and the
scientific study of human affairs. At the same time, the culture of epistemology that took shape within the Harvard complex reflected changes in
the human sciences that were spurred by the broader national and international fortunes of particular disciplines. To tell the story of the rise and fall
of scientific philosophy in the American human sciences, we must show the
weaving and reweaving of the local and the national and transnational; we
must alternate between the examination of university subcultures and the
dissemination of ideas and practices across oceans and state lines. What the
Harvard complex offers the historian is an especially useful vantage point
from which to carry out this work of historical recovery.

Working Knowledge is intended to provide a historian's antidote to the epic philosophical histories of the human sciences. I examine how the epistemological techniques of scientific philosophy, on one hand, and the malleable forms of the interstitial academy, on the other, came together in forming the human sciences at Harvard University during the decades running from World War I to the Cold War. Chapters 2 through 6 explore the ways in which the languages and practices of scientific philosophy were actually involved in Harvard's interstitial academy. Chapter 2 presents the first product of this synthesis: the cult surrounding Vilfredo Pareto's *Trattato di Sociologia Generale*. The Pareto enthusiasm at Harvard brings to light the ways in which a concern for the conditions of knowledge of human affairs was linked to discipline-building within the interstitial academy. Not only were the leading members of the Pareto circle those associated with marginal, inchoate, and interstitial enterprises at Harvard; many of them were also, for this very reason, eager to employ Pareto's systems schemas in such a way as to justify their claim to have created scientific knowledge of human phenomena. At key points, moreover, the Pareto enthusiasm dovetailed with the spread of a pedagogical tool at Harvard: the so-called case-method. The interest in Pareto and the appeal to case-based thinking went hand-in-hand with the writing and teaching of the leaders of the Pareto circle.

Another contribution to the Harvard complex during the 1930s and 1940s was the "operational analysis" championed by the experimental physicist Percy Williams Bridgman. Several members of the Pareto circle, including its founder Lawrence Joseph Henderson, drew directly on their colleague Bridgman's view of operational definition to undergird their Paretian conception of scientific knowledge. Scientific concepts, on Bridgman's account, meant no more and no less than the operations involved in applying them; the habit of action associated with a concept was thus said to constitute the meaning of that concept. Bridgman identified his operational viewpoint with the anti-metaphysical scientific philosophies of Mach, Poincaré, J. B. Stallo, and Einstein. These developments are surveyed in Chapter 3. The uptake of operationism, like the Pareto vogue at interwar Harvard, was an expression of the interplay between the epistemological arts of scientific philosophy and the work of discipline building in Harvard's interstitial academy.

Chapter 4 turns away from the proto-human sciences of the interstitial academy and toward the discipline of philosophy. The focus here is on how the work of the logical empiricists—Rudolf Carnap above all—found its way into the Harvard complex. Although philosophy had a long and distinguished tradition at Harvard, philosophers responsive to elements of scientific philosophy did not have an easy time finding support for their interests. An upstart but soon-to-be-dominant professional framework for philosophy, so-called analytic philosophy or philosophical analysis, prospered at Harvard principally in the interstitial academy. Given the centrality of scientific philosophy to the Harvard complex, it is not surprising that neophyte analytic philosophers like Willard Van Orman Quine should have found a voice there.

The final two chapters emphasize developments in the decades after World War II. In opposite ways, both assay the limits of institutionalization for those formed by the interstitial academy. Chapter 5 surveys the problems inherent in the rapid departmentalization of important portions of the interstitial academy. The Department of Social Relations (DSR) was established in 1946 under the chairmanship of Talcott Parsons—a former member of the Pareto circle. The DSR combined the disciplines of sociology, social psychology, clinical psychology, and cultural anthropology with the aim of creating what its leaders termed "basic social science." But the programmatic ambitions of the DSR masked its ad hoc, almost entirely circumstantial origins; it represented the promotion of an informal interstitial network to departmental status. The core senior faculty in the DSR could not agree on how to practice basic social science, either theoretically or experimentally. As a result, the Department failed to crystallize a discipline and became instead a kind of interstitial academy in its own right.

While the DSR showed what happened when the epistemic commitments of the interstitial academy were formalized, Thomas Kuhn's early career reveals what kinds of thinking were possible for someone formed almost exclusively by the Harvard complex. The interstitial academy provided Kuhn shelter from disciplinary monotheism. But he struggled to bring epistemology, pedagogy, and research practice into satisfactory alignment in a theory of science. As described in Chapter 6, scientific practices, pedagogical regimes, and epistemology became, in an unstable concoction, the *subject* of Kuhn's research. Kuhn was in many ways the ultimate product of the interstitial academy, yet his success also marked the decoupling of epistemology from practice and its absorption into the methodological battles of the 1960s and 1970s. The publication of Kuhn's *The Structure of Scientific Revolutions* provides us with a natural terminus for the history of scientific philosophy's encounter with the human sciences at Harvard.

Theory and Practice

When they are considered within the framework of scientific philosophy and the Harvard complex, questions about the epistemic identity of the human sciences begin to look less like problems about modernity and more like practical matters of pedagogy, research practice, and institution-building. It is important to be clear at the outset, however, what the deflationary approach to epistemology taken in this book does and—more importantly—does *not* entail. First, I will not argue that ideas about the constitution of knowledge and the proper understanding of human behavior were *merely* coded reflections on more prosaic matters of pedagogy and research technique. Instead, I want to show that it mattered to the thinkers under discussion that reflections on the conditions of knowledge in the human sciences include awareness of the kinds of practices through which the scientific disciplines were taught and conducted. It was part of their sense of what concepts such as "knowledge," "theory," and "data" entailed that they be connected to everyday academic practices of teaching, thinking, and testing. Kuhn spoke for many of his Harvard forebears when he insisted in *Structure* that a "quite different concept of science" would emerge from a full accounting of "the research activity itself."[73] This is what many ideology-focused accounts of epistemology miss in their eagerness to transform ideas about scientific knowing into cultural symptoms.

One significant consequence of examining the history of human sciences in terms of practices is that we can bring theories down from their lofty perch in the realm of ideas. Although this book is a work of intellectual history, I adopt a view of "knowledge" and "epistemology" that grasps these terms as involving an assortment of social practices: modes of self-discipline and education as well as techniques of model-building, calculation, and experimentation. I aim to portray the Harvard complex not as a set of ideas but also as an array of academic subcultures.[74] The Pareto circle, operationism, analytic philosophy, Parsonian systems theory, the history and philosophy of science—each of these parts of the Harvard complex were

scientific subcultures *in ovo*. Hence, in describing the visions of knowledge
elaborated by their membership, I will investigate the research techniques,
teaching practices, and theoretical exercises on which those visions were
founded. Of course, "ideas" and "theories"—the claims and counterclaims
set forth in the texts written by members of the Harvard complex—must
remain a principal object of concern for any history of the human sciences.
Working Knowledge is no different in this respect from other works of intellectual history. But it seeks to throw new and important light on those texts
and discourses by locating them within the academic, subcultural practices
that made them possible.

A practice-oriented perspective on claims to knowledge, of the kind advocated in this book, has long since been assimilated into one of intellectual
history's near neighbors, the history of science. Indeed, it was the turn to
the study of experimentation in the history of science in the early 1980s that
led to the "practice turn" in science studies.[75] Only in the last decade or so,
however, have these insights about practice been applied to scientific *theories* and to general epistemic ideologies such as "objectivity."[76] The present
study owes a good deal to these later developments in the history of scientific
practices, which it seeks to welcome into the fold of the history of ideas, and
especially into the history of social thought.[77]

These commitments, however, come with an important caveat. Talk
of academic practices can be misleading if it implies we have entered a
Machiavellian world of careerist strategizing. Throughout the book, I
firmly resist the idea that the relations between epistemology, research, and
pedagogy were governed by a single, utilitarian logic. I do not argue that
Harvard's marginal human scientists invoked the concepts of scientific philosophy because they believed such a strategy would have a clear institutional pay-off. In addition, I do not claim that the figures I discuss formed
the explicit plan to make their own teaching and research practices the "conditions of knowledge" for the discipline or research program they wished
to establish—as though it were understood that such a move would secure
scientific respectability. This is not to say that thinkers such as B. F. Skinner,
W. V. Quine, Talcott Parsons, and Thomas Kuhn were not interested in
building disciplines and promoting research programs or that they did not
engage in institutional politics as a means of gaining acknowledgement
and prestige. These men were indeed discipline builders and institutional

politicians. What I reject is the notion that one can identify a formula that accounts for their conceptual choices or explains whatever success they were able to achieve at the institutional level. To look for such formal explanations is to attempt to hypostatize what were in fact contingent responses to the pressures of specific, time-bound situations, imperfectly and often implicitly understood. This is where the intellectual historian's insistence on examining "ideas" in their own terms—rather than as epiphenomenal products of practices and institutions—comes into its own.

Yet the search for explicit strategies for maximizing individual cultural capital has become the leitmotif of a powerful new methodological perspective on study of intellectuals, the "new sociology of ideas."[78] I want to distinguish my own approach from this method of studying ideas in context. Inspired in large part by the writings of Pierre Bourdieu, the champions of the new sociology of ideas view academic intellectuals as locked in a perpetual, agonistic battle with their peers for recognition and social power.[79] Thus positioned in the force field of the university, sociologists of ideas argue, academicians ape *homo economicus* in their attempt to articulate projects that can secure the maximum possible professional rewards.[80] Even when they do not suggest that personal gain is at issue, the new sociologists of ideas insist that mechanisms like that of the "intellectual self-concept" can retrospectively predict the choices a given thinker made at crucial points in their research career.[81] At root, such investigations rest on the belief that there is some particular key, like self-interest or a desire for coherence with one's ideals, that accounts for the twists and turns of an intellectual life. It seems to me, however, that one can easily admit that career ambitions will often shape specific choices and even the social patterns of the academy without converting such motives into transhistorical explanatory schemes.[82] The sociologists of ideas look for causal mechanisms that "explain" why a thinker held a particular belief. I subscribe to no such program, which comes dangerously close to sociological reductionism. In practice, relations between beliefs and institutions are complex, contingent, and often counterintuitive. Casual arrows, such as there are, point in all directions. My study of the Harvard complex offers no "theory" of conceptual change, but it does seek to illuminate the reciprocal formation of theories, research practices, and training regimes.

What matters, in the end, is what we as historians and practitioners of the human sciences are to make of the hands-on epistemology developed by those formed in the Harvard complex. The book concludes on an opti-mistic note. It proposes that the Harvard complex gives us a picture of a less ideological relationship among philosophy, natural science, and the human sciences. If it is right that the balkanization of the human sciences has engendered a tendency towards adventitious philosophical history, then a historical reinterpretation of recent epistemology as a practical art may help to remove some of the philosophical and ontological baggage that weighs down communication across the dividing lines of the contemporary human sciences. I do not suppose that the example of the Harvard complex can heal all wounds: deep philosophical differences must surely remain. But by deflating some of the more expansive narratives about epistemology, and by underscoring the fact that many of the differences between behavioralists and interpretivists are very often differences of practice—of training, meth-ods, and tools—I am hopeful that this book might encourage those who seek to overcome ideological divisions in the human sciences.

1

The Interstitial Academy

Harvard and the Rise of the American University

In the late summer and early autumn of 1936, Harvard University celebrated its tercentenary.[1] The timing could hardly have been more inauspicious. Americans were mired in an unprecedented economic downturn; a fascist coup against the democratic government in Spain threatened to spark an international conflagration; Nazi Germany stood emboldened by its seizure of the Rhineland; and the grim show trials of Stalin's political opponents were opening in Moscow. Like other academic festivals of the Depression era, the Harvard tercentenary bristled with the tension between optimism about science and learning and the acknowledgment of what looked to many like the crisis of Western civilization.[2] In one example of this striking juxtaposition, on the evening of September 17, the British-based anthropologist Bronislaw Malinowski delivered a minatory oration to the Harvard Chapter of Phi Beta Kappa on the topic of "war as a menace to our immediate future." Malinowski cautioned against an American "relapse into fictitious natural pacifism" while arguing that warfare was neither a permanent feature of human societies nor, in its contemporary mechanized form, a fruitful tool of national policy. As Malinowski marshaled ethnographic data to warn of the pressing danger of armed conflict, a far larger audience—some 350,000 souls—gathered along the banks of the Charles River to watch a fireworks display. Rockets and roman candles were set alight on a moving barge that processed up the Charles, with the Harvard Band following behind, providing a musical accompaniment. When the flotilla reached the Harvard Houses near Weeks Bridge, red fire flares were ignited along the riverfront.[3]

If bathos was unavoidable in the 1930s, its presence in the precincts of the Harvard Yard was apt. In the years surrounding the tercentenary, Harvard was caught uncertainly between worlds. Since the end of the Civil War, the university had embraced some of the reforms that had reshaped American higher education in that period, notably the concepts of scientific research and the elective curriculum. Yet Harvard remained close to the traditional folkways and thoughtways of metropolitan Boston—the world in which the university had been safely ensconced since the turn of the nineteenth century. The stresses and strains produced by these competing poles of attraction were exacerbated by Harvard's institutional eclecticism. Harvard, like many other American universities, was not a single coherent entity but a matrix of distinctive, often isolated, components: departments of the liberal arts, colleges, professional schools, museums, observatories, gardens, and research laboratories.[4] The integration of these elements was never total, and at times the stitched-together whole split along the seams. This tendency was particularly characteristic of the nation's older colleges and universities, which scrambled to incorporate administrative and pedagogical reforms into preexisting— and often intractable—institutional forms. Harvard faced similar pressures during the decades spanning the two world wars. My purpose in this chapter is to assess the origins and results of this institutional fragmentation.

I shall focus on the emergence at Harvard of what I have termed the interstitial academy. A product of centrifugal forces operating on America's universities, Harvard's interstitial academy provided practitioners of nascent research programs with enclaves in which to exchange ideas and conduct inquiries outside of established departments and curricula. The growth of such interstitial spaces within emergent research universities was by no means limited to Harvard. During the middle decades of the twentieth century old and new universities alike nurtured programs and seminars that stood self-consciously outside of conventional departments and professional schools: the University of Chicago's interdisciplinary committee system for graduate training was an obvious example of an interstitial academic space, as were the "University Seminars" developed at Columbia by the historian Frank Tannenbaum.[5] More generally, interwar efforts to develop core undergraduate curricula were intended to overcome disciplinary fragmentation and professional specialization in the colleges. Harvard's interstitial academy was therefore not uncharacteristic of America's universities in this period, but it did fulfill some unique functions in its own local context. At its peak in the decades after World War I, Harvard's interstitial programs and seminars provided a unique space in which hitherto marginal projects in the human sciences were able to come to fruition. At Harvard, fields such as sociology, psychology, cultural anthropology, the history and methodology of science, social systems theory, and analytic philosophy began as fugitive professional ventures. Only slowly, given the breathing space allowed by the interstitial academy, did they acquire institutional legitimacy. It was in Harvard's interstitial academy that the tradition of scientific philosophy became useful to would-be human scientists attempting to ply their craft.

Beyond the Pioneer Paradigm

Presupposed in talk of an interstitial academy is the existence of an established institutional framework for the American university. There is certainly evidence to warrant this assumption. During the three decades between 1890 and 1920, there emerged a stable, if also diverse, organizational pattern in higher education.[6] In these years, faculty and university leaders came to hold shared commitments to the advancement of knowledge, academic freedom in teaching and research, the importance of graduate education in the liberal arts, and the confirmation of the undergraduate college as the gateway to the university. By 1900, the management of universities was sufficiently bureaucratic to warrant the publication of manuals for the new class of academic administrators.[7] Universities had become another outlet for the gospel of efficiency promoted by the business executives of the progressive era.[8] The creation of the American Association of Universities at the turn of the twentieth century marked the nationalization of this university ideal.[9]

Formative though this moment of crystallization was, the "interstitial" and "established" structures contrasted in this chapter can be characterized only in relative terms. Up to World War II, the American university was a fragile creature. As late as the 1920s, a number of academic leaders were asserting the primacy of collegiate values and warning of the dangers of university expansion and disciplinary specialization.[10] Throughout the interwar decades, advanced research in universities relied almost exclusively on tithes from private industry and philanthropic foundations.[11] Such countercurrents underline the fact that the Great Instauration of 1890–1920 embedded in American higher education not a streamlined "university," but a somewhat loose assemblage of institutions and educational goals, many of which were either distinct from, or even in tension with, one another.[12] The components welded uncertainly together in the early-twentieth-century research university had usually been advanced independently of one another: separable ideals associated with the university movement included the promotion of specialized scientific and scholarly knowledge in undergraduate education; the provision of advanced training in the liberal arts; vocational training in the professions; the preservation of cultural treasures; and the extension of the frontiers of human knowledge. In short, the formation of research universities in the United States was a piecewise, haphazard, and halting process that began in the early national period and continued on into the Cold War years.[13] Consequently, to observe that part of the institutional pattern of the university was "established" during the middle decades of the twentieth century is to imply neither permanence nor necessity.

Given the relative status of the university as an institutional form, a characterization of the interstitial academy requires a brief history of Harvard University, one that captures the contingency inherent in the development of Harvard from a liberal arts college into a research university. As well as the acknowledgment of the complex evolution of the American university, this task of historical reconstruction also entails the rejection of what we might call the pioneer paradigm. In a number of histories and biographies, the emergence of Harvard as a research university and the success of the national university movement are attributed to the leadership of Charles William Eliot and kindred spirits at new state and private universities such as Johns Hopkins, Cornell, Michigan, and Chicago.[14] Samuel Eliot Morison wrote of Eliot's election in 1869 in heroic terms: "More than at any age in our history, a leader was wanted. The leader was ready and waiting. He was thirty-five years old, and his name was Charles William Eliot."[15] "No one," observed another historian of Eliot, "did more to establish higher education in general, and Harvard in particular, as an accomplice to American greatness."[16]

These laudatory reviews are in large part the product of inherited prejudices about the academic scene in antebellum Cambridge. The prevailing image of Harvard College in a pre–Civil War state of apathy is owed above all to Henry Adams. In an oft-cited passage of his *Education,* he summarized his view of Harvard in mordant prose:

For generation after generation, Adamses and Brookses and Boylstons and Gorhams had gone to Harvard College, and although none of them had ever done any good there, or thought himself the better for it, custom, social ties, convenience, and, above all, economy, kept each generation in the track. Any other education would have required serious effort, but no one took Harvard College seriously. All went there because their friends went there, and the College was their ideal of social self-respect.[17]

Morison, confirming Adams' verdict, observed that on the eve of Eliot's election in 1869, "Harvard College was hidebound, the Harvard Law School senescent, the Medical School ineffective, and the Lawrence Scientific School 'the resort of shirks and stragglers.'"[18] The tendency to see Eliot's election as a watershed has been reinforced by the new president's attempts to clear his own path. Nowhere was this more evident than in his Delphic inaugural address of 1869, which attempted to draw a line under what Eliot suggested were petty curricular squabbles that had plagued the university in years past.[19]

The pioneer paradigm rests on a sharply dichotomous and teleological reading of the history of American higher education during the nineteenth century. This picture contrasts an antebellum epoch of falling enrolments, scholarly dissipation, and dogged commitment to the classical curriculum with the Gilded Age triumph of professional expertise and the secular university.[20] In the 1950s, many historians were deeply swayed by Donald Tewksbury's account of the college movement before the Civil War, which highlighted both the explosion in the establishment of colleges as the frontier moved rapidly west after the Louisiana Purchase and the strikingly high attrition rate of the frontier colleges.[21] Tewksbury himself described the antebellum college as a classic "frontier institution" and linked its efflorescence to Frederick Jackson Turner's vision of the frontier as a crucible of American democracy.[22] Looking back on these developments in the Eisenhower era, however, historians were less sanguine. The exponential increase in colleges, they argued, led to the dilution of standards and the sidelining of academic principles that had begun to take shape in the small cohort of Eastern colleges before and immediately after the Revolution. During the "Great Retrogression" of the antebellum decades, Richard Hofstadter argued, the forward march of secular higher education in the United States was interrupted by the Second Great Awakening, whose interdenominational rivalries led to the establishment of competing sectarian colleges across the nation.[23]

Hofstadter and others depicted the "old-time college" as a reactionary bastion in a democratic age. With the nation in need of skilled engineers and practical men of affairs, the college, they insisted, stuck doggedly to the elitist precepts of the trivium and quadrivium.[24] The stage was set, after the Civil War, for the concerted reforms of Eliot and likeminded academic pioneers such as Daniel Coit Gilman of the Johns Hopkins University, Andrew D. White of Cornell University, James Angell of the University of Michigan, and William Rainey Harper at the University of Chicago; the "old-time college" was swept aside by the "university ideal," which had succeeded by 1900 in giving birth to a small and elite set of secular, research-oriented institutions of higher education.[25]

To be sure, the story told by the consensus historians of the 1950s was neither entirely false nor irremediably simplistic. They succeeded in showing how the university movement of the 1870s and 1880s had been pieced together from a variety of sources, from the Morrill Act of 1862—the so-called Land Grant Act providing public lands to the states for the purpose of endowing technical colleges—to the professional academic associations and scientific societies founded in the late nineteenth century. Nevertheless, in the writings of these scholars, the basic contrast remained between antebellum retrogression and Gilded Age reform—what George Peterson felicitously described as "a morality play written in two acts."[26] The story of Eliot's presidency has often been assimilated to this wider narrative concerning the fortunes of higher education in the Victorian era. Beginning in the 1970s, however, a series of studies overturned prevailing stereotypes surrounding the antebellum college and questioned the discontinuity between ante- and postbellum higher learning. Charges that the denominational college was either unpopular, unresponsive to curricular innovation, or subject to inordinately high rates of failure have been called into question.[27] Likewise, the reforms of Eliot, Gilman, and others have been placed in a more nuanced context: the significance of German commitments to pure research and academic freedom have been qualified and the economic and intellectual conditions for the rise of the research university have been connected to a longer lineage that stretches back to the early national period.[28] Reassessments of this kind compel us to sharpen our grasp of complex process by which research universities were jerry-built during the nineteenth and early twentieth centuries.

Harvard University occupies a pivotal position in this reassessment of the transformation of the college and the rise of the university. In virtually every innovation in higher education that we will trace in this chapter, another institution claimed precedence over Harvard. What typified Harvard's evolution into a research university was not the boldness of its leaders, Eliot included, but rather the University's ability to finesse major pedagogical, intellectual, and organizational challenges. By hook or by crook, Harvard throughout the nineteenth century was able to remain flexible enough institutionally to respond to dramatic changes elsewhere—even if this entailed a conservative attitude toward graduate education, research specialization, and collegiate values. Harvard prospered, in other words, because it was able to sustain a certain measure of institutional "give" as a result of its size and its connections to metropolitan Boston. As World War II loomed, however, this loose organizational pattern had become so diffuse as to create a system of "interstitial" and irregular forums that existed in between, above, and across the by-then-familiar architecture of university life.

Harvard Metropolitanized

One of the shibboleths that revisionist historians of the antebellum college have sought to overturn is the claim that the pre–Civil War colleges were dominated by sectarian concerns. More recent scholarship has suggested that localism, not dogmatic denominationalism, characterized the collegiate scene in the decades after 1800.[29] As the territories west of the Mississippi were rapidly absorbed by the republic, the college carried out several important cultural, social, and economic functions in towns and cities across the expanding nation. It provided elements of practical education, teachers for local schools, revenue through student boarding fees, and exhibitions and public lectures to fill out the program of civic culture.[30] It was this multipurpose utility of the antebellum college in frontier towns and nascent urban networks that made it such a popular, and also somewhat evanescent, institution. "Physically," Richard Hofstadter observed, "the great continental settlement of the United States in the pre–Civil War era was carried out over the graves of pioneers; intellectually over the bones of dead colleges."[31]

Localism was not, however, the sole preserve of upwardly mobile Missourians or Kansans on the rambunctious frontier. In the early national and antebellum years, the longer-standing colonial colleges, too, were subject to the gravitational pull of local networks. But it was into metropolitan cultures, rather than frontier communities, that they were assimilated.[32] Of the nine colleges established in the American colonies at the time that the Declaration of Independence was issued, Harvard was among those most shaped by metropolitan forces. The "metropolitanization" of Harvard was by no means unparalleled among the colleges founded in the colonial era: during the Jacksonian period the University of Pennsylvania, Columbia University, and the University of the City of New York were likewise drawn into the orbit of increasingly assertive urban elites.[33] (On the other hand, colleges like William and Mary and Dartmouth were too far from large cities to be subject to the same gravitational pull as the colleges placed in or adjacent to the major east coast hubs.[34]) Few antebellum colleges were more thoroughly absorbed into an urban order than was Harvard. And it was from this integration into a web of urban institutions that Harvard acquired the adaptability and wealth it needed to accommodate innovations in college and university education during the nineteenth century.

The beginnings of Harvard's transformation into a bastion of metropolitan culture can be dated as far back as 1780. It was in this year that reforms to the constitution of Harvard shifted the control of the Board of Overseers—one of the governing bodies of the university—from its hitherto clerical membership to the political and commercial classes of Greater Boston. The same year witnessed the foundation of the American Academy of Arts and Sciences, which presaged the connection of Harvard with local civic culture by alternating its meetings between Harvard Yard and Boston.[35] The turning point in the emergence of metropolitan supremacy at Harvard came in the early years of the nineteenth century with two controversial appointments.[36] In 1805, a Unitarian, the Reverend Henry Ware, was elected Hollis Professor of Divinity, the senior chair in the College and one hitherto held exclusively by Calvinist divines. Just over a year later, in March 1806, another religious liberal, the Reverend Samuel Webber, was confirmed as the new president of the University. Congregationalists on the University's governing boards fiercely resisted what they (rightly) saw as a Unitarian coup, but their efforts were to no avail: the elections of Ware and Webber opened an era of Unitarian hegemony at Harvard. With the establishment of a Divinity School dominated by Unitarians, the University became the lodestone of a post-Calvinist religious movement in the Massachusetts Bay.[37]

Liberal Protestantism in early-nineteenth-century Boston was intimately connected to the emergent commercial order; the appointments of Ware and

Webber signaled the ascendancy of urban, maritime Massachusetts over the Calvinist backcountry of New England in the affairs of Harvard College. Appalled by this turn of events, the well-to-do living outside of metropolitan Boston began to send their sons elsewhere for college. From 1810 to 1820, the portion of students coming to Harvard from outside New England more than doubled.[38] The Yankee majority that remained came increasingly from Boston and the coastal towns, and they embraced Unitarianism as a Christian ethos more congenial to "prosperity, urbanity, and worldliness."[39] In particular, the wealthy shippers, textile magnates, and financiers of Boston "preferred a religious stance looking out on wide contemporary horizons rather than back to the old Puritan ideals."[40] Unitarians replaced orthodox convictions regarding predestination, conversion, and a wrathful God with reasoned appeals to a benevolent divinity, the ethics of Christian life, and religious tolerance that drew on the heritage of the Christian Enlightenment and Scottish common-sense philosophy.[41]

Boston's connection to Harvard through Unitarianism became, in due course, one of several ways in which the rising business and professional families of Massachusetts came to overlap and interpenetrate all levels of University life. For the "Brahmin" upper class that flourished so visibly after the Civil War, Harvard occupied the center of a web of cultural, medical, and educational institutions in greater Boston that conferred class status and social authority. This process of class formation through the patronage of higher education, hospitals, and museums was repeated in several other rising American metropolises in the nineteenth century; Philadelphia, New York, and Chicago nurtured similar expressions of the cultural capital of their elite residents. But none were as insular or as deeply interconnected as those of Boston, with Harvard the central node in the metropolitan network.[42] "The great families built Harvard," one historian has written, "while building themselves."[43]

The metropolitanization of Harvard was therefore driven by a more general process of class formation. Historians have often followed Alexis de Tocqueville in describing Jacksonian America as a rough-and-tumble egalitarian society steered by a preponderance of "middling sorts."[44] While the antebellum period may indeed have laid the foundations of mass democracy, it has become clear that the "Age of the Common Man" was at least as much the "Age of the Bourgeois."[45] The political culture of the republic was increasingly affected by the noisy clamor of smallholding farmers and petty traders in the years before the Civil War, but its urban institutions and social wealth were consolidated in the hands of fewer and fewer families. In 1780, the wealthiest 10 percent of Americans owned somewhere in the region of half of the property in the United States.[46] On the eve of the Civil War, the richest 10 percent claimed approximately 70 percent of all property, a shift in the concentration of capital that amounted to "an economic and social revolution."[47]

Boston and Massachusetts followed a similar trajectory. The top 1 percent of Bostonians owned a quarter of the taxable wealth of the city in 1800 and about 40 percent in 1860, while the top 1 percent of the population of Massachusetts accounted for half of all taxable wealth in 1850.[48] The moneyed classes of Boston were distinguished by their stability and endurance. In New York, for example, the volatility of the market and high-risk investment ensured a rapid turnover of the wealthiest families; with some notable exceptions, the Knickerbockers did not reproduce themselves down the generations. By contrast, the Bostonian elite in the antebellum period made social reproduction into an art. This success was due in part to the relatively slow cycles of the New England economy, which encouraged the preponderance of a small set of commercial dynasties. Maritime commerce was the bedrock of the New England economy throughout the eighteenth and nineteenth centuries and was dominated by merchant families such as the Amorys and Cabots. The Embargo Acts of 1807–1808 and the War of 1812 sharply reduced trade in the Atlantic and induced a turn toward industrial production in the shape of textile manufacturing, which blossomed in eastern Massachusetts under the stewardship of Jacksons, Appletons, and Lawrences. Finally, as the mercantile economy recovered and textiles flourished, banking and other financial enterprises grew in Boston in the hands of the Phillipses, Higginsons, and Coolidges.[49]

These birds of a feather certainly flocked together. The top Boston families frequently intermarried, and sometimes intramarried. Typically, they worshipped in Unitarian churches and, in politics, were drawn to the Federalists (and later the Whigs and conservative Republicans).[50] But the real crucible of class-consciousness was the interlocking network of societies, academies, and philanthropic institutions patronized by the rising Boston elite. The Boston Athenaeum, the Massachusetts General Hospital, the American Academy of Arts and Sciences, the Public Library—each of these formed a node in the metropolitan nexus. No civic institution received more patronage, or became more important to the projection of cultural authority, than Harvard. Between 1800 and 1850, the number of endowed professorships increased from six to twenty-one; the College library more than quadrupled in size; and three new professional schools were added to the university. Brahmin wealth underpinned this expansion of staff and resources. The price for Harvard was near complete control by the commercial classes. Boston businessmen, drawn from a very small circle of elite families, came to dominate the Harvard Corporation. Their daughters began to marry members of the faculty; their sons populated more and more of each graduating class. Elite student clubs for scions of the Boston aristocracy flourished. For these reasons, Harvard's tercentenary biographer was moved to call the 1820s the University's "Augustan Age."[51] During the antebellum decades, Harvard became more expensive to attend, more Unitarian in its faith, and more exclusive in the schools and academies from which it recruited. Membership in the Corporation or the holding of a professorship often went hand in hand with trusteeships or memberships in a web of metropolitan institutions. Harvard became the socioeconomic doppelganger of the Bostonian power elite.[52]

Josiah Quincy's Compromise

What were the effects of localism upon the curriculum in Cambridge, and especially upon the formation of the rudiments of "university" education? Metropolitan ties certainly influenced pedagogical reforms before the Civil War, but they did so in complex and often indirect ways. Most obviously, the Unitarian ascendency shaped the clerical training provided at the new Divinity School; liberal Protestantism acquired at Harvard a systematic theological and pedagogical foundation.[53] Yet, perhaps because of the efforts directed at the intellectual legitimation of Unitarianism, Harvard was not central to debates about the viability and reform of the classical curriculum during the 1820s and 1830s. It was left to the trustees and faculty of Yale to issue in 1828 the most influential statement of the virtues of a classical education in the liberal arts;[54] Jefferson's University of Virginia, on the other hand, led the way in bringing the voluntary system into the college curriculum.[55] In the restive 1840s and 1850s, moreover, it was Francis Wayland

of Brown and Henry Tappan of the University of Michigan who sounded
the call for practical instruction and advanced research in American higher
education.[56] Harvard, meanwhile, was caught flatfooted by the infusion of
German principles of academic freedom and pure research. When a quar-
tet of notable scholars arrived at Harvard between 1819 and 1822 after
advanced studies in Europe, they found themselves stymied in their attempt
to introduce the new academic ethos in Cambridge. Edward Everett, George
Ticknor, Joseph Green Cogswell, and George Bancroft experienced varying
degrees of disillusionment with what they perceived as the stubborn back-
wardness of Harvard's faculty and student body and sooner or later directed
their energies elsewhere.[57]

If these details suggest the outlines of the pioneer paradigm—with
Charles William Eliot waiting in the wings to realize ambitions thwarted
in the 1820s—the already intricate connections between Harvard Yard and
State Street in Boston point toward a more complex narrative of Harvard's
transformation. It was Harvard's ongoing involvement with the economic,
professional, and cultural world of metropolitan Boston that led it, during
the 1830s and 1840s, to accumulate the trappings of a modern university.
Financial concerns were the decisive factor. When George Ticknor succeeded in having a tranche of curricular and pastoral reforms written into
the 1825 *Statutes of Harvard University,* the put-upon faculty thwarted most
of Ticknor's plans.[58] Yet, by the close of the decade, the University could
scarcely afford the luxury of dissent. As early as 1824, the political weather
had changed in Boston. That year, the Massachusetts General Court failed
to renew a ten-year grant to Harvard of $10,000 per annum, a move that
reflected the waxing of Democratic forces in the state. Federalist-Unitarian
Harvard could no longer count on being the favored son in Massachusetts.
Meanwhile, student unrest catalyzed a decline in enrollments, which continued through to the end of the 1830s. President John Thornton Kirkland,
in office since 1810, proved himself singularly unable to deal with the crisis
engulfing the college and stepped down in acrimonious circumstances in
the spring of 1828. After a lengthy search for a replacement, the Corporation
eschewed the appointment of another minister in favor of the former mayor
of Boston, Josiah Quincy. Greater Boston's men of affairs were taking
more direct hand in the running of the College; a "new era" had begun.[59]
Nonetheless, the early years of Quincy's presidency were marked by low

enrollments, ongoing financial instability, and student unrest so disruptive that it almost unseated the new president.[60]

Quincy's initial inclination was to resist any significant moves to revise Harvard's curriculum in ways that would accommodate ever-more-vociferous demands for "utility" and "research" in higher education. The president was a staunch believer in the virtues of the prescribed curriculum, especially its emphasis upon facility in ancient Greek and Latin; standards of admission to Harvard should be raised, not lowered, he argued. At the request of the Overseers, Quincy in 1830 outlined a new "General Plan of Studies" that aimed "chiefly to effect 'a more thorough education in the Greek and Latin Languages, the Mathematics, and Rhetoric'" than hitherto available.[61] The continuing decline of the University, however, forced Quincy to moderate his strict humanism. Throughout the 1830s, Harvard remained mired in financial woes, with enrollments showing no sign of budging upward. Principles had to be sacrificed if the future of the university was to be secured. President Quincy soon conceded his fight to raise admission standards and turned himself into an able fundraiser.[62] More importantly, professionalizing, entrepreneurial forces within the academy, and on Harvard's faculty, would give Quincy a powerful pedagogical rationale for yielding to the vulgar needs of the Jacksonian Age.

Where Ticknor had failed, Benjamin Peirce succeeded, at least in part. Returning to Harvard in 1831 as a tutor after graduating two years earlier, Peirce was an ambitious mathematical astronomer. He was appointed to the Hollis Professorship in Mathematics in 1835 and later served as the first Perkins Professor of Astronomy at the Harvard Astronomical Observatory.[63] During the 1830s, Peirce repeatedly argued against the restrictions of the prescribed curriculum and in favor of a voluntary system in which advanced mathematics courses would be optional. In pressing this case, Peirce's main goal was to spare himself the burden of teaching students who were either uninterested in his subject or incapable of grasping it; by letting only the most able students take advanced courses after their freshman year and giving the rest the opportunity to avoid compulsory studies in the calculus, Peirce hoped to free his time and energy for the publication of scientific papers, the execution of government contracts, and the leadership of professional organizations. When Peirce first proposed making mathematics courses voluntary in 1835, Quincy, wedded to an elitist vision of his own, was not persuaded. Three years later, however, with Harvard's outlook remaining bleak, he was ready to be convinced. In June 1838, with Quincy's backing, the Corporation agreed to a trial run of Peirce's elective system in the mathematics department. The reform allowed Harvard to demonstrate its responsiveness to the "spirit of the age" by opening the curriculum to the winds of choice and specialization, while at the same time allowing for advanced courses that could produce a new intellectual and cultural elite.[64]

Soon after the elective system was adopted in mathematics, the Latin and Greek departments followed suit. By the end of Quincy's tenure in 1846, most Harvard College courses save those for freshmen were optional. Moreover, entrepreneurial professors like Peirce began to proliferate. Asa Gray, appointed to the Fisher Professorship of Natural History in 1842; Louis Agassiz, a professor of geology and zoology who joined the faculty in 1847; and the historian Jared Sparks shared Peirce's concern for scholarship, publication, and professional engagement in academic life. Sociologically, these men looked less like the much-bemoaned timeserving clerics of the 1820s faculty and more like the enterprising lawyers and doctors with whom the new professoriate socialized in Boston.[65] The success of these figures reflected the growing organization and professionalization of science in the antebellum republic.[66] An important mark of these changes was the founding in 1847 of the Lawrence Scientific School, which offered the opportunities of "advanced study" to the burgeoning community of experimental scientists in Cambridge.[67] This professional, entrepreneurial academic ethos was another consequence of Harvard's metropolitanization.

The Eliot Complex

By 1850, Harvard's encounter with metropolitan Boston had endowed it with the rudimentary elements of a university culture. Burgeoning investment on the part of the "leading gentlemen" of Boston had left Harvard with a growing number of professorial chairs, a bountiful library, and the means to create new schools and research units, notably the Astronomical Observatory and the Lawrence Scientific School. Quincy's compulsion to bow to the demands of his practical-minded constituency had placed the elective system firmly on the agenda of curriculum reform. Finally, Harvard's professors, led by the trio of Peirce, Gray, and Agassiz, had made the academic life into a vocation akin to the respectable callings of the law, medicine, commerce, and the ministry—the professions that occupied the well-to-do families in Boston. Metropolitan Massachusetts had served as the crucible of what we would now consider a "modern" university.[68]

Once in place, however, these elements did not coalesce into the canonical institutional form that would characterize research universities across the United States by 1920. In fact, for the better part of half a century after Quincy's fateful backing of Peirce's plan for electives in mathematics, each component of Harvard's university culture—academic professionalism, the elective system, and graduate and professional education—endured vacillating fortune. The early years of Charles William Eliot's presidency saw not a qualitative break with the past but yet another attempt to fit the various pieces of the University together. Institutional crystallization, when it came toward the end of Eliot's reign, was driven largely from without: university reforms elsewhere in Gilded Age America, coupled with the emergence of national associations for the various arts and sciences and the settling-in of professional identities, more or less forced Eliot's hand. But by the time Harvard began to take its place among the set of leading research universities, the legacy of the earlier attempts at university reform had made Harvard a sprawling and fragmented place.

We can begin to grasp the continuing dialectic of reformism and conservatism in the period running from 1840 to 1880 by examining Eliot's early views of university administration in the context of his predecessors. After Josiah Quincy stepped down in 1846, he was replaced by Edward Everett, the former Eliot Professor of Greek Literature who had left Harvard in exasperation shortly after his return from Europe in the 1820s. Everett had designs to establish in Cambridge faculties of philosophy and science similar to those he had seen at Göttingen. Although he oversaw the foundation of the Lawrence Scientific School, his grander plans went unrealized, in part because he was, throughout his tenure, deeply ambivalent about the job he had reluctantly accepted.[69] With respect to Harvard College, Everett actively rolled back the move toward the voluntary system, citing the dispersion and lack of discipline inherent in the elective model. When he departed in 1849, Jared Sparks, himself a dedicated scholar, continued the hemming-in of the elective component of the college curriculum and clashed with Peirce and his supporters on the faculty.[70] Sparks's successors—James Walker (1853–1860),

Cornelius C. Felton (1860–1862), and Thomas Hill (1862–1868)—were by no means hostile to university reforms. Walker, for example, anticipated some of Eliot's (and, later, James Bryant Conant's) gestures toward meritocracy. He also sensed the growing importance of expertise and specialization in the liberal arts and saw the university moving toward division into departments of knowledge.[71] Like Walker, Hill was a man of the cloth and thus a throwback to the minister-presidents who had led the university up to Quincy's appointment, but he too supported the promotion of scholarship among the faculty.[72] In general, however, the Civil War era witnessed backsliding among Harvard's presidents from Quincy's commitment to electives.

Eliot's path to the presidency was far from smooth. When the call came from Harvard in 1869, Eliot was Professor of Analytical Chemistry at the Massachusetts Institute of Technology, where he had been teaching since 1865. Five years earlier, Eliot had left the Harvard faculty in acrimonious circumstances after the chair he coveted was awarded to Wolcott Gibbs. MIT, meanwhile, was associated with a turn toward vocational education in the pure and applied sciences that many Brahmins found distasteful.[73] Eliot's publication in 1869 of a pair of articles in the *Atlantic Monthly* on the virtues of technical education only confirmed the impression many at Harvard held concerning his wanton utilitarianism.[74] Yet if Eliot was not an entirely comfortable choice for the Corporation, his election was hardly a radical departure. Eliot was a solid representative of metropolitan Boston. His father was a successful merchant, and Eliot had seriously considered following in his footsteps before joining other Brahmin scions on the Harvard faculty.[75] Despite his differences with some senior faculty, he shared their professional concerns and in due course became the most effective advocate of the elective system that Benjamin Peirce had known. Moreover, the institution he inherited was almost living up to the ideal of a university: at the time of Eliot's election, Harvard boasted professional or quasi-professional schools of medicine, divinity, law, science, and mining and geology, as well as an observatory, a botanic garden, and a zoological museum.

Harvard's trustees and faculty were ready to hear what Eliot told them in his inaugural address; indeed, they had heard quite a lot of it already. As one historian has acutely noted, "The address was well received not because what it said was new but because it encapsulated much that [others associated with Harvard] had recommended. . . . [Eliot] restated in an incisive way much that had been said before."[76] Certainly, the menu of administrative concerns the new president considered in his inaugural remarks was familiar: the question of the proper place of the elective system had been on the table since the 1830s; Eliot's interest in improving scientific instruction had been shared by Everett; Walker had turned his attention to the development of the talents of individual students long before Eliot. Throughout the first two decades of his presidency, moreover, Eliot was rather less the progressive educator he had seemed at MIT and more a regal president in the mold of Quincy and Everett.[77] Owing in part to financial constraints, Eliot did not expect to appoint to the faculty a new generation of Peirces and Grays; in the severe economic climate of the Reconstruction era, Harvard simply could not afford to pay the rising cohort of professional scholars a comfortable salary. But Eliot also believed that faculty should be appointed at the president's discretion and according to personal criteria such as loyalty, trust, and character. In one example of this attitude, he coaxed Christopher Columbus Langdell from his law practice in New York into the Dane Professorship in the Law School on the basis of estimations of the latter's "genius" made when Eliot was an eighteen-year-old junior at Harvard College.[78] Professional standards and disciplinary self-governance, then, played little role in Eliot's initial approach to the management of the Harvard faculty. That some of his appointments turned out to be distinguished, as was the case with Charles Eliot Norton—Eliot's cousin—and Henry Adams, was a matter of luck rather than judgment. Even in the closing years of the nineteenth century, Eliot took such a personal hand in appointments and course allocations that, as a bewildered George Santayana was to learn, he was not averse to dropping in to see his staff personally when a class needed to be reassigned.[79]

Bureaucratic governance this was not. Nonetheless, in regard to one key pedagogical issue, Eliot did offer a bracing departure. If for more than thirty years the fate of the elective system had been in the balance, Eliot removed any doubt about both its importance and its scope. "The College," he told the audience at his inaugural address, "proposes to persevere in its efforts to establish, improve, and extend the elective system. Its administrative difficulties, which seem formidable at first, vanish before a brief experience."[80] This was a hard line, which Eliot buttressed with two forceful arguments. For the student, the elective system encouraged the cultivation of advanced skills in a particular field, which would in turn provide the republic with a variety of expert men. The United States, Eliot intoned, had learned the hard way that professionalism was no materialistic contrivance. "Only after the years of the bitterest experience [in the Civil War] did we come to believe the professional training of a soldier to be of value in war."[81] Yet there was also the expansion and growing maturity of the student body to contend with. The close class ties and large admixture of adolescents of the antebellum period had become a thing of the past. These two demographic factors alone "would compel the adoption of methods of instruction different from the old, if there were no better motive for such change."[82] And so, finally, the "system of liberty" was firmly ensconced. Undergraduates could choose whichever courses they liked provided they had completed by graduation a mandatory quota of classes. Classmates could follow almost entirely different curricula: one might take all their courses in one field, another no more than one in any; some might be tested to the limits of their abilities in rigorous subjects, while others would coast through choosing only unchallenging courses.

But for developments outside of greater Boston, Eliot's Harvard might well have continued into the twentieth century as an institutional oddity, shaped above all by its elective system and otherwise by a jerry-built structure of professional schools, museums, and observatories, with a semiprofessional faculty and a sprawling college at its center. Indeed, the rolling out of the elective system turned out to be a long and arduous process: not until the end of the 1870s did Eliot succeed in removing prescribed courses from the junior and senior years, and only in the mid-1880s did the President introduce an elective component into the freshman curriculum.[83] His one major innovation at university level was the creation, in 1872, of the Graduate School of Arts and Sciences; but, as Eliot himself later would admit, his aim in doing so was not to make Harvard the new Göttingen. It is better to think of the Graduate School as an extension of the elective system, for it offered specialist instruction both to graduates and to undergraduates, who were (as—with special permission—they still are) allowed to take graduate seminars.[84] What ultimately pulled these initiatives into a generic institutional pattern were two national developments: the ongoing professionalization of the liberal arts and the emergence of competing research universities.

After the Civil War, and with gathering pace in the 1870s and 1880s, more and more Americans were choosing to become professional academics.[85] When Harvard was undergoing metropolitanization, teaching positions in colleges and universities were considered fallbacks for those who had failed to gain a footing in the ministry or in the legal and medical professions, or as stopgaps for those between engagements in public affairs. By the middle decades of the nineteenth century, however, an academic career had become a more respectable, and for some a very desirable, option.[86] The trickle of Americans travelling to Europe to study in the universities of Germany became, by the 1880s, a flood.[87] And when these professional scholars and scientists returned to the United States, many sought employment in colleges, especially those, like Harvard, that offered the possibility of endowed chairs and resources for advanced research. Under Eliot's nose, a crowd of ambitious young academics had gathered in Cambridge. The elective system demanded teachers specialized in a given field of learning, and so Eliot was obliged to populate his faculty with new strata of tutors, instructors, and assistant professors, as well as full professors. Many of these career academics took junior and poorly paid positions in the hope that they would be promoted up through the ranks into one of the increasing number of professorships in the University.

It was the need to rationalize appointments that compelled Eliot to find new mechanisms for determining promotions. He found such mechanisms in the rising professional academic associations, which expressed the growing power of peer-group identification and the imposition of disciplinary standards across a range of academic fields. The founding of the American Philological Association in 1869 presaged a slew of similar initiatives: these included the Modern Language Association (1883), the American Historical Association (1884), the American Economic Association (1885), and the American Chemical Society (1892).[88] University presidents like Eliot now had to ask, in assessing the qualities of a potential professor, whether the man in question was a good historian, or chemist, or classicist, according to the standards of the field in which they worked. It also made sense to group professionals within these fields into departments within the expanding liberal arts faculties: such divisions could then be responsible for overseeing instruction and promoting research in a given field as well as helping to regulate hiring and promotion. Thus departments, rather than chairs, became the basic organizational units for disciplinary teaching and research in American universities.[89] At Harvard, the boundaries of these nascent organizational forms remained porous into the 1940s.

Alongside the emergence, at the national level, of a professional-departmental system, there unfolded a new university movement. Although distinct, these two developments were intimately connected. What drew them together was the foundation of universities whose primary function was graduate education and advanced research. We have seen that elements of graduate teaching and scientific research had been acquired by some of the colonial colleges in the 1840s and 1850s. Yale was more forward-looking than Harvard in this respect: the Sheffield Scientific School preceded the Lawrence Scientific School, and Yale awarded the first American PhD in 1861. But these initiatives did not trigger a wider move toward graduate education. The Morrill Act of 1862, which awarded federal land grants to states to establish technical colleges, precipitated the foundation of a new generation of state universities. These universities brought much-needed vocational and scientific training to the young states of the Middle and Far West and provided the framework within which a national system of universities could emerge. But the involvement of the "land grant" universities with technical and professional education meant they did not provoke the colonial colleges into feats of emulation. Only when Johns Hopkins opened its doors in 1876 did Harvard and its peers begin to take the new academic movement seriously.[90] Daniel Coit Gilman, the first president of Hopkins, moved Eliot to action by courting several of his most prized faculty members. A market for academic talent quickly emerged when Eliot was compelled to offer distinguished faculty inducements to resist Gilman's overtures, and in time he himself began to poach top prospects from Hopkins and elsewhere. When additional private universities came on the scene in the closing years of the nineteenth century—Chicago, Clark, Stanford—the pattern was set. Professional academic specialists would circulate among a set of competing institutions across the nation, and these universities would come to resemble one another organizationally as they sought to keep up with their rivals.[91]

Emulation did not, however, entail homogenization for Eliot's Harvard. The compartmentalization at work in the remaking of the faculty, along with the consolidation of graduate teaching and advanced research, were compounded by the dispersive logic of the elective system and the expansion of Harvard's student body. A vertiginous increase in the number of student societies, elite "final clubs," and athletics teams during the later years of Eliot's reign funneled undergraduates into discrete groups as never before.[92]

Exponentially increasing enrollments compounded the fragmentation of University life. George Santayana echoed the impressions of many students and faculty at the turn of the century when he made the following observations in his memoirs:

> Harvard, in those waning days of Eliot's administration, was getting out of hand. Instruction was every day more multifarious and more chaotic; athletics and college life developed vigorously as they chose, yet not always pleasantly; and the Graduate and associated Schools worked each in its own way, with only nominal or financial relations with Harvard College. . . . Eliot, autocrat that he was, depended on the Fellows, half a dozen business men in Boston who were the legal proprietors of Harvard. . . . All this formed an immense tangle of disconnected activities: the President was driving not a four but a forty-in-hand.[93]

A. L. Lowell and the Politics of Retrenchment

The chief legacy of Eliot's reign was a University of many parts. The Brahmins largely maintained their grip on the Corporation, as the choice of the ur-Bostonian Abbott Lawrence Lowell as Eliot's successor made clear. But now the castle of metropolitan power in the University was surrounded by a network of hamlets in which specialized teaching, professional research, and student factionalism thrived. Lowell sought to bring order and a strong dose of aristocratic cultural values to this chaotic scene.[94] So he did, often by imposing restrictive quotas on racial and ethnic groups—notably Jews and blacks—that he considered obstacles to the natural unity of Anglo-Saxon gentlemen.[95] Yet rather than changing the fundamental structure imposed by Eliot, Lowell's reforms, while in certain respects far-reaching, kept the core academic organization of the university in place; Lowell ultimately came to accept the basic tenets of an elective college curriculum and the establishment of professional schools. His task, as he saw it, was to resist further fragmentation and to bridge where he could the gulf between the College and the other parts of the University.[96] The crosshatched institutional pattern he created in pursuit of these goals allowed the interstitial academy to take shape during the interwar decades.

When it came to graduate education, Lowell, along with the leaders of other research universities, was faced with a fait accompli. By no later than the second decade of the twentieth century, it was possible to follow a relatively clear path of training and accreditation through to an MD, a JD, or a PhD.[97] American research universities, so long in the shadow of their German paragons, could now produce accredited practitioners of discrete professions and academic disciplines. Professional lawyers and doctors had been in existence since the late eighteenth century, but now new and exotic professional species bloomed forth: philosophers, managers, physicists, and historians came on the scene. Harvard was at the forefront of these developments.[98] In philosophy, in business, and in education, the University led the march toward professionalization. Such was the efficiency of the academic machine that late-Victorian intellectuals like William James worried about the *strength* of disciplinarity, not—as is now common—its contingency.[99]

Lowell shared James's misgivings about the "PhD octopus." "No one," he remarked in lament of what had become of graduate education, "goes into a graduate school in order to acquire a love of learning."[100] Advanced study in the liberal arts had become, for Lowell, another form of vocational drudgery. Disillusioned, Lowell focussed his efforts as president squarely on the College. His reforms were part of a national revival of collegiate values, a movement that reflected Progressive-Era concerns about social cohesion and common culture. "Counterreform" is how one scholar has described Lowell's project, and the term captures Lowell's attempt to curb the fragmentation of academic life at the undergraduate level.[101] In his own, inveterately elitist, fashion, Lowell was trying to restore a measure of meritocracy and collective identity to what had become a dispersed and largely unregulated student body. He thought it disastrous that, under the elective system, there was so little regulation of each undergraduate's course of studies. How could it be wise to let a certain portion of undergraduates coast through the College by picking basic courses, without facing a rigorous set of examinations? Between 1902 and 1909, Lowell sat on a series of committees charged with oversight of undergraduate instruction, and during this time he became a leading spokesman for greater order and stricter standards of assessment in the curriculum. As president, he introduced concentration and distribution requirements in the college, so as to ensure that the energies of each individual student were focused on a more comprehensive and systematic course of study. Lowell's interest in the use of general examinations to identify the "best men" stemmed from his comparative analysis of the recruitment and training of colonial civil servants in Great Britain, the Netherlands, and France.[102] Examinations appeared to provide the key to making and selecting able men; Lowell applied the same principle to college education, which would provide future statesmen, jurists, and businessmen for the republic.[103]

There was also a sociological dimension to Lowell's reforms. Like Woodrow Wilson, who as president of Princeton University endeavored to combat the power of exclusive dining clubs and private housing, Lowell recognized a class system at work within the Harvard student body.[104] While members of the Boston upper classes ensconced themselves in private housing near the Yard and separated themselves off in final clubs, a larger constituency of undergraduates, drawn from the middling classes of greater Boston, commuted to Harvard or lived in less salubrious residences in Cambridge. Combined with the partial anarchy of electives, this clubbiness and exclusivity produced little sense of shared identity in the student body, nor a clear hierarchy of intelligence or character—attributes that Lowell valued highly. In the 1920s, Lowell began to enact a plan whereby undergraduates would be lodged in a series of University residential houses. Only with the Harkness donation to Harvard in 1927, which totaled $10,000,000, was Lowell able fully to realize his vision: a set of houses along the north bank of the Charles River—Eliot, Adams, Leverett. Lowell added to this emerging undergraduate world a set of tutors to engage the students directly in their studies, in much the same way as the preceptor system worked at Princeton. No longer would collegians be left to their own devices: examinations, concentration and distribution requirements, prescribed accommodation, and tutorials would pull them together and set them in dialogue and competition with their peers.[105]

Despite its meritocratic trappings, Lowell's reforms of the college were deeply exclusive. Those who could get inside the gates of the Yard would be given a testing and character-forming education, but there were many whom Lowell felt would have to be left outside in order for his experiment to work. Rising enrollments of Jewish students were a particular concern of Lowell's, for he believed such students to be inimical to the gentlemanly culture he was seeking to foster. When his attempts to impose a strict quota on the admission of Jews failed, he was nevertheless able to use informal means to bring enrollments of Warburgs, Lehmans, and Loebs down to an acceptable minimum.[106] African Americans fared even worse. As for second- or third-generation European immigrants, the laissez-faire policy of the Eliot administration was, as in so many other respects, reined in.

Having focused so much of his effort on the College, Lowell was less an interventionist than an obstructionist at the faculty and postgraduate levels. As someone who had taken up a chair in Harvard's Government department on the basis of some well-received books on British politics and a mediocre career at the Massachusetts bar, he despaired of the specialization and the recondite jargons of the professional schools and academic departments. He was not, for instance, in any sense a supporter of the Business School, which opened its doors in 1908. A notional "political scientist," Lowell was nonetheless ill disposed toward the more recently established social sciences.[107] The departments of History, Economics, and Government met with Lowell's approval, but he took no steps at all to move with the scientific times when it came to sociology, anthropology, and psychology, which were then rising to prominence in the nation's universities. Lowell starved the Department of Social Ethics of funds—it existed only as the result of a private benefaction—and presided without regret over its disbandment in 1930. Only with reluctance did he replace it, in 1931, with a Department of Sociology.[108] Anthropology remained tied to the Peabody Museum, despite the discipline's autonomous departmental status elsewhere in the country. Psychology had a laboratory, but would be yoked institutionally to the Department of Philosophy until James Bryant Conant assumed power in 1933.[109] Under Lowell and his predecessors, there was a clear tutelage system for the new social sciences: political science with history, sociology with economics, and psychology with philosophy, with anthropology bracketed with the Peabody. Consequently, during the 1930s and 1940s, the story of the human sciences at Harvard involved "as much a search for institutional standing as a saga of disciplinary development."[110]

Lowell's hostility toward the professionalization of what he viewed as the gentlemanly enterprises of business and liberal education would have important knock-on effects for how inchoate disciplines and research programs took shape at mid-twentieth-century Harvard. For a corollary of Lowell's opposition to specialization was his support for initiatives that promised either common education for students across departments or the unification of interests among the various departments of the University. One act serves as symbol of Lowell's commitments on this score: his astonishingly generous personal donation of $2,000,000 to the Society of Fellows, which was established at the end of his tenure. The Society, as will be documented at length in the chapters that follow, would become one member of a set of interstitial clubs, societies, seminars, and teaching programs that were, in part, the product of Lowell's resistance to departmentalization.

Harvard at Three Hundred: The Higher Disorganization

The tercentenary celebrations of 1936, in combination with James Bryant Conant's accession to the presidency three years earlier, made acute the vexing question of Harvard's identity as a place of teaching and research.[111] As Harvard approached its three-hundredth birthday, questions of common purpose asserted themselves in the design of the undergraduate curriculum, the best divisions between departments, and the relationships among the professional schools, museums, libraries, and the faculty of arts and sciences. These Depression-Era tensions were expressed in some of the books produced to mark Harvard's birthday festivities. In the introduction to one of these tercentennial volumes, the organizational structure of the University was depicted as a great wheel. At the core of the University was the Harvard Corporation, composed of the president and fellows; these men were responsible for the formation and implementation of policy and the supervision of the University. If (to change metaphors) the Corporation was the yolk, then Harvard College was the white.[112] Connected to the college, but in a yet larger stratum of university organization, was the Graduate School of Arts and Sciences. The Faculty of Arts and Sciences, with its various academic departments, straddled both the College and the Graduate School. On the same administrative level as the Graduate School of Arts and Sciences and the Graduate Department of the Engineering School were ten professional schools. The College Library and an accretion of museums, gardens, and observatories completed the graduate-professional tier of the university. This concentric, segmental structure was portrayed as resting on the stabilizing ballasts of, first, the Board of Overseers, which had oversight over the Corporation, and, second, underpinning the whole system, the alumni of the University.[113]

If this graphic representation of the University resembled a carefully balanced wheel, it was just as easy, and indeed more accurate, to view the picture as a centrifuge, whose elements were fanning outwards and pushing further away from one another. This was how the newly installed president, James Bryant Conant, saw matters. The Harvard he inherited from Lowell was ungainly and fissiparous, as Conant quickly learned upon assuming office. Two years into his administration, he noted with alarm the "tremendous subdivision of the fields of learning which has occurred in the past thirty-five years." "Every day," Conant lamented, "the expansion of the academic universe separates still further one specialist from another. . . . The situation will soon become so desperate that the process must be reversed and we shall enter a period when the stars in the university firmament will slowly tend to coalesce." The president went so far as to predict that "a marked fall in the birth rate of special vested academic interests will be the first sign of the approaching reform. The present inchoate situation is the product of a necessary transitional era."[114]

What Conant objected to was not the specialized pursuit of knowledge but the enervating effects of unchecked intellectual and administrative compartmentalization. His views were connected to widespread concerns about a lack of direction and distinction at the faculty and postgraduate level— a malaise many blamed on Lowell's attempts to revive the College at the expense of advanced research and professional training. Lowell's retirement in 1933 released pent-up dissatisfaction with the national standing of the University. In the *Harvard Graduates' Magazine,* Bernard DeVoto (then a part-time English instructor in the College) voiced the common view that "the faculty of Harvard is not, on the whole, as good as it used to be. . . . One hears that the distinction of this department has lapsed, or that there is much dead wood in another one, or that the best students of a given field now go to Chicago or Columbia or Wisconsin, whereas once they came to Harvard."[115] Conant certainly shared these concerns. When members of the Harvard Corporation first sounded out the young chemistry professor about his objectives should he become president, Conant expressed his frustration with the standards of faculty promotion. All too often, he complained to one Corporation member, professors were appointed because of their "devotion to Harvard" rather than for the promise of their research or their eminence as scholars.[116] One of his most important reforms as president was to institute a new procedure for faculty promotions, whereby junior faculty, after a period of eight years, were moved "up or out." Conant's aim was to ensure that the "best men" were given tenure and unproductive or undistinguished scholars shown the door.[117]

The effects of Conant's reforms would take some time to register. Even when his meritocratic vision of the University was partially realized in the years after World War II, it did not create a uniformity of organizational structure or intellectual purpose. Interwar Harvard, meanwhile, was a study in uneven development, especially where newer fields of learning were concerned. Symptomatic of the University's uncertain direction was the state of the human sciences. While sociology, anthropology, psychology, and political science thrived as autonomous enterprises at the University of Chicago, Columbia University, and Johns Hopkins, none, as noted above, enjoyed high status at Harvard before World War II.[118] Even where Harvard seemed to be moving with the social-scientific times, its claims to innovation were weak. The Department of Government did not offer the training in policymaking that the first graduate programs in political science provided at Johns Hopkins and Columbia.[119] Likewise, Economics enjoyed at Harvard a stable existence over many years but, as a stronghold of neoclassical orthodoxy, it remained blissfully unmoved by the popularity of "institutionalism," which energized American economics after World War I.[120]

The tercentenary of 1936 made manifest the precarious existence of the Harvard human sciences. The centerpiece of the celebrations was the two-week-long Conference of Arts and Sciences, attended by scholars from across the globe. In sections devoted to mathematics, astronomy, physics, and the biological sciences, the program took shape around the dynamic research fronts of the day: cosmic radiation, parasitism, industrial chemistry, and communications engineering.[121] The three wide-ranging symposia dedicated to the social sciences and humanities proved harder to organize, for there seemed to be no common area of agreement among the participants. The official summary of these sessions hedged to the point of unintelligibility its description of symposia. They had showcased "a series of typical methods and outlooks which jointly play their part in approaching the great problem of seeing human life in an ever truer perspective."[122] At all events, the volumes that emerged from the symposia scarcely concealed the disparity of interests on display; in few other books could a disquisition on modern mathematical logic by the German philosopher Rudolf Carnap have been grouped with a ponderous meditation on constitutional history by ex-President Lowell.[123]

Although the University had professional sociologists, anthropologists, and psychologists on its payroll, they carried little institutional weight; the suggestions for speakers and themes they had made were far from uncritically accepted by the faculty powerbrokers who sat on the Executive Committee of the Tercentenary Conference. Among these men were the professor of biochemistry Lawrence Joseph Henderson, Edwin Wilson from the School of Public Health, the astronomer Harlow Shapley, and the historian of religion Arthur Darby Nock. Wilson in particular took a dim view of the recommendations offered by his colleagues in the social sciences, which seemed to him to have been offered on the basis of tribal loyalties rather than scientific standing. The psychologists, he noted, had ignored a number of eminent educational psychologists, while the anthropologists' list was rated "very poor."[124] Early in 1936, Wilson also reported to Henderson that he had been told by two eminent sociologists outside Harvard that the list of world-leading sociologists provided by Harvard's own Professor of Sociology Pitirim Sorokin "could hardly be worse."[125] Henderson, meanwhile, saw fit to embroider his invitations to those participating in the symposia with suggestions for topics based on his own understanding of the current state of social knowledge. Neither Wilson nor Henderson questioned their right to "add names" to lists outside their field or to "sit in" with the social science groups responsible for inviting the leaders of their disciplines.[126]

Cynicism toward the human sciences started at the top. Although Conant was not in principle hostile to the human disciplines, he maintained the natural scientist's wariness of the scientific pose struck by many social scientists. Shortly before the tercentenary, he admitted in private to doubts as to "whether or not the social sciences are the modern equivalents of astrology or of medicine"—did they amount to prophecy or science?[127] Conant did not discount the possibility that social inquiry could be scientific but, when judged by the chemist's yardstick of prediction and control, "social science" barely deserved its name. For Conant, the present state of social science posed the question whether it was "important to have university prophets as well as scientists; or, to put it another way, propagandists anxious to convert as well as social scientists eager to examine."[128]

Members of the Tercentenary Executive Committee shared Conant's suspicions. Indicting "teachers of political science, economics, sociology, etc.," Henderson told an audience in 1930 that there was "probably at least as much nonsense taught in the colleges and universities of this world in those subjects as there is sense."[129] Henderson's rhetorical stock-in-trade was an invidious comparison between those disciplines that studied human relations, on the one hand, and, on the other, the exact or empirical sciences, from mathematics to biology. Whereas the latter had predictive power or deductive completeness, the former were afflicted by "myths" and "superstition." There was "at least as much philosophy mingled with our current social sciences as there was at any time in the medical doctrines of the Greeks." As such, "a great part of social science today consists of elaborate systemization on a very insufficient foundation of fact."[130] In correspondence with one another, Henderson and Wilson were apt to be scathing about practicing social scientists. "There isn't any use trying to please these beggars," Wilson wrote Henderson in the spring of 1935. Of his stint as president of the Social Science Research Council, Wilson recalled that "[o]ne merely refrains from doing things he likes to do so as to be more pleasing to them and they are so stupid anyhow and so completely hostile in intellectual processes that they can't be persuaded."[131]

Neither Conant nor the organizers of the Tercentenary Conference knew exactly what to do with the human sciences; their marginal, unstable place on Harvard's organizational chart was mirrored by the eclectic manner in which they were bunched together in the Tercentenary Conference and its published proceedings. It is striking that, at this relatively late moment in their emergence in the United States, the human sciences remained an object of debate, diagnosis, and intervention on the part of faculty power-brokers at Harvard, some of whom had tenuous or nonexistent professional ties to the fields in question. This insight is crucial to understanding the interstitial academy that had come into being at Harvard over the course of the preceding century. The conversation that Henderson, Wilson, and their colleagues held under the auspices of the Executive Committee was part of an ongoing discussion about the nature of the human sciences that had been underway in Cambridge since the 1920s. Here we find the most consequential feature of the tercentenary debates: the institutional fate, scientific problematics, and disciplinary practices of the human sciences at Harvard were being decided in a cluster of more or less ad hoc or irregular spaces that existed alongside the well-established departments of the University.

In the first half of the twentieth century, a cluster of interfaculty discussion groups and clubs provided the setting for these developments. We shall encounter these interstitial forums repeatedly in the chapters that follow: the Royce Club, the Shop Club, the "Levellers" cohort of social scientists, the Science of Science Discussion Group, the Inter-Science Discussion Group, and the Institute for the Unity of Science. Then there were departments and schools that became subject to extraprofessional control or influence owing to their unstaffed or weakly institutionalized faculties: the Business School, the Department of Sociology, the Department of Philosophy, the Department of Psychology, and the Department of Social Relations. These interprofessional or interdisciplinary meeting grounds in turn sheltered a variety of heterodox research centers, course offerings, and discussion groups, notably Henderson's seminars on Pareto and "Concrete Sociology" in the Department of Sociology. Finally, there were organizational innovations that cut orthogonally across the tiers of the College and the University: the Society of Fellows and the post–World War II General Education program were the foremost examples of such interstitial structures.

It is this collection of clubs, discussion groups, societies, pedagogical programs, seminars, and marginal department and schools that I term "the interstitial academy." This label picks out not an enduring institutional nexus that can be easily distinguished from the firmly set organization of the university, but the foregoing account of Harvard's development can help us to see how interstitial academic spaces emerged and how they were navigated. First, most of the interstitial organizations were both made possible by the ad hoc and largely unplanned accretions of university departments begun under Josiah Quincy and brought to a head by Charles William Eliot. If the competition for faculty, students, and money touched off by Daniel Coit Gilman at Hopkins led to the crystallization of an institutional pattern for American universities, Eliot and later Lowell grafted this pattern onto an already eclectic organizational plan. While the elective system provided Eliot with the flexibility to accommodate the new age of a research-focused faculty, strong departments, and dynamic professional schools, the reign of free choice exacerbated the intellectual and institutional fragmentation that was being driven by expanding student numbers and the rise of collegiate athletics and student societies. Since the time of Quincy, indeed, the elective curriculum had served Harvard presidents as a valve through which to release the pressures brought to bear on the University by the demands of utility, academic professionalization, and student dissatisfaction.

The story of Harvard from the Webber to the Eliot administrations is the story of how problems of this nature were endlessly finessed rather than systematically addressed—this being, if not Harvard's "genius," then certainly the explanation for its place among the top rank of higher education institutions during the transformative nineteenth century. Virtually all the major innovations in the American college and university happened elsewhere: technical education was nurtured in the land grant universities and down Massachusetts Avenue at MIT; Cornell, Johns Hopkins, and Clark led the way in graduate training and advanced research; Princeton and Yale ushered the college into the age of the university; Chicago promoted departments for the social sciences. Bold plans were enacted elsewhere; Harvard trimmed as necessary as it made its way across these choppy educational waters. Although other distinguished colleges—Columbia being the most notable example—had to navigate similar challenges and later sprouted their own interstitial programs and seminars, Harvard responded to the new environment in ways that were especially consequential for the human sciences.

The spectacle offered by this maneuvering was not always edifying: Eliot, for example, was sure that a university devoted to graduate education and advanced research could not succeed and was forced to adapt quickly when Hopkins flourished and began to covet his best faculty members. Even more critically, Harvard hedged on the human sciences, favoring the evidently more cultivated pursuits of history, government, political economy, and philosophy over sociology, political science, anthropology, and psychology. Something of this ambivalent stance toward the social sciences could be glimpsed as late as the mid-1950s, when a faculty report on the development of the "behavioral sciences" at Harvard offered the following gloss on the University's complex relationship with intellectual change:

> All through the years, Harvard has had its adventurous minds . . . who pioneered in the development of their disciplines. If newer institutions like Johns Hopkins or Chicago were at certain periods even more daring than Harvard in innovations, it must be remembered that Harvard had a responsibility not only to venture into the new and uncharted, but also to preserve the best already existing in its heritage from many generations of scholarship. This is not a responsibility lightly to be regarded.[132]

As these words were published, however, the attitudes toward the social sciences that they expressed were going through a drawn-out process of revision, which spanned the middle decades of the twentieth century. Both scientific philosophy and the interstitial academy would be centrally involved in these developments. What emerged from this synthesis was an intellectual culture in which the human sciences engaged epistemology and the natural sciences in ways deeply mediated by institutional reform, pedagogical experimentation, and a sensitivity to the knowledge-making practices of working scientists.

2

Making a Case

The Harvard Pareto Circle

In the autumn of 1932, a group of Harvard faculty and students began to meet for a weekly seminar, which ran under the title of "Pareto and Methods of Scientific Investigation."[1] Discussions were led by the biochemist Lawrence Joseph Henderson, who organized the course with the assistance of the prominent Boston lawyer and member of the Harvard Corporation Charles Pelham Curtis, Jr. Participants had one assignment: to read the French translation of Vilfredo Pareto's *Trattato di Sociologia Generale*.[2] An Italian nobleman, engineer, and economist, Pareto had turned to sociology when it became clear that the calculative rationality of agents assumed in his economics fell short of social reality. The *Trattato*, published in 1916, was Pareto's anatomy of the "non-logical" features of human action that tended to foul up the economist's theories.

Numbered among those who read the Italian's treatise of "general sociology" in Henderson's seminar were present and future luminaries of the academy. From the Harvard faculty came Joseph Schumpeter, Pitirim Sorokin, Thomas North Whitehead, Fritz Roethlisberger, Hans Zinsser, Clyde Kluckhohn, W. Lloyd Warner, Henry Murray, Elton Mayo, Crane Brinton, Bernard DeVoto and Talcott Parsons. Student members included Robert K. Merton, William Foote Whyte, Kingsley Davis, George C. Homans, and James Grier Miller.[3] In addition to the leaders of the Hawthorne experiments at the Western Electric Company (Mayo, Roethlisberger, and Whitehead) and one of America's rising men of letters (DeVoto), the group included five future presidents of the American Sociological Association, and one each of the American Historical Association, the American Anthropological Association, and the American Economic Association.[4] Many of the works that emerged from the Pareto seminar would go on to become classics of mid-century social thought: Parsons's *The Structure of Social Action* (1937), Brinton's *The Anatomy of Revolution* (1939), and Dickson and Roethlisberger's *Management and the Worker* (1939).

Given these associations, it is not surprising that Henderson's seminar rapidly acquired a certain mystique. As early as 1934, the *New Republic* drew its readers' attention to "the first seminar on Pareto at Harvard," and noted that the recently published *Introduction to Pareto* (1934), written by Curtis and Homans, had the "semi-episcopal *Imprimatur*" of Henderson and "Pareto's chief defender and angry-letter writer," DeVoto.[5] When the English translation of the *Trattato* was published the following year, its editor genuflected toward "Professor Henderson's epoch-making seminar in Harvard."[6] Three-quarters of a century later, the "Harvard Pareto circle" is a key point of reference in the history of American social thought. Alvin Gouldner set the tone for much of this commentary in his 1971 polemic, *The Coming Crisis of Western Sociology*. Included in his long bill of indictment against Talcott Parsons's functionalist paradigm for sociology was the youthful Parsons's involvement in the "politically conservative, anti-Marxist" Pareto clique.[7] Subsequent assessments of Parsons and American social theory have likewise seen the Pareto circle as the source of postwar "structural-functional" sociology.[8] Today, the Harvard Pareto circle figures in accounts of the emergence of organizational sociology, general systems theory, management science, and the history, sociology, and philosophy of science.[9] Historians have explained the embrace of Pareto at Harvard as a response to the fevered politics of the Depression years. Pareto, in this reading, was Harvard's answer to Marx during the high tide of socialist activism on college and university campuses. Some scholars go further: European social theorists like Pareto, Weber, and Durkheim, they claim, were cited and discussed by conservative Harvard faculty as an intellectual counterweight to New Deal social democracy.[10]

Such interpretations miss two crucial features of the Pareto enthusiasm, which become visible once we are attuned to the salience at Harvard of the tradition of scientific philosophy. It was Pareto's theory of scientific knowledge, not his critique of social democracy, that mattered to Henderson and his associates. Moreover, the Pareto vogue found its strongest influence and support within the interstitial academy. Pareto was a key figure for Henderson because the former showed the social sciences how they could be scientific, and Henderson, in turn, mattered to interstitial sociologists, anthropologists, and management scientists because he was "the arch exponent at Harvard of what made science science."[11] In the Pareto circle, the technical understanding of epistemology made available by scientific philosophy was put to work in attempts to legitimize, in the chilly intellectual atmosphere of Harvard, the practices of the human sciences.

We find in the Pareto circle our first example of the Harvard complex—the combination of scientific philosophy, the interstitial academy, and the marginal human sciences. Crucially, however, Henderson and his colleagues also relied on one additional notion with a long Harvard pedigree. The case method, as a mode of instruction and as a form of inquiry, traced its roots back to the 1870s, when Christopher Columbus Langdell introduced it into his teaching at the Harvard Law School. In the notion of the case, as it was invoked by Langdell and his successors, pedagogy, research practice, and epistemology became reflections of one another. What scholars did in acquiring knowledge, what it was they claimed to know, and how they taught their subject were merged. This was a reflexive style of thinking that Harvard's psychologists, philosophers, sociologists, anthropologists, management theorists, and methodologists of science would make their own.

Henderson's Pareto

During the 1930s, the name of Pareto was spoken in many academic tongues. Under the influence of Pareto's *Manual of Political Economy,* the fields of general equilibrium theory and welfare economics were transformed by a new generation of mathematically sophisticated economic theorists.[12] Pareto's reception in social and political theory, meanwhile, took a number of unexpected turns. In interwar Europe, Pareto attracted the attention of leftist intellectuals. The relationship was a mixture of repulsion and attraction. For a young Raymond Aron, still on the moderate left at this stage of his career, Pareto was "a fascist thinker whose sociology represented little more than a Marxian deviation turned to conservative purposes." Aron recognized that Pareto's theory of the circulation of elites was an attempt to portray class struggle as an eternal feature of human society. He read Pareto's account of the skillful manipulation of mass sentiments by political elites as an accurate description of fascist rule.[13] At the Institute of Social Research in Frankfurt, Franz Borkenau integrated Pareto's sociology of elites into a Marxist theory of totalitarianism.[14]

In the United States, responses to Pareto's social theory were less refined. The general attitude among American liberals was that Pareto's doctrines of the circulation of elites and of the manipulability of the democratic masses were realized in "Hitler's totalitarian state."[15] It was well known that Pareto had accepted Mussolini's invitation to become a life senator in the Italian parliament, and that he had been lauded by the Duce as the intellectual godfather of the fascist movement.[16] The typical American reaction to the *Trattato* in the 1930s was one of puzzlement, coupled with a curiosity about Pareto's tough-minded, scientific-sounding language of "sentiments," "systems" and "equilibrium."[17] At all events, no American thinker in the age of the New Deal possessed an explanation of totalitarianism to rival those of Aron and Borkenau.[18] In this uncertain scene, Henderson was able to define Pareto's social thought for a generation of American human scientists.

The Harvard biochemist was primed to react to very specific parts of Pareto's sprawling treatise. Henderson's principal achievement in his scientific work was to generalize Josiah Willard Gibbs's model of the physicochemical system across several new fields of inquiry, first in biological chemistry and then in human biology.[19] While this approach yielded several important technical results relating to acidosis and role of the blood in maintaining physiological equilibrium, it also engendered in Henderson an acute methodological and, at times, metaphysical self-consciousness. More than a decade before the outbreak of World War I, he became transfixed by a cluster of philosophical problems emerging from his scientific research. At this stage of his career, Henderson was especially interested in explicating the uniquely favorable properties of the critical elements of living organisms—carbon, oxygen, and hydrogen (the importance of nitrogen was not then known)—in the promotion of organic life.[20]

As a champion of Gibbs, moreover, Henderson was led to ponder the explicitly conventional nature of the systems studied in the physical and biological sciences: the systems framework developed by Gibbs was an abstraction, which scientists brought to bear on the phenomena at hand. Yet, rather than functioning as a mere aesthetic indulgence or arbitrary hypothesis, this framework made possible the collection and interpretation of facts. Familiar with the writings of philosopher-scientists like Henri Poincaré and Ernst Mach, Henderson was aware that man-made symbolic conventions and a priori frameworks were elsewhere being acknowledged as respectable elements of science.[21] Henderson's investment in the concepts of system and organization placed him at the leading edge of the philosophical turn against mechanical explanations of nature. In physics, the mechanical world picture had been toppled by relativity theory, after a long crisis during the nineteenth century; the concept of the organized *system,* spurred by developments in biology, was now the order of the day.[22]

Over the course of the 1920s, Henderson's concerns regarding environmental fitness dropped away, but his interest in the nature of scientific reasoning and the phenomenon of organization grew yet more intense. It was against this background that he encountered Pareto, whom he first read in 1926 at the suggestion of the entomologist William Morton Wheeler. In its reflections on scientific method, on the explanatory scheme of mutual dependence, and on the conditions of equilibrium within systems, the *Trattato* treated ideas already of central concern to Henderson.[23] Pareto's sociology, Henderson later recalled, "crystallized certain half formed ideas and conclusions in my mind and led me at once to realize that, while I had formerly not seen how it was possible to think about problems of the kind here in question, I could see how to think a little about them."[24] Using the *Trattato* as his guide, Henderson outlined his vision of an authentic social science in his 1935 study *Pareto's General Sociology: A Physiologist's Interpretation.*

Henderson applauded Pareto for his clear-eyed view of the place of "non-logical" motivations in social and political life. Not since Machiavelli, Henderson insisted, had the sources of human conduct in the sentiments been dissected with the dispassion and precision shown in the *Trattato.*[25] Pareto had demonstrated that "no other elements of the social system are more important than the sentiments."[26] In their interactions with one another, social actors usually did not behave in what Pareto called a "logico-experimental" fashion—the manner of the experimental scientist, the engineer, or the calculating entrepreneur. When they did, the disciplines of economics and scientific methodology could account for their behavior. But Pareto, frustrated by the limits of his economic theory in the real world, felt compelled to insist that social agents most often acted on deeply held, nonrational sentiments, which manifested themselves either as "residues"—core values visible to the sociologist—or as verbal rationalizations, which Pareto termed "derivations."[27]

Henderson considered Pareto's analysis of the non-logical elements of human social life a scientific contribution of the first order. Nonetheless, he bought the Italian's theory retail, not wholesale. Even as he worked his way toward the publication of *Pareto's General Sociology,* Henderson refused to become bogged down in the *Trattato*'s densely nested subdivisions among the residues and derivations, and he ignored altogether Pareto's exhaustive listing of the species of logical and non-logical conduct.[28] Henderson seized instead upon the most general categories of Pareto's social theory: sentiments, residues, and derivations. These concepts appealed to Henderson because they functioned in the same way as notions of temperature, pressure, and concentration in the physicochemical theory outlined by Gibbs: they were attributes of a system, and thus furnished the core concepts of a general social theory. It was vital for Henderson that the general elements of Pareto's scheme allowed for holistic explanations in terms of systems of mutually dependent variables. Although Henderson was quick to acknowledge "that there is not the slightest reason to believe that Pareto was led to his theory by a consideration of the properties of a physicochemical system," he understood the *Trattato*'s discussion of social interaction as a system to be yet another vindication of the organismic model of explanation. "There seems to be a psychological, if not a logical, advantage," he wrote, "when conceptual schemes consist of *things* which have *properties* (or attributes) and *relations*."[29] Pareto's sociology had just this advantage: its basic entities or "components" were persons; these components were heterogeneous, for persons related to each other in various ways according to class, occupation, kinship, and so forth; and, finally, the components had certain attributes, notably the residues, the derivations, and "economic interests." Here was a system of mutually dependent elements existing in dynamic equilibrium much like those Henderson had studied in other departments of science. Pareto had shown how to bring sociology into line with "the type of description previously found necessary for dynamical, thermodynamical, physiological, and economic systems."[30]

In matters of epistemology, Henderson combined the scientist's robust empiricism with an unapologetic defense of theoretical constructivism. Here we see the heritage of scientific philosophy making its presence felt: the sharp separation of constructed, analytic frameworks and unreconstructed sense experience was the primary result of the nineteenth- and early twentieth-century attempts to find a way beyond Kant's synthetic a priori. Henderson emphasized the practical skills scientists demonstrated when they employed "conceptual schemes" or a priori frameworks to explain the phenomena they confronted. On his view, abstract conceptual schemes like Gibb's physico-chemical system were certainly human constructs, but just in the sense that a tool was a construct. Ideally, the conceptual scheme of the practicing scientist should become absorbed into the process of scientific inquiry:

> The effective use of [Pareto's] scheme is like swimming in that, as a rule, it is more or less unconscious. Nobody was ever the master of the use of a tool, physical or mental, who constantly asked himself, "What am I doing?" or "What shall I do next?" . . . There are few differences in the behavior and procedure of investigators more interesting and more important than that between half unconscious, habitual use of a clear, simple, convenient, highly generalized conceptual scheme, and any other method of work. This is, in one respect, the significant differ-ence between the work of almost all natural scientists, on the one hand, and the work of most social scientists, on the other.[31]

This view allowed Henderson to eschew questions about the truth of con-ceptual schemes for the same reason that one did not ask about the truth of a walking stick: both were convenient instruments for getting on terms with the world.[32] His principle authority for this view was Henri Poincaré, whose writings on scientific methodology Henderson had encountered at the turn of the century. Henderson especially admired Poincaré's pragmatic dictum regarding statements about the existence of the world: "These two proposi-tions, 'the external world exists,' or 'it is more convenient to suppose that it exists' have one and the same meaning." Notions of human convenience and convention were at the heart of Henderson's epistemology.

Mapping the Pareto Network

Henderson disseminated his Paretian social theory through several chan-nels. For some commentators, the Pareto circle begins and ends with the seminar on Pareto and Methods of Scientific Investigation.[33] The life of the circle was not, however, limited to the Pareto seminar. Three forums, in addi-tion to the seminar, stand out: Henderson's course on "Concrete Sociology," the Society of Fellows, and the Graduate School of Business Administration.

Henderson's leadership in all three of these enterprises indicates a broader conclusion: the biochemist's advocacy on behalf of Pareto was continuous with a career devoted to academic entrepreneurship along the institutional fault lines of Harvard University. The four clearest embodiments of Henderson's Pareto circle—the seminar on Pareto, Concrete Sociology, the Society of Fellows, and the Business School—were a critical part of the interstitial academy that took shape during the interwar years.

By the mid-1930s, Henderson had embedded the study of Pareto in the undergraduate curriculum of Harvard College. In the academic year of 1935–36, with the Pareto seminar still running, Henderson began teaching Sociology 23, a course initially entitled "Seminary in Methods and Results of Certain Sociological Investigations," and later renamed "Concrete Sociology: A Study of Cases."[34] Concrete Sociology and the Pareto seminar were complimentary enterprises: Sociology 23 was Henderson's attempt to demonstrate the applicability of the Paretian framework to the study of human relations in a variety of cases, each provided by a speaker acquainted with "practical affairs" in medicine, business, and the law. Several of those asked to lecture on the course had participated in the Pareto seminar. Brinton, Homans, Parsons, Mayo, Roethlisberger, Whitehead, DeVoto, and Kluckhohn all presented cases in Sociology 23. In addition, Henderson invited a number of close associates to teach in the course. These included Wallace Donham, Melvin Copeland, David Dill, Arlie Bock, and Dean Wilson, all from the Business School, along with the young anthropologists Conrad Arensberg and Eliot Chapple. Henderson also recruited senior figures to his cause: Harvard professors Arthur Darby Nock and Edwin B. Wilson (with whom he had sat on the Executive Committee of the Tercentenary Conference); the president of the New Jersey Bell Telephone Company Chester I. Barnard; and the president emeritus of Harvard, Abbott Lawrence Lowell.[35]

The contributors to Sociology 23 were a heterogeneous bunch but, as Roethlisberger later recalled, they were united, in Henderson's mind at least, by a common orientation: "the property which determined who was included in this set of persons was 'Did you get the point of Henderson's seminar on Pareto?' If you had, you were in; if you had not, you were out."[36] Henderson did not soft-peddle these commitments. Each year, he circulated to his lecturers his opening remarks for Sociology 23 on Pareto's "conceptual scheme"—a tactic designed to solicit comments from, and to encourage doctrinal consistency among, the speakers.[37] For the guest lecturers, participation in the course was therefore not a light undertaking. Student attendance, meanwhile, could be slight—a young Bernard Barber was one of the few undergraduates to take Sociology 23 in the spring of 1938.[38] Henderson's intention was that the presenters themselves (and their wives) should attend the lectures.[39]

Even before Concrete Sociology appeared in the Harvard Course Catalogue, Henderson extended his proselytizing on behalf of the *Trattato* into the newly established Society of Fellows. As chairman from the election of the first Junior Fellows in 1933 to his death in 1942, Henderson was, in the words of William Foote Whyte, "the dominant figure" in the first decade of the Society.[40] Henderson's Society—and no fellows from the period dispute that it was indeed Henderson's Society, from the choice of the wine served at the Monday dinners to the common topics of conversation—was a vital link in the Pareto network at Harvard. Several former fellows remembered Henderson holding forth on Pareto and scientific method during the Society's social gatherings.[41] And he brought into the life of the Society his closest associates from the Pareto courses. Homans was elected a Junior Fellow in 1934 largely on the basis of his coauthored book on Pareto (his collaborator Curtis was one of the first cohort of Senior Fellows of the Society).[42] Brinton was a frequent guest at the Monday dinners of the Society, and in 1939 was named Senior Fellow—very likely at his friend Henderson's instigation. Parsons, too, attended several dinners of the Society as Henderson's guest and protégé.[43]

Ties between the Society and Henderson's wider Pareto network thickened as the years passed by. One of Henderson's lecturers in Sociology 23, the anthropologist Conrad Arensberg, was elected to the Society, as was one of the few undergraduates to participate in the seminar on Pareto, James Grier Miller. The sociologist William Foote Whyte, elected in 1936, studied Pareto under Henderson while a Junior Fellow. Ultimately, Arensberg and Whyte were drawn more to the functionalism of the social anthropologists Bronislaw Malinowski and A. R. Radcliffe-Brown than to Pareto's concepts of social system and equilibrium, but the influence of Henderson's Pareto was nonetheless visible in their research as Junior Fellows.[44] James Grier Miller, meanwhile, dedicated his life's work in general systems theory to the precepts laid down, in part, by Henderson's account of organic systems.[45]

Henderson's preoccupations were all but impossible for Junior Fellows to ignore. Homans, who had concentrated in English literature as an undergraduate and aspired to be a poet, became a sociologist largely under the direction of Henderson. Whyte notes that he and other social scientists in the Society were virtually obliged to take Henderson's seminar.[46] Yet more telling are the relatively trivial expressions of Henderson's influence. One of his coinages, the phrase "conceptual scheme," became ubiquitous in American philosophy, history, and social science.[47] Strikingly, more than one of the early Junior Fellows turned his interests toward those consonant with Henderson's upon election to the Society. Henry Guerlac, for instance, abandoned biochemistry for history of science—a subject taught by Henderson in the College— soon after his election. The physicist John Howard, to take another example, became a sociologist as a Junior Fellow in the late 1930s. Whyte eschewed an early interest in economics for social anthropology.[48] Many human scientists in the Society treated Henderson as a tribunal for their research projects: both Arensberg and Whyte viewed Henderson's approval of a research project or theory as the surest mark of its scientific standing.[49] Always an energetic evangelist, the chairman made a point of circulating his programmatic lectures for Sociology 23 among the Junior Fellows.

The web of associations evident in the Pareto seminar, Sociology 23, and the Society was further strengthened by Henderson's involvement with the Graduate School of Business Administration. Although Henderson held appointments in the College and the Medical School, from 1927 he worked out of an office on the Business School campus, where he served as director of the Fatigue Laboratory. The Fatigue Laboratory was itself one wing of a dual program at the Business School devoted to studying the physiological and psychopathological aspects of industrial labor—what management experts of the time liked to call the "human problems" of business and industry. Henderson led biological research into worker fatigue and related nonpathological factors impacting upon productivity and workplace efficiency.[50] The Australian social psychologist Elton Mayo, meanwhile, orchestrated complementary field studies of morale and working conditions. Mayo's team at the Business School was responsible for the collection and interpretation of data gathered at the Hawthorne Works of the Western Electric Company in Cicero, Illinois. In 1930, Henderson's and Mayo's programs merged to form the Industrial Hazards Project, which received a generous grant from

the Rockefeller Foundation.[51] Henderson sat on the administrative committee of the Hazards Project, and chaired the Committee on Industrial
Physiology, which funneled the Rockefeller money to Mayo and the Fatigue
Laboratory.[52] His ties to the Business School, and to the field of business
administration, were deep.

Henderson became involved in the affairs of the Business School through
its dean, Wallace Donham, who had been a classmate of Henderson's at
Harvard College. In the years after World War I, Donham came to believe
that business education, at that time barely a professional enterprise and, at
best, a form of applied economics, had failed to tackle the complex human
dimension of industrial capitalism. For Donham, these challenges were eminently practical: competent managers needed to have technical knowledge
of how to deal efficiently with the physiological, social, and psychological
factors that affected the productivity of their workers. The problem was one
of "human relations," and the business of the business professor was to make
that problem into the foundation of a science.[53] It was in this connection
that Donham first began talking to Henderson about his academic concerns.
These discussions helped prepare the way for the Industrial Hazards Project,
which became in due course the centerpiece of Donham's attempts to set the
professional training of modern business executives on firmer ground.

Henderson's enthusiastic embrace of Pareto as the Gibbs of sociology was
a response to the problems of human relations posed by Donham, Mayo,
and like-minded researchers in the Business School. Indeed, Henderson was
much less involved in the administration of the Fatigue Laboratory (which
he left largely to his trusted lieutenant, David Bruce Dill) than he was in setting out the conceptual framework and methodological prerequisites for the
management technology of human relations.[54] That was the goal Henderson
kept in view in his fatigue research, in the Pareto seminar, in Sociology 23,
and to some extent in the Society of Fellows. Pareto, he believed, gave the
science of human relations things it otherwise lacked: a set of objects of
study, a repertory of explanatory strategies, and a method of inquiry.[55]

The Business School contingent in both the Pareto seminar and Sociology
23 was larger even than that of the Society of Fellows. Mayo, the guiding
spirit of the latter phases of the Hawthorne experiments, attended the seminar on Pareto and Methods of Scientific Investigation and gave a case study
on the Hawthorne data to Sociology 23. So too did Fritz Roethlisberger,

who with William J. Dickson of Western Electric supervised tests of worker motivation and carried out a comprehensive program of interviews at the Hawthorne Works. Roethlisberger's work on industrial relations, notably his seminal *Management and the Worker* (1939), co-authored with Dickson, bore the indelible stamp of Henderson's Pareto.[56] Thomas North Whitehead, like Mayo, Dickson, and Roethlisberger, was also involved in the Hawthorne research. He attended the Pareto seminar, presented two lectures in Sociology 23 during its first two years, and manifested the influence of Henderson in his published writings of the period.[57] The connections between the Pareto seminar and the interpretation of the data collected in Cicero, Illinois, were not surprising. In September 1932, as Henderson's Pareto seminar was held for the first time, the records of the Hawthorne experiments were deposited in Morgan Hall at the Business School, the same building where Henderson had his office. Pareto's concept of the social system marked some of the Hawthorne researchers' attempts to interpret the interview records and group test data they had gathered. Henderson himself seems to have encouraged such applications.[58]

Far from being confined to the seminar on Pareto, then, the Harvard Pareto circle was institutionally eclectic: it connected various schools, departments, and research facilities. This characteristic of the circle reflected Henderson's position within the University. He was one of President Lowell's most trusted lieutenants, and was placed in positions of authority over fields far outside the usual domain of a chemist. Nor did Lowell's retirement in 1933 diminish his status. Henderson was the uncle by marriage of Lowell's successor James Bryant Conant.[59] The new president deepened Henderson's involvement in the social sciences.[60] Not only was Henderson allowed to teach his courses in the Department of Sociology, despite his lack of training in the field; he was placed on the Executive Committee of the Harvard Tercentenary Conference that selected speakers for a special interdisciplinary symposium on "Factors Determining Human Behavior."[61] The biochemist moved freely through the weakly professionalized human sciences at Harvard, gathering around him as he went a heterodox band of anthropologists, sociologists, and management scientists.

Henderson was thus a prime mover in Harvard's interstitial academy, a scientist without portfolio. In fact, his activities extended beyond the Pareto network and into the wider interstitial culture of scientific philosophy at

Harvard. In 1939 he participated in the Fifth International Congress of the Unity of Science, which had been organized at Harvard by the former Junior Fellow Willard Van Orman Quine and the physicist Percy Williams Bridgman.[62] The following year, he attended the meetings of the Harvard psychologist Stanley Smith Stevens's Science of Science Discussion Group, which recruited from across the Faculty of Arts and Sciences.[63] These meetings, which we will examine in greater detail in the next two chapters, continued Henderson's engagement with issues of scientific method and epistemology. It would thus be misleading sharply to distinguish the seminar on Pareto, Sociology 23, the Society of Fellows, and the Industrial Hazards Project from Henderson's administrative positions, teaching interests, research activities, and interdisciplinary discussions.

C. C. Langdell and the Harvard Case Method

Armed with this institutional map of the Pareto circle, how are we to explain the connections between the Pareto network and the pursuit of the human sciences in the interstitial academy? What was it that made the ideas of Pareto resonate so powerfully among Harvard's students of human affairs? The manner in which Henderson taught Sociology 23 points to the answer to these questions. Henderson's sociology and theory of science presupposed an understanding of scientific philosophy in which the study of *cases* was central. For many Harvard Paretians, Henderson's use of cases in his account of scientific knowledge had the advantage of bundling teaching, research, and epistemology together in ways that made the human sciences a more legitimate enterprise. In order to grasp this key point, we need to go back to the origins of the case method of teaching in post–Civil War Harvard, and then trace its later development. By filling in this history, we will see that a particular understanding of thinking in cases chimed in with the culture of scientific philosophy that absorbed Henderson and other members of the Pareto circle.

The case method of teaching was introduced into professional education in the Law School of Charles William Eliot. Its champion was Eliot's handpicked dean, Christopher Columbus Langdell, who found in the case method a solution to a set of challenges surrounding the professionalization of the study of law. Some of these challenges centered on academic standards in

American law schools. When Langdell arrived at Harvard in 1870, incoming students at the Harvard Law School were not required to hold a bachelor's degree. Once enrolled, moreover, they could proceed to the bar examination without facing formal examinations or following a systematic curriculum. The prevailing modes of instruction in the Law School were the lecture and textbook recitation, neither of which added a significant measure of rigor to what was already a desultory teaching program. Langdell's institution of the case method, beginning with his own courses on contract law, was one component of a broader set of reforms that included compulsory annual examinations, a new three-year curriculum, and the imposition of stricter standards for entry into the School.[64]

Langdell's turn to the case method was part of a wider movement after the Civil War to rationalize the canons of legal knowledge and the training of lawyers.[65] As would later hold for Henderson's scientific methodology, matters of pedagogy were not side issues for Langdell but central to the problem of knowledge. Where was legal knowledge to be found? On the European continent, the reforms of the Enlightenment had codified large swathes of jurisprudence; in Britain and America, however, the common law tradition had persisted. In place of legal codes, the law in Britain and the United States was embodied in written case decisions.[66] According to the doctrine of *stare decisis*, these adjudged cases formed precedents to which future decisions appealed for their legitimacy: the principles of the law in the Anglo-American tradition were thus to be extracted not from explicit codes but from the accretion of written precedents.[67] During the nineteenth century, however, the combined force of the Continental example and the professionalizing imperative in American higher education led legal educators like Langdell to look for a more durable epistemological foundation for the law than the informalities of *stare decisis*.[68] The movements known as formalism and legal positivism were responses to this demand for the determination of judgments on the basis of universal principles that could be methodically applied in individual cases.

Langdell's justifications for the case method echoed the concerns of the legal positivists, but it is wrong to range him among the formalists.[69] Indisputably, Langdell saw the law as a science and treated adjudged cases as the empirical materials from which general principles were to be extracted. The oft-cited preface to *A Selection of Cases on the Law of Contracts* (1871) outlined his commitment to the notion that knowledge of the law was knowledge not of precedents per se but of "certain principles or doctrines" implicit in those precedents. Langdell further contended that "if these doctrines could be so classified and arranged that each should be found in their proper place, and nowhere else," then the vast archive of written decisions making up the common law could be reduced to a set of formative cases in which the cardinal principles of a given branch of the law were expressed.[70] In this view, the professional study of law involved (a) the classification of general principles and (b) the identification of a set of seminal cases from the annals of the common law. Each activity, in a circular fashion, aided the other: the better one understood the general principles, the more skilled one became at finding the seminal cases; conversely, the more cogent one's set of cases, the easier it was to extract the subtending doctrines. Langdell pushed these claims further in an address to the Harvard Law School Association in 1886. If law was indeed a science, he told his listeners, then it treated the printed record of cases much as botanists treated their plant specimens or as chemists drew conclusions from the results of their experiments: "we have . . . constantly inculcated the idea that the library is the proper workshop of professors and students alike; that it is to us all that the laboratories of the university are to the chemists and physicists, all that the museum of natural history is to the zoologists, all that the botanical garden is to the botanists."[71]

Langdell did not think that cases related to general doctrines as specimens related to the classes or species to which they belonged: they were not mere instances whose sole purpose was to give access to universal truths. According to Langdell, each general doctrine was "a growth, extending in many cases through centuries." The use of cases, given this evolutionary character of doctrines, had a twofold importance: a chronological series of cases relevant to the understanding of a doctrine traced an arc of development, thereby allowing the lawyer-in-training to grasp the doctrine as a "growth," not a fixed form. By revealing in this manner what a doctrine was, a series of cases allowed the neophyte jurist to master the appropriate use of the doctrine in legal judgment by showing how particular crucial decisions "embodied" its application. Understanding the doctrine—as either a student or a scholar of the law—and learning how to apply it in skilled legal judgment were one and the same thing. Teaching, research, and professional judgment converged on the handling of cases.

This adaptability of the case method—its wearing of three hats, as it were—was, in the Harvard of Eliot, Lowell, and Conant, a source of considerable strength. Judging from the record of its use across different schools and departments of the University, it would seem that the case method in teaching, research, and epistemology helped to stabilize otherwise marginal professional ventures. Langdell introduced case teaching in the Law School as part of his attempt to move legal education away from amateurism and toward academic respectability. Next, around 1900, the case method cropped up in the Medical School, where its use was designed to professionalize clinical training. The situation in the Medical School was both similar and strikingly different to that of the Law School. Like their colleagues in the Law School, the champions of pedagogical reform in the Medical School wished to break away from the classical model of rote learning and recitation. When Medical School student W. B. Cannon wrote in support of "The Case System of Teaching Systematic Medicine" in 1900, he was inspired by "the eagerness and zest with which [a roommate studying at the Law School] and his fellow students discussed cases and their implications."[72] This was, on the face of it, an odd source of inspiration. The notion of the case traced its origin to Greek medicine—the doctors had got there first. Nevertheless, the reason why the Langdell's case method provided inspiration was that the "case" which attracted Cannon's attention was not the sort of descriptive, open case described by Hippocrates, but the "printed clinical records" produced in modern clinical assessments—the case file in which examination, diagnosis, and the results of treatment were logged for each individual patient.[73] These clinical cases resembled more closely the written decisions of judges than the descriptive cases of Hippocrates.

Soon after the Graduate School of Business Administration opened its doors in 1908, it too adopted the case method, with similar professionalizing motives in mind. As early as 1915, a member of the Business School faculty declared that the "profession of business cannot be taught from text-books. Actual business problems, as the business executive has to meet and deal with them, are as unlike any purely text-book presentation of them as the sick person calling at the young doctor's office is unlike the 'symptoms' in the medical text-book."[74] Teaching by cases was nurtured at the Business School under the deanship of Wallace Donham, and began in 1919. Donham had studied in the Harvard Law School under Langdell and his disciple James Barr Ames;

he became in due course the foremost defender of the case method in the Business School.[75] Under Donham, Harvard's Bureau of Business Research was transformed from a data collection agency, devoted to the compilation of business statistics, into an organization that collected and published "management precedents" to be used in the training of executives.[76] The fruits of this enterprise were made public in 1925 with the publication of the first volume of the *Harvard Business Reports*.[77] These *Reports* were intended to function pedagogically in much the same way as Langdell's casebooks. Although the series was abandoned by the end of the decade, use of and reflection on the case method continued at the Business School into the 1950s.[78]

Henderson, a product of the Medical School and a leading figure in the Business School, carried the case method into Harvard College when he began to teach Sociology 23. We have already noted how many of the rising sociologists, anthropologists, and management scientists at Harvard presented cases in Henderson's course. Immediately after World War II, President Conant himself, cognizant of his uncle Henderson's emphasis upon case-based learning and conceptual schemes in science, deployed the same technique in teaching the history of science in the General Education program. Only the study of cases, Conant argued, could provide undergraduates not specializing in science with the expert's sense of the "tactics and strategies" by which scientists generated knowledge.[79] Finally, while a member of the Society of Fellows and then a professor in the Gen. Ed. program, Thomas Kuhn made reasoning by cases or "paradigms" the key feature of "normal" scientific inquiry and training.

At each stage of the career of the case method at Harvard, we see the method adopted in interstitial enterprises: preprofessional "academic" subjects like law or business; uncertainly placed fields like sociology and history of science; and cross-departmental programs like General Education. The case method even gained traction in the one place one might have expected it to be unnecessary, the Medical School, precisely at the point when the teaching of clinical medicine had fallen into disrepute. One explanation for this striking institutional profile of the case method brings us back to the blurring of pedagogy, inquiry, and epistemology that we first encountered in Langdell's view of legal training. The proponents of thinking in cases were able to place their own practices of research and teaching into epistemic roles. In this way, claims to knowledge in these fields were buttressed by the research practices that produced those claims (i.e., case-based empirical generalizations) as well as by the mode in which the same claims were taught (the case method of teaching). Although there is no evidence for the view that simply adopting the case method set one on the road to scientific respectability, it surely aided the attempt to legitimize otherwise marginal or preprofessional disciplines, for Langdell's model of legitimate knowledge-making was just what legal scholars *did* in libraries, and, also, how lawyers-in-training *learned* in the classroom.

Scientific Philosophy and the Hendersonian Case

Having seen that the case method aided professionalizing projects at Harvard for the better part of the century, we need to explore further how and why it did so for Henderson and the Pareto circle. Crucial to Henderson's institutional success was the conjoining of the case-based epistemology popular at Harvard since Langdell to newer currents in scientific philosophy. Even as an undergraduate in Harvard College, Henderson felt that traditional philosophy had less insight to offer than its champions supposed. In a series of autobiographical reflections recorded in 1936, Henderson remembered recoiling immediately from a popular course on ethics taught by George Herbert Palmer, a stalwart of the Department of Philosophy. After hearing Palmer's first lecture, Henderson "went straight from the lecture room to the college office and dropped the course." "The lecture seemed to me meaningless," he recalled; "in one way or another I did reach a conclusion that Palmer was thoroughly unscientific and therefore, to me untrustworthy as a guide to objective thinking." This act was "the first manifestation of a certain kind of intelligence I can remember in myself." In retrospect, that intelligence belonged to the skilled clinician, who valued practical effects and intuitive familiarity with concrete affairs over grand abstractions and "the tricks of the word-game."[80]

The young Henderson felt able to turn from philosophy because of what he later characterized as an unreflective commitment to the robustness of scientific knowledge. During his time at Harvard College, at the Harvard Medical School, and then at the University of Strasbourg conducting post-doctoral research, Henderson was, by his own admission, a "naïve realist": "throughout my student days and for a good many years longer I felt no doubt about the general lines of the universe. It was real, it was certainly Newtonian, and that was good enough for me." Moreover, Henderson "did not regard this conclusion as philosophical at all." It was simply the viewpoint of someone raised before the transcendental status of the a priori was called into question: "I am just old enough to have experienced the old comfortable assurance that the world of science is stable, true, and real; that not only the facts and uniformities but also the theories and conceptual schemes are on the whole such that they will endure with nothing more than improvements, refinements, and occasional corrections, and that they will be marred only by an occasional minor catastrophe."[81]

Soon after his return to Harvard in 1904, when he took up a position at the Medical School, Henderson's views began to change. Scientific philosophy and scientific practice both played their part. Sometime after completing his medical degree, and most likely during his stay in Europe, Henderson read Karl Pearson's *The Grammar of Science* (1892) and Poincaré's *The Value of Science* (1905) without yet fully grasping their instrumental, practitioner's view of scientific concepts. At the same time, after two years of toil in his Strasbourg laboratory, Henderson "had come to feel and . . . to see that biological chemistry and the sciences in general are what I should now be inclined to call social processes rather than abstraction." He soon became "much occupied with unmethodical reading and reflection about scientific method and philosophy." The works of Bacon, Descartes, and Kant grabbed his attention, and he continued his engagement with Poincaré, Pearson, and also Ernst Mach.[82] These efforts became more formal in 1908, when Henderson began to attend philosopher Josiah Royce's Seminar on Logic, which "might better have been [called] 'everything else and logic', for topics discussed ranged over many fields of knowledge and speculation and the discussion led to endless and extraordinarily varied consideration."[83]

Henderson's involvement in Royce's seminar was not a concession to the power of philosophy, but rather a symptom of how reflection on epistemology and the methodology of science were, in the early years of the twentieth century, escaping traditional philosophical boundaries. If Henderson treated the Seminar on Logic as a "separate compartment of my life" and "an amusement," it undoubtedly gave a spur to his ruminations on scientific method and the epistemic significance of the processes of inquiry.[84] As Henderson admitted, the Seminar offered two "concrete advantages, which

even came within the range of my professional interests": "a better understanding of the history of science" and an introduction to "at least the elements of symbolic logic . . . as well as the general question of postulates and
postulation, and there was some talk of the work of Peano, of Dedekind, of
Cantor, and of Whitehead and Russell."[85] These latter figures were key contributors to debates held around the turn of the twentieth century about the
ontological and epistemological foundations of mathematics. Under such
influences, Henderson's realism rapidly gave way to an interest in the actual
practices, techniques, methodology, and psychological demands of science.
In 1911, feeling stymied at the Medical School, Henderson secured Lowell's
permission to teach a course on the history of science, a change of direction
he credited in part to his "association with Royce, and my long, if irregular
membership in his seminar on logic." Henderson spent much of the next two
years preparing History of Science 1, which led him to a sustained engagement with Ernst Mach's *The Science of Mechanics* (1883).[86]

Although Henderson's move toward history was an avocation of sorts, it
further extended his interests in scientific philosophy. Indeed, the tradition
of scientific philosophy encompassed a set of concerns about the social study
of science. The great Victorian philosopher-scientists Ernst Mach, Henri
Poincaré, and Pierre Duhem welcomed historical investigations of science,
but—and this was a crucial caveat—only insofar as history was conceived as
a medium for principles of scientific reasoning. Mach, for instance, opened
his *Science of Mechanics* with the following bold claim: "The history of the
development of mechanics is quite indispensible to a full comprehension of
the science in its present condition. It also affords a simple and instructive
example of the processes by which natural science generally is developed."[87]
For Mach, history was valuable only insofar as it could be used to explicate
scientific concepts and to show how the natural sciences had emerged over
time as part of the more general evolution of the human race.[88] As far as
Mach was concerned, giving an account of the genesis of a concept was
a means of clearing away the metaphysical baggage that had accumulated
around it during the course of actual social history—thereby allowing its
transhistorical, logical structure to stand out in full relief. Pierre Duhem's
The Aim and Structure of Physical Theory (1914) likewise invoked the history
of physics in order to lay bare its formal conceptual structure and theoretical mutability.[89]

Some of these commitments carried over into Henderson's engagement with the history of science, albeit in a different pedagogical context.
Teaching in the history of science at Harvard had begun in earnest in the
early 1890s, when the chemist Theodore William Richards (Henderson's
brother-in-law) began to deliver a regular series of lectures on the history of
chemistry.[90] This was part of a national trend, with courses in the history of
chemistry, biology, and the physical sciences established from 1887 onwards
at the Massachusetts Institute of Technology, Stanford, the University of
Chicago, the University of Pennsylvania, the University of California at
Berkeley, Cornell, Johns Hopkins, Michigan, Illinois, Yale, North Carolina,
and Northwestern.[91] Such courses had an ideological purpose. Around this
time, American institutions began producing their first generation of doctors of philosophy—a cohort that included T. W. Richards. College courses
on the history of science were designed to lend authority to the rapid diffusion of scientific research under the auspices of the university and its
wealthy patrons. In this early phase of the field there was little concern for
"the history of science as an academic discipline or of the problems to be
dealt with in the application of historical methods to the natural sciences."[92]
Much like the philosophical historiography of science of Mach and Duhem,
the "history" in American historiography of science was a vehicle for more
than purely scholarly ends. But in these early years of American history of
science, pedagogy, not epistemology, was the overriding concern.

Henderson brought epistemology and the theory of research practice
back into the picture in his teaching of the history of science. In History
of Science 1, he had established the first course at Harvard devoted to the
history of *science,* rather than to a particular science. Henderson asked his
undergraduate tutees in History of Science 1 to grasp the variety of scientific methods involved in the solution of empirical problems. Using recently
published original sources from Harvard Classics and Everyman editions
of canonical scientific texts, Henderson assigned excerpts from Aristotle,
Hippocrates, and Galileo.[93] At the outset of the course, he had students reading "Harvey on the circulation of the blood, as a prime example of scientific
thinking and discovery."[94] Henderson well understood that his approach to
these texts did not rely on any deep understanding of, or aspiration to, the
methods of historical inquiry. He was long convinced that the "cultivation of
the history of science" as it stood in the 1930s "may profitably be undertaken

by one who lacks a training in the professional methods of the historian." In what could have become a motto for those Harvard historians of science who followed in his footsteps, Henderson was confident that "acquaintance with the sciences, with the literature of scientific method, and with other topics may afford advantages that largely offset the disadvantages that arise from this lack technical proficiency" in the historian's craft.[95]

The purpose of Henderson's course was not to elicit new historical knowledge or lay the foundation stones of a new domain of historical scholarship; nor, on the other hand, was it designed to contribute to the toolkit of the research scientist. Rather, in keeping with the mission of fin-de-siècle historians of science, Henderson's intention was to cultivate an appreciation of the skilled and professional nature of scientific reasoning, presumably because in a modern social order this would be required of the well-rounded men of affairs whom Harvard could be expected to produce. Notably, few among the small number of those who attended Henderson's classes were scientists-in-training, as Henderson himself was well aware. Moreover, the demands of the course were not as stringent as those of the more exacting disciplines—History of Science 1, as Bernard Cohen later recalled, was a well-known "easy A."[96] Hence the most notable products of the early years of the course were men of letters such as Bernard DeVoto and John Dos Passos.[97] Later auditors included the instigators of novel research programs in the human sciences: Cohen, Robert Merton, and B. F. Skinner.[98]

By 1914, Henderson's cluster of concerns in the nature of scientific method encouraged him to establish with Royce "an informal club, to be made up of some twenty of our colleagues in the Faculty of Arts and Sciences and the Faculty of the Medical School, and including those interested in the more general problems of a scientific nature that were then under discussion."[99] Known first as the New Club, and later, after Royce's death in 1916, the Royce Club, Henderson's elite interfaculty discussion circle met outside of Cambridge, in the Harvard Club of Boston, with Henderson presiding as secretary. It lasted for roughly a decade. The Royce Club, said Henderson, "helped me to find out how many of my most intelligent colleagues thought about general problems of scientific method and philosophy of science, and that was something I very much wanted to know at the time."[100] Both the constitution and the physical location of the Royce Club indicated the direction in which Henderson's thought and academic practice were heading during the interwar years. Questions of epistemology and methodology had become, for Henderson, pressing problems within scientific practice, but outside of the narrow parameters of philosophy and, indeed, the scientific disciplines themselves.

It was in this context that Henderson came to embrace the notion of the case as the key to his epistemological concerns. For Henderson, the handling of cases encapsulated the techniques of the skilled research scientist and provided a model for training historians and neophytes in an understanding of the research process. This perspective combined elements of scientific philosophy with the pedagogical vision outlined by Langdell. But it also drew on classical understandings of the case dating back to the casebooks of Hippocrates. In his history of science course, Henderson celebrated the insights provided by the Hippocratic case, and his other offerings in the subject involved selected examples of good and bad scientific reasoning from the ancients down to the writings of Mach.[101] Henderson's unification of inquiry, epistemology, and pedagogy proved especially potent in the interstitial academy.

Henderson added to Langdell's formula a subtle equation of cases and systems. This equation took shape in Henderson's appeal to the Hippocratic method—an appeal that characterized his work in history and the methodology of science. Although indebted to Pareto and Poincaré, Henderson's epistemology placed both men in a tradition of scientific methodology that originated with Hippocrates. The primary lesson Henderson drew from Hippocrates was that all sound reasoning in science started with actual sets of recorded empirical circumstances. "In the beginning," he told his audience in Sociology 23, "are the cases."[102] The classical Hippocratic case history, Henderson insisted, possessed two singular virtues. First, it consisted of "bare observations of bare facts, uncolored by theory or presupposition and condensed to the very limit of possible condensation."[103] All relevant details of the patient's condition were to be recorded, however obvious they might be. Second, cataloguing the progress of an illness in this manner opened the way to identifying uniformities as the sickness developed. Cases provided the investigator with a wealth of concrete details through which they could enhance their familiarity with the phenomena at hand. Henderson underscored the inductive sources of general categories and classifications: they were to be derived only by recording "the recurrence again and again in different cases . . . of single events or of the uniformity observed in a single case."[104] Classifications based on uniformities identified in concrete cases were, in turn, to lead the investigator to provisional conceptual schemes, whose purpose was to employ empirical generalizations in diagnosis. All three levels of investigation—the recording of the salient details of a particular case, the classification of the phenomena, and the theoretical deployment of the classifications in diagnosis—would be most effective when the diagnostician acquired, through immersion in cases, "intuitive familiarity" with the relevant scientific functions: recording, classifying, and theorizing.[105]

The diligent handling of cases, according to Henderson, was the sine qua non of effective scientific thinking. It followed that particular conceptual schemes, such as that of the system of mutually dependent variables, were to be considered working hypotheses whose value stemmed exclusively from their utility in placing observed uniformities and battle-tested generalizations in a fruitful explanatory framework. But Henderson's agnosticism about conceptual schemes was not quite as strict as it appeared. In practice, he endowed the conceptual scheme of the system with special significance, and in doing so made the notion of the case as much a product of his conception of science as its first principle: "case" and "system," that is to say, were mutually defining concepts in Henderson's epistemology.

Model Societies

In his opening lectures for Sociology 23, Henderson came close to suggesting that the concept of a system was the most general and effective of all empirical conceptual schemes, the holy grail of scientific conceptualization. He did this indirectly, through a discussion of the concept of equilibrium. The "widest of all generalizations in the work of Hippocrates," Henderson told his listeners, was "that as a rule sick people recover without treatment."[106] Systems tended to toward equilibrium: a shift in the status of one element would produce corresponding effects in the others, leading to the restoration of the status quo. On this point, Henderson found Hippocrates and Pareto in agreement. The concept of equilibrium was therefore a cardinal component of scientific knowledge:

> This definition [of equilibrium] applies to many phenomena and processes, both static and dynamic. It applies not only in the fields of pathology and sociology but very generally in the description of almost

all kinds of phenomena and processes. It is indeed a statement of one of the most general aspects of our experience, a recognition of one of the commonest aspects of things and events. For example: (1) a ball which is in the cup, and which is struck a blow that is not too hard, will return to its original position; (2) a candle flame which is deflected by a draft that is not too strong will resume its original form; (3) a trout brook that is "fished out" will, if carefully protected, regain its former population of fish; and (4) to take a Hippocratic instance, an infant after a disease that is not too severe will gain in weight until that weight is reached which is approximately what would have been reached if there had been no sickness.[107]

If the notion of equilibrium was a generalization derived from careful observation of cases in each of these fields of knowledge, then the conceptual scheme of the system was needed to explain the efficacy of the generalization. Equilibrium was always the property of a system, not of a single, isolated element. For Henderson, the notion of equilibrium and the conceptual scheme of the system were the telos of scientific inquiry. At the very least, he was prepared to short circuit the careful description and cataloguing of human affairs (an activity that occupied none of his own research time) in order to reach the commanding heights of equilibrium theory:

So in February 1937 the people of Louisville, driven away by a flood, returned to their homes when the flood receded; and a few weeks later life was going on with little change. So within a decade the traces of the earthquake and fire in San Francisco could hardly be seen, or the devastation of the war of 1914–1918 along the battlefront in Northern France. In such cases the "forces" that tended to produce "the conditions that would have existed if the modification had not been impressed" are what we describe as habits, sentiments, and economic interests.[108]

In effect, Henderson assumed that a system was what needed to be identified in social science. As such, all of Henderson's cases were descriptions of systems, and descriptions of social systems were—for Henderson himself—exemplary case studies. Members of the Pareto circle learned to exploit this circular logic. Henderson's social systems theory offered an interpretive framework in which Harvard's marginal human scientists could fashion cases and thereby attempt to gain institutional recognition for their research projects.

An inspection of the scholarly fruits of Sociology 23 brings this conclusion home. Several case studies of Paretian social systems for Sociology 23 were written up and published.[109] In an early prospectus for the course, Henderson stated that it had been planned "by a number of persons . . . who are concerned to see to it that something like the Hippocratic method shall be employed in teaching the most general aspects of social phenomena." A little further on, Henderson made explicit the status of his scientific epistemology as a form of pedagogy: he would "proceed . . . according to the method that is implicit in the . . . works of Hippocrates and that has been followed in many circumstances by others since the fourth century BC, though not often in the field of the social sciences." Henderson further noted that "every person presenting a case will be asked to consider how far the interaction of sentiments, prejudices, and passions are the principal features in any concrete situation."[110]

Under Henderson's influence, and within the parameters of the interstitial academy, the production of case studies of social systems became a cottage industry. George Homans's *English Villagers of the Thirteenth Century* (1942) was a particularly effective mobilization of the Hendersonian conceptual scheme. Ostensibly a historical study, *English Villagers* was conceived as an essay in social anthropology and systems theory conducted by means of archival research. Homans used the manorial records and court rolls of thirteenth-century England to "describe as a whole a social order of the past."[111] This was, he reported in his memoirs, an application of "the ideas of modern social anthropology to an historical society."[112] "Some of the most fruitful work in the social sciences" he noted in *English Villagers,* "has been done by anthropologists studying so-called primitive societies, that is, societies that are sufficiently small, simple, and isolated to make possible consideration of every aspect of each society in relation to all the other aspects."[113] The village folk of lowland England were Homans's Andaman and Trobriand Islanders. Equally important was Homans's implicit claim about typicality: the thirteenth-century villagers were important because they could be treated as a case study in the general phenomenon of social order.

We can gauge the extent of Henderson's and the Pareto circle's influence on Homans's early work if we compare it to kindred community surveys

produced in the United States during the interwar decades. *Middletown,*
a study of Muncie, Indiana, published by Robert and Helen Lynd in 1929,
along with the new Gallup polls measuring public opinion, were popular
examples of this new literature. What marked out these new surveys was
their preoccupation not with the deviant or the marginal, as had been the
case with the major social surveys of the Victorian era, but with the average
and the typical.[114] The Lynds, for example, employed the analytical armory
of statistics and social anthropology to investigate the common cultural
habits and folkways of a "normal" American community. Homans, in turn,
sought nothing less than an "anatomy" of a medieval peasant society—its
customs, folkways, and so forth. The chief novelty of *English Villagers* within
the interwar social sciences was Homans's invocation of the logic of cases
and systems. His immersion in manuscript sources implicitly chimed in
with Henderson's rhetoric of "intimate familiarity."[115] More important still
was Homans's appropriation of the conceptual scheme of the social system. He took certain terms directly from Henderson's Pareto, in particular "sentiments," "non-logical actions," and "mutual dependence."[116] But he
put them to work in a broadly functionalist account of the customs of the
English villagers. In doing so, he made his subjects into a representative case
of a social system.

Homans's insistence on working with "cases" of social systems continued
beyond the publication of *English Villagers.* Although it was his behaviorist account of the foundations of social organization that would ultimately
define his reputation as a sociologist, Homans spent the first two decades of
his academic career attempting to draw a general theory of social systems
from a carefully constructed set of exemplary cases.[117] Adumbrated in a pilot
essay in the late 1940s, Homans's interest in the universal properties shared
among discrete communities culminated in *The Human Group* (1950).[118] The
book had two purposes, both of which betrayed the influence of Henderson
and the Harvard Pareto network. First, Homans sought to outline and
put to the test a "conceptual scheme" for the study of his titular "human
groups." Not only was this scheme rooted in Hendersonian-Paretian ideas
about sentiments and other non-logical factors in social behavior; many of
Homans's representative cases had been common currency in the Pareto
circle. Among the model societies Homans assessed were the assembly lines
of the Hawthorne experiments and William Foote Whyte's North End of

Street Corner Society (1943). *The Human Group* was therefore a testament to the enduring influence of the social thought instigated by Henderson. The second purpose of Homans's book further confirmed its Hendersonian heritage. As Homans saw it, the purpose of his case studies was to provide "drills" for social-theorists-in-training. *The Human Group* had a pedagogical mission not dissimilar to Sociology 23: it was designed to show sociologists how to apply a conceptual scheme in a given range of carefully selected cases, the better to facilitate a general approach to understanding the principles of social order.[119]

The case studies on which Homans relied were themselves model social systems that emerged from within the Pareto network. A Depression-era survey akin to the Lynds' *Middletown,* W. Lloyd Warner's *Yankee City* studies, published in five volumes between 1941 and 1959, took shape while Warner held a joint appointment in the Department of Anthropology and the Business School. His research assistants on the project included two members of Henderson's Pareto seminar, Eliot Chapple and Conrad Arensberg, each of whom also presented case studies in Sociology 23. Arensberg's *Family and Community in Ireland* (1940) was in many ways akin to Homans's combination of history and anthropology in *English Villagers;* Homans, in turn, devoted chapters of *The Human Group* to the Warner and the Arensberg studies. Chapple, by contrast, pushed the systems model to a scientific extreme in his creation of the "interaction chronograph," a modeling tool designed to represent in a purely formal and uninterpreted manner patterns of social interaction in small groups. William Foote Whyte followed the approaches of the Lynds and of Henderson in *Street Corner Society,* a book on which Whyte began work as a Junior Fellow in the Society of Fellows during the 1930s. Whyte's anonymization of Boston's North End (which he referred to as "Cornerville") was motivated in part by his desire to screen out particularities so as to isolate the elements of a total social system.[120] This was also true of the anthropologist Clyde Kluckhohn's treatment of the Ramah Navajo, another of the Sociology 23 case studies later published. Kluckhohn was explicit that the combination of reservation order and the sparse population of the Rimrock region made the Navajo a physically realized "ideal type" of a social system. The Rimrock Navajo, as Kluckhohn presented them, offered the perfect case study for a theoretical exploration of the functioning of non-logical elements in the constitution of a social order.[121]

The most prodigious source of model societies at 1930s Harvard was the Hawthorne data. Early in the life of the Pareto seminar, Roethlisberger, Dickson, and Whitehead embraced Henderson's social systems theory as offering the best framework for understanding the problem of human relations. In *Management and the Worker* (1939), Roethlisberger and Dickson applied Henderson's concepts. Industrial organizations, they averred, had to be understood as "social systems." Citing Henderson's *Pareto's General Sociology*, they defined a "system" as "something which must be considered as a whole because each part bears a relation of interdependence to every other part."[122] They paid particular attention to the "logic of sentiments" within the system.[123] The task of personnel management or "human resources" was to ensure equilibrium among the various elements of the organization. As Roethlisberger and Dickson presented their data, the experimental situations created by the Hawthorne investigators in the Relay Assembly Test Room, and later in the Bank Wiring Observation Room, embodied "representative" instances of social systems.

Some of these studies would go on to become classics of American social science. They also helped to stabilize the careers and projects of their authors. Homans was given tenure in the Department of Sociology after World War II; Roethlisberger, an institutional wanderer, became a leading professor in the Business School; Kluckhohn was eventually made Director of the Russian Research Center; Brinton became Henderson's successor as chairman of the Society of Fellows. The provision of Paretian case studies was one way in which the human sciences were institutionalized at mid-twentieth-century Harvard. However, the production of cases did not professionalize the human sciences as successfully as it had the study of law. Instead, an offshoot of the Pareto network soon became the focal point for the new social sciences at Harvard: the Department of Social Relations. Before we address the history of Social Relations, however, we must explore two further projects in which scientific practices were tied to epistemological concerns within the interstitial academy.

3

What Do the Science-Makers Do?

Migrations of Operationism

Notably underrepresented in the deliberations of the Pareto circle were Harvard's psychologists. When Henderson's seminar commenced in 1932, the institutional status of psychology at Harvard was just as uncertain as that of sociology, anthropology, and business administration. Since the 1870s, when William James established in Cambridge the nation's first psychological laboratory, psychology at Harvard had existed in "forced cohabitation" with philosophy.[1] Very much the junior partner in this alliance, psychology would not attain independent departmental standing until 1934. Yet, marginal figures though they were, Harvard psychologists of the interwar decades—especially Edwin G. Boring, Karl Lashley, B. F. Skinner, and Stanley Smith Stevens—evinced no interest in either the Paretian theory of social systems or the case method.[2] Only Henry A. Murray, the untenured director of the Harvard Psychological Clinic, took part in his friend Henderson's courses, but Murray's intellectual sympathies lay squarely with Jung and psychoanalysis, not with Pareto.[3] Ongoing research at the Fatigue Laboratory and at the Hawthorne Works was certainly pertinent to the biological and sociological wings of professional psychology, but neither venture provided a direct connection between psychologists and the Pareto circle. That the Hawthorne experiments were on their way to becoming "a creation myth in *social* psychology" was unlikely to please Harvard's experimental psychologists, who were committed to laboratory-based measurement techniques in the tradition of Wilhelm Wundt and E. B. Titchener.[4] For his part, L. J. Henderson insisted that Pareto's key notion of sentiment "implies nothing concerning psychological processes." Steering shy of the murky deeps of psychology, and especially of psychoanalysis, Henderson regarded sentiments as necessary "assumptions" or Poincaré-style conventions within Pareto's scheme.[5]

Why this mutual indifference? The answer is that Harvard's psychologists, especially those who worked in the experimental branch of the discipline, had developed their own "indigenous" epistemologies, which answered to the organizational context in which psychological investigations were conducted at interwar Harvard.[6] The interstitial existence of Harvard's experimental psychologists engendered a distinctive form of epistemological self-consciousness. Instead of thinking in terms of systems and cases, Boring, Stevens, and Skinner articulated a position soon to be baptized "operationism." The foundations of this perspective on meaning, method, and legitimate knowledge were laid down by the physicist and Harvard faculty member Percy Bridgman in his 1927 treatise *The Logic of Modern Physics*. Bridgman's directives were intended to make plain the meaning of concepts in the experimental natural sciences. Within a decade of the appearance of Bridgman's *Logic,* however, the operational stance was being adopted with increasing ardor by experimental psychologists, many of whom taught, or were trained, at Harvard.

Like the Pareto vogue, then, the story of operationism in American psychology is in large part a Harvard story. The prime movers in the dissemination of operationism into psychology had ties to Cambridge: the émigré philosopher of science Herbert Feigl introduced Boring to Bridgman's ideas during postdoctoral studies at Harvard in the early 1930s; Boring's students and protégés B. F. Skinner and Stanley Smith Stevens adapted operational analysis to their own ends soon after, while another leading advocate of the operational stance, Edward Tolman, gained his PhD at Harvard under the behaviorist philosopher Edwin Bissell Holt. Other contributors to professional debates about operationism, notably Carroll Pratt, also traced their intellectual roots back to Harvard Yard. Many non-psychologists who invoked Bridgman's dictums in the 1930s and 1940s likewise had ties to Harvard. Finally, a series of major exchanges of the 1940s and early 1950s, which capped a cross-disciplinary engagement with operationism across the human sciences, emerged from a discourse about the "science of science" that took shape in Harvard's interstitial academy. Operationism became, in time, part of a flourishing cross-disciplinary intellectual culture that spread across the organizational fault lines of the university.

This chapter explains why Harvard should have expedited the migration of operationism into psychology. The intellectual and institutional sources of the rise of operationism in psychology are similar to those that produced the Pareto circle. As with Henderson's network, we find in operationism an epistemological and methodological doctrine that captured the attention of practitioners of weakly institutionalized human sciences. And, as with Henderson's case-based conceptual scheme for social theory, we can identify in operationism an epistemological conceit that placed the practices of research—in this case, the procedures of the experimental psychologist— into epistemic roles. To insist that scientific concepts be synonymous with the operations through which they were applied was to say that the scientific procedures of the laboratory were the *content* of empirical concepts. What made this blurring of the line between epistemology and the techniques of inquiry possible was, once again, the license of the tradition of scientific philosophy. Having assessed how these features of the Harvard complex came together in Henderson's theory of social systems, our purpose here is to examine a parallel case.

Technoscience: Landscapes of Interwar Psychology

An account of the reception and transformation of operationism by Harvard's experimental psychologists requires, as a first step, a sense of the state of American psychology during the interwar years. National and local developments are equally important in grasping how and why operationism resonated in the discipline. At the national level, the 1930s were a period of expansion and fragmentation. Even at this relatively late stage of its emergence, professional psychology in the United States was not a strictly defined scientific field but a loose assemblage of technoscientific practices.[7] The notion of technoscience, Bruno Latour has remarked, highlights the sheer range of social actors involved in the creation, legitimization, and dissemination of scientific knowledge; it is often invoked in order to undermine the distinction between "pure" science and its technological "application."[8] From its inception as a professional field, psychology in the United States was a sprawling technoscientific enterprise in Latour's sense: when the so-called new psychology took shape during the last third of the nineteenth century, it combined laboratory experimentation in physiological psychology—pioneered by Wilhelm Wundt at the University of Leipzig—with an enduring engagement with "philosophy, medicine, positivistic social reform, educational interests, and those secularized 'social science' aspects of academic moralism concerned with the development of an appropriate political and social order."[9]

In practice this meant that, at the turn of the twentieth century, American psychology organized itself around two poles. On one hand, the emergence of the new psychology was marked by the proliferation of university-based laboratories modeled on Wundt's Institute of Experimental Psychology. After James had pieced together the Psychological Laboratory at Harvard in 1876, and his student Granville Stanley Hall had followed in his footsteps (after postdoctoral work with Wundt) at Clark University in 1883, the 1890s saw the consolidation of experimental psychology under Hugo Münsterberg at Harvard, Edmund Clark Sanford at Clark, Edward Wheeler Scripture at Yale, and, most influentially, Edward Bradford Titchener at Cornell.[10] From these experimental programs there emerged, before World War I, the earliest doctrinal "schools" in American psychology: structuralism, associated above all with Titchener; functionalism, which derived from James's adaptationist view of consciousness and took root in the work of John Dewey and James R. Angell at the University of Chicago; and behaviorism, the controversial view put forward by the Johns Hopkins animal psychologist John Broadus Watson.[11]

On the other hand, the enduring moral and technocratic commitments of the new psychology quickly produced an expansive domain of applied psychology, which was to shape the profession as deeply as the academic instauration of the psychological laboratory. James, Hall, Münsterberg, Watson, and other founding figures encouraged the application of psychological knowledge to the social challenges of the day.[12] In 1921, an influential group of experimentalists banded together to form the Psychological Corporation, a consulting organization dedicated to the promotion of psychological expertise.[13] The public consumption of applied psychology had in fact begun two decades earlier. During the progressive era, the prophets of social efficiency had latched onto intelligence tests and statistical studies of the distribution of mental capacities among populations as a means of measuring and improving the performance of education and labor. Henry Herbert Godard, Lewis Terman, and Edward L. Thorndike adapted intelligence scales developed in France and used them to launch a mental testing movement. By the early years of the twentieth century the testing movement had unleashed an "information fever" in American public life.[14] A wave of school surveys beginning in 1909 was one of the earliest expressions of the desire for quantitative data on pupil and institutional performance.[15] The vision of a statistics-based social policy, to which the testing movement had given rise, was enthusiastically voiced in *The Wisconsin Idea* (1912), a manifesto in which the Progressive Party's candidate for president, Theodore Roosevelt, insisted that "measurable results" were the proper criterion of "social progress."[16] Intelligence and aptitude tests came into their own, however, only during World War I, when the Army adopted the Stanford-Binet scale and various other measures in selecting and managing cannon fodder for the Western Front.[17] At roughly the same time, and with growing intensity during the 1920s and 1930s, psychological research into motivation, energy, and fatigue was sponsored by corporate interests.[18] Psychology in America was, from its earliest professional incarnation, never only a "pure" discipline whose knowledge was gradually "applied": it was a technoscientific enterprise that encompassed experimental research into the physiological basis of consciousness, at one end of the spectrum, and personnel management, at the other.

Over the course of the 1920s and into the 1930s, however, the multidimensional presence of psychology in American intellectual and public life began to pose severe methodological, epistemological, and professional problems for academic psychologists. In addition to the divisions between "pure" and "applied" psychology, there were intramural fissures between experimentalists, clinicians, and social or "personality" psychologists, as well as ongoing turf wars with philosophy, biology, and medicine. Experimentalists were set apart from one another by the investigative traditions established in the university laboratories: psychophysics, neurophysiology, and animal psychology possessed footholds in different labs and were frequently justified by appeals to conflicting ideologies of method and scientific epistemology. At the same time that psychologists sought to mark themselves off from one another, they also had to defend their expertise not only from the claims of faith-healers, spiritualists, and assorted quacks, but also from laypersons who, prima facie, had no reason to suppose that an academician could claim more authoritative knowledge about the contents their minds than their own folk wisdom.[19] Two developments of the 1920s made these conflicts especially acute. The first was the rise of psychoanalysis.[20] By invoking a set of prurient driving forces in mental life—sex, death, desire—and promising therapeutic control of common neuroses and anxieties, Freud's theory of the unconscious threatened to upend the carefully constructed scientific authority of university-based psychologists. As psychoanalysis made its way into popular magazines and psychiatric medicine during the 1910s and 1920s, professional psychologists sought to repudiate Freud's picture of the mind and reassert their claim to objective knowledge of mental life. This move instigated a contest and gradual accommodation with psychoanalysis on the part of academic psychologists that ran through the middle decades of the twentieth century.[21] The second key development involved a combination of professional expansion and the contraction of the academic job market during the Great Depression. In the two decades after World War I, the number of doctorates awarded in psychology shot up. Between 1910 and 1914, 111 PhDs were awarded in psychology in the United States; in the five years after 1929, roughly five times that number were handed out by American universities. As employment opportunities dried up in the academy during the Depression, the glut of psychology PhDs began to move into applied sectors, a migration that consolidated the testing movement in public institutions.[22]

By the time Bridgman's "operational viewpoint" was being adopted by psychologists in the early 1930s, these centrifugal pressures on the practice of psychology had created a balkanized discipline. To be sure, "pluralism" was a feature of other human sciences in this period, notably economics and sociology.[23] But, by 1930, psychologists seemed especially conscious of the division of their field into competing "schools."[24] The tumult of the 1920s and early 1930s had produced a range of doctrines, traditions, and methodological visions that extended well beyond the parameters of the new psychology. Columbia University's Robert Woodworth, in his 1931 textbook *Contemporary Schools of Psychology,* listed the following positions in the field: introspective psychology (in the tradition of Wundt and Titchener), behaviorism, Gestalt psychology, psychoanalysis, and purposivism (latter-day functionalism).[25] The presence of the Gestaltists signified another European school, in addition to psychoanalysis, that had moved into the United States during the 1920s.[26] By 1933, Edna Heidbredder could identify "seven psychologies" vying for primacy.[27] Even existing traditions were revised in the 1930s, with Watsonian behaviorism giving way to the "neo-behaviorism" of Yale's Clark L. Hull, Berkeley's Edward Tolman, and

Harvard's (then Minnesota's) B. F. Skinner. Diversity did not necessarily entail the absence of synthetic ambitions. On the contrary, Yale's Institute of Human Relations, which coalesced around Hull's neo-behaviorist program of a deductive theory of behavior, evidenced the vitality of systematic psychological research programs in the 1930s.[28] Nonetheless, the challenge for individual psychologists, laboratories, departments, and subfields was to mark out and defend a legitimate domain of expert practices in a densely populated technoscientific landscape.

Harvard's Psychological Complex

At the local level, these struggles produced strikingly different results, depending on the balance of power among research programs. Yale had its Institute of Human Relations, but functionalism persisted at Chicago, while Columbia became notable for its strength in dynamic psychology.[29] Harvard, meanwhile, was a peculiar case. The defining feature of Harvard's role in the university movement was that it was flexible enough to absorb innovations in research, administration, and pedagogy while remaining conservative in its policies, organization, and hiring practices; it was this characteristic of nineteenth-century Harvard that produced, by the interwar years, the host of marginal, weakly institutionalized disciplines and research projects that defined the Harvard complex. Few fields followed this line of development more closely than psychology. When James Bryant Conant assumed the presidency in 1933, psychology at Harvard was dominated by the experimentalists, yet experimental psychology was under severe pressure from three sources. It was undermined by the philosophers who controlled the department in which it was housed; it was challenged by a cohort of clinical psychologists who embraced heterodox currents in abnormal, personality, and social psychology; and it was divided increasingly within itself by various institutional and methodological differences. It was this precarious institutional position that encouraged a number of Harvard's experimental psychologists turn to operationism during the 1930s.

In its familiar dual role of laggard and pioneer, Harvard could lay claim to the status of founder in the history of American psychology. After encountering the research of Wundt and Helmholtz in Germany, William James offered in 1875 the first course in the United States on physiological psychology and psychophysics. Around the same time, he also established the nation's first psychological laboratory, albeit one limited in its early years to a small space under the staircase of the Agassiz Museum and to a collection of crude instruments for classroom demonstrations that included "a metronome, a device for whirling a frog, a horopter chart and one or two bits of apparatus."[30] Experimental psychology, with a distinctly biological edge, was thereby confirmed at Harvard many years before figures such as Titchener and Sanford took up their posts; but, crucially, James's experimental research program was institutionalized at Harvard under the auspices of philosophy, where it would remain (with increasing rancor on the part of the experimentalists) until the mid-1930s. In the short term, Harvard psychology undoubtedly prospered under the philosophical aegis. William James trained the first American PhDs in psychology, including a young Stanley Hall. James also produced a uniquely American form of scientific psychology in his *Principles of Psychology* (1890).[31] Furthermore, he saw to it that the Psychological Laboratory secured funds and professional leadership in the shape of Münsterberg, whom James lured from Wundt's Institute in Leipzig.[32] Empirical psychology found a further place in Harvard's Department of Philosophy—soon rebaptized the Department of Philosophy and Psychology—in the absolute idealism of James's colleague Josiah Royce. Royce welcomed new insights in logic, mathematics, and scientific philosophy; his engagement with experimental psychology in the 1890s aimed to illuminate the connections between the two dimensions in which Royce's "absolute self" was known, the "world of appreciation" and the "world of description."[33] In the first decade of the twentieth century, three more psychologists, Edwin Bissell Holt, Robert Yerkes, and Herbert Langfeld, were added to the Department.[34]

The apparent prodigality of experimental psychology in late-Victorian Harvard was just that: apparent. Its prosperity depended largely upon the enthusiasm of the philosophers, and behind them the administration. When the ardor of both began to cool on the eve of World War I, the vocational psychologists found themselves in a precarious position. James's dedication to experimental psychology had always been more a matter of principle than practice; his backing for the appointment of Münsterberg was motivated principally by his desire to make someone else responsible for the Laboratory, so that he could devote himself to his philosophical, psychical, and metaphysical

interests.[35] Even Münsterberg was no experimental purist in the mold of
Titchener: during the early 1900s he was as concerned with therapeutic and
applied aspects of psychology as with laboratory research.[36] The crunch came
when the leading lights of Harvard philosophy's "Golden Age" began to pass
away—James in 1910, Royce and Münsterberg in 1916. Holt abruptly resigned
in 1918 in disgust at the professionalization and specialization of academic
life.[37] These events left a rump of junior psychologists and no full professors
in the discipline. To make matters worse, Abbott Lawrence Lowell, as Edwin
Boring would later laconically note, "had not cared much about psychology."[38]
The president, indeed, was content to leave the Psychological Laboratory in
the hands of junior figures like Yerkes and Langfeld, with the latter acting
as caretaker after Münsterberg's death. Such was Lowell's indifference to
psychology that he allowed Yerkes to depart for Wisconsin after the latter
was told that psychology would not be given departmental status at Harvard.
An invitation to Titchener to fill Münsterberg's shoes was withdrawn when
the Cornell experimentalist stated that he would come to Cambridge only if
independent departmental rank for psychology was assured. Lowell allowed
the philosophers to sway him in the eventual appointment of a senior psychologist, the British scholar William McDougall, in 1920.[39] McDougall
harbored a special animus against American currents of experimentalism,
a sentiment that was more than reciprocated by the experimentalists, who
regarded McDougall's work in social psychology, personality theory, and
parapsychology as unscientific at best.[40]

    None of this was a sign of rude disciplinary health. Between 1920 and
1927 there was one full professor in psychology, one assistant professor
(Langfeld and Boring, who arrived from Clark in 1922, overlapped for one
year at the same rank before Langfeld left for Princeton), and a handful
of instructors, notably Leonard Troland, Carroll Pratt, and John Gilbert
Beebe-Center.[41] After McDougall hastily exited for Duke in 1927, and before
Boring secured a permanency two years later, there was no full professor for
psychology at all. With falling graduate enrolments and no distinguished
representatives in the 1920s, "the very existence of psychology at Harvard
[was] in serious jeopardy."[42]

    Examples of psychology's uncertain, unhappy status at Harvard con-
tinued to multiply. Up to 1926, one could at least say that, whatever there
was of psychology at Harvard, it was experimental. To be sure, the social

psychologist Gordon Allport had been appointed in 1930 at the insistence of the philosophers, but it soon became evident that Allport's humanistic vision of the discipline did not jibe with that of the experimentalists.[43] While the Psychological Laboratory remained the focus of psychological research in the interwar years, its lack of institutional traction was made plain by the foundation in 1926, within the Department of Philosophy and Psychology, of a psychological clinic administered not by experimental psychologists, but physicians interested in abnormal psychology and the theories of Freud and Jung. The establishment of the Clinic was connected to a bequest from an anonymous donor, and it raised the problem of experimental psychology's relation to psychoanalysis, social psychology, and personality theory at just the time when debates about this issue were raging at the national level. The initiative came from Lowell; the rump of psychologists had simply to adapt to the new circumstances as best they could.[44] A New York physician, Henry Murray, soon became the leader of the Clinic. The consequences of this decision would not come to a head until the mid-1930s, when an ad hoc committee had to decide on whether to grant Murray tenure. By that stage, psychology had succeeded, finally, in separating itself from philosophy, but this, as it turned out was a decidedly mixed blessing, for it exacerbated rather than reduced the fissiparous tendencies of psychology at Harvard.

In contrast to Lowell, Conant was well disposed toward the discipline and quickly allowed the division of philosophy and psychology into separate departments. He also formed a committee to identify and appoint "the best psychologist in the world," who turned out to be the University of Chicago physiological psychologist Karl Lashley, an ardent neo-behaviorist.[45] Then the problems began. Lashley was so dismissive of Allport, Murray, and the work of the Clinic that he demanded the title of his chair be changed "to professor of neuropsychology, or to physiological psychology, to distinguish himself from what he regarded as his less scientific clinical colleagues."[46] Moreover, Lashley conducted his research not in the rooms given to the Psychological Laboratory in Emerson Hall, but in the newly opened Biological Laboratories; rarely did he make his way over to the luncheons Boring had organized for the Department.[47] Lashley's aloofness from his colleagues did not, however, prevent him from insisting that Murray's application for tenure in 1936 be emphatically rejected, a position that Boring too came to share. Much of this maneuvering was due to the still uncertain future of Harvard psychology: Conant's early enthusiasm for the subject had diminished, and Lashley and Boring were no doubt aware of the need to present their field as a legitimate empirical science. In the end, the tenure committee was split and the decision was made to guarantee Murray two five-year appointments, but without tenure.[48] This clash turned out to be the opening round of a battle between Harvard psychology's "biotropes" and "sociotropes" that would culminate in the departure of Allport and Murray from Psychology into the Department of Social Relations in 1946. We shall survey these developments in Chapter 5. For now, however, we must focus on the scene in which a number of Harvard's psychologists came to embrace the theory of knowledge known as operationism.

P. W. Bridgman and the Operational Attitude

All accounts of the history of operationism must begin with Percy Bridgman and *The Logic of Modern Physics* (1927). "In general," ran Bridgman's surmise of the operational viewpoint, "we mean by any concept nothing more than a set of operations; *the concept is synonymous with the corresponding set of operations*."[49] Bridgman's exposition of this principle singled out Einstein's critique, in his special theory of relativity, of the notion of simultaneity.[50] In Newtonian physics, simultaneity was a derivative notion, explained in terms of an absolute frame of temporal reference: simultaneity was a property of two events, such that they occurred at the same time. Faced with an array of puzzles in the theory of electromagnetism and optics, Einstein broke with Newtonian absolutes and reversed the order of explanation, defining time in terms of simultaneity. Insisting that both simultaneity and sequences of events in time could intelligibly be gauged only with cog-and-spring clocks, rather than with reference to the schema of Absolute Time, Einstein observed: "all our judgments in which time plays a role are always judgments of *simultaneous events*. If, for instance, I say, 'That train arrives here at 7 o'clock', I mean something like this: 'The point of the small hand of my watch to 7 and the arrival of the train are simultaneous events'."[51] But how, for instance, could one know that the pointing of the small hand of one's watch to 7 and the arrival of a train 300 miles away were "simultaneous events"? Einstein inspired Bridgman's operational attitude by outlining an ideal procedure for determining simultaneity at a distance, the upshot of which was the redefinition of "time" as a purely mechanical problem of synchronizing measuring devices.

Einstein reduced the problem of testing the synchrony of events-at-a-distance to a matter of observers coordinating clocks by means of light flashes and auxiliary calculations. The meaning of simultaneity and, a fortiori, of time, was thereby operationalized: the concept of "occurring at the same time" had been made synonymous with a concrete set of operations—coordinating clocks—carried out by scientific observers. "Einstein," concluded Bridgman, "in thus analyzing what is involved in making a judgment of simultaneity, and in seizing on the act of the observer as the essence of the situation, is actually adopting a new point of view as to what the concepts of physics should be, namely, the operational view."[52] Bridgman's account of the "operational attitude" focused on such practical understandings of conceptual definition as those described in Einstein's remarks on the concept of simultaneity. This was the greater part of the attraction of operationism to scientists outside of physics. But we have to probe deeper to discern in what Bridgman thought an "operation" ought to consist.

Born in 1882 in Cambridge, Massachusetts, to pious Congregationalist parents, Bridgman rejected at an early age the claims of religion "on the basis of inherent improbability, because of their failure to jibe with the external world as I found it."[53] Thus committed, in both his personal and later in his scientific life, to the single tribunal of "the external world" as he personally encountered it, Bridgman was subjecting himself to a stringent demand. Scientific knowledge could rest only on personal empirical confirmation of empirical claims. This privatistic requirement entailed ruthless self-scrutiny. In the words of one of his graduate students, Gerald Holton, Bridgman "had an ethos of lucidity and candor of the most difficult kind: with himself."[54] Yet, if one's criterion of correct belief was its "jibing" with the external world as registered in private experience, one was always vulnerable to the demonstration that one's methods of assessment were inadequate or misleading. This was exactly how Bridgman experienced the relativity and quantum revolutions in physics.

"It was a great shock," he wrote in *The Logic of Modern Physics,* "to discover that classical concepts, accepted unquestioningly, were inadequate to meet the actual situation."[55] Einstein's special theory of relativity was, for Bridgman, a very personal embarrassment, and he responded to this revelation with the ingenuous excuse that "physics is a young science, and physicists have been very busy." It would be nothing less than a "reproach," he wrote, if the change in the understanding of the nature of physical concepts precipitated by Einstein "should ever prove necessary again."[56] These strikingly personalized responses to scientific-philosophical revelations—as if they put in question the scientist's honor—were the direct product of Bridgman's privatistic vision of science. His development of the operational perspective on concepts was intended to contain the empirical content of scientific concepts within the horizon of those actions for which a single individual could be alone responsible: testing, calculating, speaking. In his Preface to the *Logic*, he noted that this move had been made, albeit before the relativity revolution, by philosopher-scientists such as J. B. Stallo, Ernst Mach, and Henri Poincaré.[57] Bridgman took their central lesson to be that even the abstract conceptual element in knowledge had to be identified with actual practices of calculation and experimental observation.

But what was the character of the "operations" appealed to as the literal meaning of scientific concepts? Bridgman's answer to this question would be crucial to the reception of operationism by the psychologists. There would, Bridgman admitted, be ambiguity in any scientific definitions, "for no operation can be uniquely specified in all of its details"; and experience, even as it yielded to operational description, would also teach that some of those details "which were at one time unthinkingly assumed to be without importance are actually important when our range of experience is extended or the accuracy of our measurements is improved."[58] The critical point was that the operational attitude would prevent scientists from attributing to concepts more than they could rationally or experientially warrant, thereby forestalling the possibility of future lapses into metaphysics. In terms that smacked of hubris, but which spoke instead to an acute sense of the fragility of empirical knowledge, Bridgman counseled his fellow scientists to "make it our business to understand so thoroughly the character of our mental relations to nature that another change in our attitude, such as that due to Einstein, shall be forever impossible."[59] Tellingly, Bridgman repudiated the widely held view that his intention was to "set up a philosophical system and a theory of the properties that any method *must* have if it hopes to be successful."[60] He aimed, on the contrary, to return scientific concepts to the arena of technical practices where they were actually put to work. "What we are here concerned with," he wrote in a 1938 restatement of the operational stance, "is an observation and description of methods which at least some physicists had already, perhaps unconsciously, adopted and found successful—the practice of the methods [i.e. operational methods] already existed."[61]

This qualification was important. Several commentators have taken Bridgman to mean by "operation" a publicly observable set of actions entailed by the concept in question—ideally, and paradigmatically, a laboratory procedure.[62] For example, the meaning of "length" was determined in the case of stationary objects visible to the naked eye as follows:

> What we do is sufficiently indicated by the following rough description. We start with a measuring rod, lay it on the object so that one of its ends coincides with one end of the object, mark on the object the position of the other end of the rod, then move the rod along in a straight line extension of its previous position until the first end coincides with the previous position of the second end, repeat this process as often as we can, and call the length the total number of times the rod was applied.[63]

The concept of length, on this view of Bridgman's operationism, meant just these layings of rods, markings of ends, and toting up of rod applications, along with any other auxiliary actions actually undertaken to measure the object in question. But it is vital to grasp what such a view did and did not entail. Operations, for Bridgman, were avowedly not a mechanical form of behavior. To understand this, we have to pay particular attention to Bridgman's repeated insistence that he was not offering a normative account of how to construct and validate scientific knowledge, but was rather seeking to understand physical concepts in terms of ongoing practices of experimentation and theoretical understanding in the natural sciences. "Operations" for Bridgman were defined by a phenomenological appreciation of what it was like to carry out experiments in a laboratory; knowledge making was all about the craft skills of the scientist. Operationism may have seemed to some an epistemology for robots, with its naïve emphasis upon physical movements and rule following; but it was in fact intended as a characterization of the lifeworld of the practicing scientist.

In a letter to Arthur Bentley written in 1932, Bridgman highlighted the importance of nonintentional, nonmechanical dimensions of "thinking" in his laboratory: "A good deal of this cerebration . . . I find by analysis of my own activity, is apparently divorced from any verbal element, but is almost entirely motor and visual in its character."[64] What has often appeared in Bridgman's discussion of the operational attitude to be an evasive or insufficiently philosophical position on the nature of operations stems from his claim that the salient operations to which the content of scientific concepts must ultimately be reduced could only be grasped by one who had the expert's comprehension of the necessary technical skills and dispositions involved in being a particular kind scientist. A given action—the shifting of a measuring rod along a prescribed path, the use of the calculus to determine the orbit of a planet, etc.—could only count as an operation on Bridgman's definition if "our experience has shown us that certain sorts of operation are good for certain purposes." This kind of experience was not delivered to the intellect by the senses, as the naïve empiricist would have it; it was built up from prolonged exposure to the relevant experimental and theoretical situations.[65]

This view was the product of Bridgman's experience of research in high-pressure physics. His earliest contributions to the field involved improvements to experimental equipment. Inspired by the work of the French physicists Louis Paul Cailletet and Emile-Hilaire Amagat, Bridgman began his doctoral research in 1905 with the intention of measuring the effects of volume on the refractive index of light. Such research depended upon the production of high pressures using specially designed experimental apparatus. Early in his research, Bridgman invented more or less by accident a self-tightening, leak-proof joint that allowed for the exertion of pressures far beyond existing limits. The only checks on the new instrument were the tensile strength of the steel employed and the design of the compressors, which were not as efficient as they might have been. In addition, the entry into untested levels of pressure also required the creation of new pressure gauges sufficiently robust to measure up to the new standards. These were engineering problems, and Bridgman applied himself to their solution with relish: he solicited materials from steel companies, designed pistons, and assessed the resistance profile of mercury to determine its suitability as a pressure gauge. Soon enough, the young physicist was able to produce and measure pressures beyond 20,000 Kg/cm^2—a near four-fold increase from those attained by Amagat. Bridgman submitted his dissertation in 1908 under the title "Mercury Resistance as a Pressure Gauge."[66]

Bridgman's early scientific career was thus dominated by grimy material concerns with the tensile strength of steel and the operation of pistons, seals, and measuring devices. These involvements with engineering techniques did not merely help to solve existing empirical problems; by opening up a whole range of new pressures, they led experimentation by the nose and allowed for the recording of hitherto unobserved, because unobservable, phenomena. In the words of Bridgman's biographer, Maila L. Walter, "technique preceded theory . . . . Bridgman did not set out to test theory, because there was no theory to test."[67] As Bridgman himself put it, he "endeavored to use this new technique to the limit, attacking with it any problems in which the information to be expected from the behavior under high pressures seemed likely to be of significance. This is not the usual procedure in scientific work, in which the problem usually presents itself, and the suitable technique discovered." The research almost wrote itself; results flowed from Bridgman's skill in running tests on a wide range of materials at increasing pressures. Between 1911 and 1915, he published his results on the thermodynamic properties of various liquids up to 12,000 Kg/cm^2. Bridgman next deployed his apparatus on solids, a task that involved the assessment of melting phenomena and compressibility at increasing pressures. By the 1920s, he was measuring the electrical resistance of metals and alloys at high pressure. In each case, new and important information was gleaned and unexpected phenomena encountered.

Given this background, we can see why, for Bridgman, operational analysis was "a technique of analysis which endeavors to attain the greatest possible awareness of everything involved in a situation by bringing out into the light of day all our activity or operations when confronted with the situation, whether the operations are manual in the laboratory or otherwise 'mental'."[68] By definition, in order to attain this reflexive awareness one had already to be immersed in a scientific practice, and the salient operations in which one's concepts were bodied forth could only ever by grasped against the background of a partially implicit understanding the situation itself— those "motor responses" and minute gradations of different practice that the expert alone could detect.

The Migration of Operationism

Bridgman wrote *The Logic of Modern Physics* because he wished to bring clarity to the increasingly unruly concepts of the exact sciences. But it is not hard to see why his strict identification of meanings with the procedures of the scientist might have resonated with the Harvard human scientists who sought to reform the murky ideas of their home disciplines. Bridgman's lexicon flashed briefly across the Pareto network. In 1932, Henderson invoked the operational criterion of meaning in a discussion of the formulation of scientific facts.[69] Another alumnus of the Pareto circle, the sociologist Talcott Parsons, managed to conjure the spirits of Bridgman and Pareto in his suggestion that the concept of "residues" could be defined by a set of complex classificatory operations.[70] In his *Foundations of Economic Analysis,* published in 1947, the economist Paul Samuelson gave Bridgman a Popperian accent by claiming to offer *"operationally meaningful* theorems" that presented "a hypothesis about empirical data which could conceivably be refuted, if only under ideal conditions."[71] Samuelson had written most of the papers that composed the *Foundations* while a Junior Fellow in the Society of Fellows, having been brought in by Henderson as an economist working in the Paretian vein.[72] Evidently, Bridgman's artisanal view of the content of scientific concepts chimed in with Henderson's equally craft-oriented view of scientific epistemology. Nevertheless, the reception of operationism by the Harvard Paretians remained superficial. The operational stance was avowedly that of the laboratory scientist. While this orientation kept the social theorists at a safe distance, it proved crucial to the resonance of Bridgman's *Logic* among Harvard's experimental psychologists.

Unlike the Pareto enthusiasm, the story of the operationist vogue is national in scope, but it is a tale that returns often to Harvard Yard. It is difficult to pinpoint the moment when Bridgman's strictures concerning the meaning of scientific concepts began to register among psychologists. With some justification, several commentators have argued that the basic commitment to definition in terms of technical operations was discernable as early as the 1920s in the pronouncements of prominent neo-behaviorists like Berkeley's Edward C. Tolman, as well as in the writings of less doctrinal experimentalists such as Edwin Boring.[73] Both Boring's peremptory claim, in the pages of the *New Republic* in 1923, that "intelligence is what the [intelligence] tests test," and Tolman's suggestion, three years later, that the best answer to those who could not understand psychological concepts was to take them to "a really good laboratory and show them a really good rat in a really good maze" spoke to what one historian has termed a "proto-operational" style of thinking.[74] Clark Hull's work at the Institute of Human

Relations, along with Boring's *The Physical Dimensions of Consciousness* (1933) and Tolman's *Purposive Behavior in Animals and Men* (1932), are also often seen as early operationist projects in psychology.[75]

Identifying the moment when psychologists began to read and cite *The Logic of Modern Physics* is no more straightforward. Some have suggested that the principal conduit for Bridgman's teachings was the German philosopher Herbert Feigl, who, it is said, introduced Boring and his students B. F. Skinner and Stanley Smith Stevens to the *Logic of Modern Physics* while at Harvard on a Rockefeller Foundation scholarship in 1930.[76] It was Feigl, observed Boring in the second edition of his *History of Experimental Psychology* (1950), "who introduced the Harvard psychologists to the ideas of their own colleague, Bridgman, to the work of the Vienna Circle, to logical positivism and to operational procedures in general."[77] The historical record, however, presents a more complex picture. It is true that Feigl applied for a fellowship at Harvard after reading, with the Vienna Circle, the *Logic,* and that he spent enough time with the antisocial Bridgman to secure a letter of reference from him when he left Harvard for a position at the University of Iowa in 1931.[78] Perhaps Feigl did bring Boring's attention to the operational viewpoint, but the former's role in alerting Skinner and Stevens is less clear. It seems likely that the "friend" who pointed Skinner toward Bridgman's *Logic* was not Feigl.[79] Moreover, it would appear that Skinner read Bridgman's operationism as a natural extension of the work of his most formative intellectual influence, Ernst Mach.[80] Skinner's citation of the *Logic* in his 1930 PhD dissertation, "The Concept of the Reflex in the Description of Behavior," was one of the first references to Bridgman in the psychological literature.[81] Yet both the priority of the citation and its significance for Skinner's Machian philosophy of science are questionable.[82] Meanwhile, Stevens came to Harvard as a graduate student after Feigl departed for Iowa, and the connections of Stevens's view of operationism with that of his teacher Boring are tenuous at best. Boring had it right when he surmised that the historian "can not say that operationism began at any point."[83]

In spite of these uncertain beginnings, the course of explicit debate about operationism in psychology is not difficult track. 1935 was the turning point.[84] In this year, Stevens published two articles announcing the emergence of an operationist movement: "The Operational Basis of Psychology" in the *American Journal of Psychology,* and "The Operational Definition of

Psychological Concepts" in the *Psychological Review*.[85] These two pieces, derived from a single extended manuscript, followed the model of Bridgman's *Logic* in outlining the nature of operational definition and providing a series of operational reinterpretations of core psychological concepts, including experience and the attributes of sensation.[86] Stevens followed up his initial statements with two further interventions: an essay situating the operational view of psychology within the burgeoning movement of scientific philosophy, published as "Psychology: The Propaedeutic Science" in 1936; and a 1939 literature survey of the new "science of science."[87]

In the meantime, other psychologists had sought to align their research with operational procedure. At around the same time that Stevens's articles were published in 1935, the Wesleyan University psychologist John A. McGeoch presented a paper to the annual meeting of the American Psychological Association on "Learning as an Operationally Defined Concept" in which he contended that "learning can be more adequately defined in terms of the operations of measurement than in terms of phenomenal (i.e. subjective, private) properties."[88] He offered a more critical, but still sympathetic, take on operationism in a paper delivered to the APA two years later.[89] By 1936, two heavyweights of the profession had weighed in. Stevens's teacher Edwin Boring tried operationism on for size in a note published in the *American Journal of Psychology*.[90] Bringing the procedures of operational definition to bear on the question, "Can *duration* be an *immediate experience?*" Boring asserted that this "classical problem" turned out to be "specious when it is considered with respect to the operational procedures by which perception is known." Perceptions of temporal patterns, for the operational psychologist, could only be "inferred from and defined by certain operations of introspective report, which adequately imply the differentiation of perception": perception of time, as of anything else, was, for the purposes of scientific inquiry, a publically observable report in the presence of a controlled stimulus. This "point of view toward experience," Boring noted, made "perception, not a private immediate experience, but a psychological *construct* which (in a rat or a person or myself) is just as public as is any other convincing inference from data."[91] Boring attempted a similar "'behavioral' inversion," which substituted the publicity of testing procedures for private introspective experience, in a 1941 operational translation of the German psychologist G. E. Müller's psychophysical axioms, according to which psychological and neurophysiological events were identical.[92] Ever the historian of the experimental tradition, Boring soon contrasted the new positivism in psychology with the humanist approach of his Harvard predecessor William James.[93] Edward Tolman, meanwhile, was prima facie an even more eager convert to operationism. Mirroring the efforts of Stevens, and behind him Bridgman, Tolman published a program statement on "operational behaviorism" and offered an emblematic operational redefinition of the psychological concept of "demands" in the journal *Erkenntnis*.[94]

After the initial spate of psychological translations of Bridgman's ideas between 1935 and 1936, there commenced a discussion of the nature, proper applications, and limits of operational analysis that lasted until the early 1950s. This was a conversation that included psychologists, philosophers, and scientific methodologists. Bridgman himself came back on the scene, seeking to clarify the psychological and epistemological aspects of his position in *The Nature of Physical Theory* (1936) and in a subsequent essay in the periodical *Philosophy of Science*.[95] As a matter of principle, Bridgman never used the term "operationism" or "operationalism," because he intended his remarks to constitute not a doctrine or philosophical thesis but a scientific laborer's reflection on the procedures associated with the use of scientific concepts in the context of the laboratory.[96] But an "ism" is what his ideas became. In the wake of the proselytizing work carried out by Stevens, Boring, and Tolman, operationism became a "position" to be discussed, elaborated upon, and contested. Among those seeking to defend elements of the operationist program were the personality psychologists R. H. Seashore and B. Katz, the philosopher A. C. Benjamin, the former Harvard instructor Carroll Pratt, the sociologist George Lundberg, and the experimental psychologists J. R. Kantor and Sigmund Koch.[97] Even Clark Hull, whose epistemological views mirrored those of Mach and Pearson, felt compelled to pay obeisance to operational analysis in his *Principles of Behavior* (1943).[98] Several critics also appeared on the scene. William Malisoff, editor of *Philosophy of Science*, lit the way for the critique of operationism with his searching review of *The Nature of Physical Theory*.[99]

Indeed, it was on the home ground of physics that operational analysis began to encounter philosophical difficulties. The physicist R. B. Lindsay was especially trenchant in pointing out the troublesome implications of the operational stance as a means of understanding and conducting research in

the physical sciences.[100] Soon enough, the pages of the psychological journals began to fill up with caveats and critiques of operational definition. Two
articles published in the *Psychological Review* in 1938 marked the beginning
of a backlash against operationism, which according to R. H. Waters and
L. A. Pennington of the University of Arkansas, was neither "the key to the
final solution of psychological problems" nor "the shibboleth that will unite
all psychologists."[101] A more systematic attack came in 1944, when two Smith
College psychologists, reviewing the literature to date, demonstrated that
what psychologists meant by "operationism" was something quite different
from the methodological and semantic rules laid down by Bridgman.[102] This
article prompted Boring to put together a symposium for the *Psychological
Review* in which Bridgman and a select group of almost exclusively Harvard-
trained psychologists and philosophers—in this case, Boring, Pratt, Skinner,
and Feigl—might work out their differences. Instead, the symposium
revealed just how deep the divisions were among the leading operationists.[103]

By the early 1950s, another turning point was reached. In 1951 Bridgman
published his most extensive engagement with epistemology, methodology,
and semantics in physical theory since the *Logic*.[104] The strictures of operational definition, meanwhile, began to be treated in psychological textbooks
as the sine qua non of scientific method. All psychologists, it seemed, had
to define their variables "operationally."[105] Yet, at least among philosophers,
the credibility of operationism was more or less destroyed. A roundtable on
"The Present State of Operationalism" held in Boston at the annual meeting of the American Association for the Advancement of Science in 1953
revealed how philosophically naïve Bridgman's position had been. In particular, philosophers of science associated with the Vienna Circle such as
Gustav Bergmann and Carl Hempel showed the operational viewpoint to
be fundamentally confused about the nature of meaning and the rational
reconstruction of scientific inquiry.[106]

Operationism was no longer a serious position in the philosophy of science, but it lived on in textbook psychology. Many historians of psychology
have wondered, with some exasperation, why a discredited position in scientific philosophy should have been so potent and enduring in psychology.
Operationism, they suggest, must have found rich soil in psychology because
psychologists were desperate for an objective methodology that could legitimate the only practical achievements the discipline could claim, that is, the

ability to test, measure, and calculate. Operational definition, in this view, allowed psychologists to avoid coming to terms with the actual content of the concepts they purported to be measuring.[107] Now they could define those concepts, which were presupposed in the application of tests and measurements of various sorts, as the application of tests and measurements. With the Harvard complex in mind, however, we must read the migration of operationism in a different way. We can explain its fortunes in psychology by grasping the interstitial situation of psychology at Harvard.

Contesting Operationism

The Harvard community of psychologists, methodologists, and philosophers was the setting for each of the major turning points in the diffusion of operationism. It was at Harvard that the initial connections between Bridgman's *Logic* and the practices of experimental psychology were forged around 1930 by Feigl, Boring, Skinner, and Stevens; it was a Harvard-trained psychologist, Stevens, who propelled the operationist vogue to the forefront of the psychological profession in 1935 with his seminal articles on operational definition in psychology; it was the Harvard contingent, notably Boring, Skinner, and Bridgman—with former Harvard scholars Feigl and Pratt—who tested operationism to its breaking point in the defining 1945 symposium in the *Psychological Review;* and, finally, it was a Harvard interfaculty scientific discussion group, led by the émigré physicist and philosopher of science Philipp Frank, that revived the operational debate in the early 1950s. Although it would be shortsighted to treat the operationist movement as a purely internal Harvard conversation, the centrality of Harvard networks to the dispersion of operationism is clear. Why should this have been so?

By the mid-1930s, Harvard's psychologists were in a difficult spot. The simultaneous expansion and fragmentation of psychology at the national level was compounded at Harvard by the prolonged philosophical tutelage of the discipline during the Lowell administration. This extended stay in organizational limbo, in turn, rendered the post-1934 Department of Psychology an institutional rump consisting of the Laboratory, the Clinic, and sundry experimental projects located in neighboring departments. Social and clinical psychologists like Allport and Murray lived in uneasy cohabitation with experimentalists such as Boring and, later, Lashley and Stevens, while the experimentalists were themselves divided between exponents of neurophysiological and psychophysical methodologies. Against the background of the disciplinary flux of the 1930s, members of Harvard's psychological community began to encounter Bridgman's *Logic,* and the operational attitude.

What Bridgman's operational viewpoint sparked among the psychologists was the desire for the possibility of conceptual clarity in which the various factions could agree on the foundations of their discipline. In the early operationist writing of Skinner, Stevens, and Boring, the clarifying powers of Bridgman's definitional procedures were repeatedly invoked. Skinner's dissertation on the concept of the reflex, for example, defended its engagement with the history of the concept on the scientific ground marked out by Mach, Poincaré, and Bridgman. The reflex concept, Skinner contended, had been "deeply marked" by "extrinsic interpretations," which had accumulated over a long history of the usage of the term; because these accretions "now appear to embarrass the extension [of the concept of the reflex] to total behavior," it was necessary to undertake the "historical method of criticism" through which merely "incidental interpretations" of the reflex could be cleared away, so that its practical meaning for empirical scientists could stand forth. Bridgman's *Logic* offered an "excellent application of the method to modern concepts" in physics.[108]

A similar compulsion to clean up the vocabulary of psychology was evident in Boring's two most explicit papers on operationism. Boring's concern was to show that troublesome mentalistic terms like "immediacy" were meaningless and could be substituted for experimentally observable terms like "physiological *continuity.*"[109] His 1941 paper on Müller's psychophysical axioms, meanwhile, pushed these definitional imperatives further by offering straight operational "translations" of Müller's claims about mind-brain identity.[110] It was Stevens's twin 1935 papers, however, that spoke most directly to Harvard psychologists' anxieties about the possibility of fruitful intradisciplinary conversation. "It is clear," Stevens declaimed in "The Operational Definition of Psychological Concepts," "that the examination of psychology's conceptual heritage under the search-light of operationism needs to be undertaken seriously if we are to be rid of the hazy ambiguities which result in ceaseless argument and dissension."[111] For Stevens, "hazy ambiguities" and "ceaseless argumentation" were abiding preoccupations. Into the "most fundamental notions" of psychology had "seeped a variable mixture of initiation, a priori postulations, reified entities, and empirical fact." Only the definitional procedures of operationism "insures us against hazy, ambiguous, and contradictory notions and provides the rigor of definition which silences useless controversy." Stevens looked forward to a psychology "stabilized on the operational basis" and thereby "fortified against meaningless concepts."[112]

Undoubtedly, these linguistic concerns about the possibility of a purified lexicon in psychology reflected concerns about the technoscientific fragmentation of psychology at the national level. But they also spoke to the divisions among experimentalists, and between social and clinical psychologists and experimentalists, at Harvard. The operationist vogue was a sign of psychology's heterodox, interstitial condition. Faced with an acute lack of linguistic resources for talking with one another and an uncertain institutional future, many of Harvard's psychologists embraced a procedure that promised to limit conceptual ambiguity and—at least in principle— to encourage methodological unity in the discipline. In practice, however, centrifugal forces operating within Harvard psychology proved too strong and produced competing versions of operationism—a babelic splitting of tongues that was made plain in the 1945 *Psychological Review* symposium.

One of the problems faced by interwar operationists in psychology can be traced back to the rise of the new psychology. The researches of Wundt and his followers in Germany, along with the statistical methodology of Francis Galton in Britain and the testing movement in the United States, rested on technical and instrumental innovations, not theoretical synthesis. Whatever social and political esteem the new psychology possessed in the North Atlantic world derived from the utility of its technological applications, which ranged from the brass instruments of the Wundtian laboratory to the therapies psychologists offered to psychiatric patients. To be sure, psychology, like other disciplines, was established in the American university through ideological battles over authority on topics of mind and behavior.[113] Nevertheless, the psychological profession in the United States acquired scientific authority in large part through the efficacy of its technoscientific practices. Questions of a unified theory emerged with force only during the 1920s and 1930s.

This was the context in which the interest in operationism emerged. What was needed was a methodological and epistemological position that could justify what it was that psychologists already did with some success: namely, test mental aptitude, measure psychophysical responses, and describe the mechanisms of behavior. Before Bridgman published the *Logic* in 1927, Boring and Tolman had tested the water with their proto-operationist arguments about the meaning of intelligence being just whatever it was the intelligence test tested and about the content of concepts being identical to what happened under the right experimental conditions. The historian of psychology Tim Rogers has suggested that Bridgman's *Logic* helped to square this circle by providing psychologists with the methodological license to identify the meaning of their concepts with the practices of testing and experimentation.[114] Certainly, Bridgman's claim that operational definitions only made sense within an ongoing experimental practice chimed in with the psychologists' wish to legitimize their own practices. The *Logic* was therefore the right book at the right time as far as professional psychologists were concerned. On the other hand, operationism had to do more for the psychologists than Bridgman needed it to do for Percy Bridgman. And it was here that the centrifugal forces at work in the expansive field of scientific practices that constituted American psychology began to counteract the unifying potential of the operational attitude. The landscape of practices covered by the estate of psychology proved too diverse, at Harvard and across the United States, to give operationism the univocal quality that Boring and Stevens hoped it would have.

We can begin to fill in the details of this general picture by drawing out a lesson contained in the historiography of operationism. The lesson is that, even at Harvard—or rather, especially at Harvard—operationism was seldom invoked in exactly the same way by Bridgman, Skinner, Stevens, and Boring. Despite their mutual admiration for Mach's insistence on stripping away the metaphysical growths from scientific concepts so as to uncover their practical, empirical meaning, Bridgman and Skinner never saw eye-to-eye on the nature of the operational stance.[115] Bridgman, as we have seen, relied on a private, phenomenological sense of the experimentalist's craft and insisted throughout his career on the absolute distinction between private experience and public facts, or what he described as the "patent operational differences between my feelings and your feelings, between my thought and your thought."[116] Committed as he was to a fully objective science of behavior, Skinner's account of the operational stance had no place for the lingering "mentalisms" of Bridgman's position. He regarded operationism as, at best, a methodological gloss on what the behaviorist approach in psychology was already doing. The two positions were, as far as Skinner was concerned, part of the same scientific attitude: "behaviorism has been (at least to most behaviorists) nothing more than a thoroughgoing operational analysis of traditional mentalistic concepts."[117] Skinner sought, unsuccessfully, to persuade Bridgman that the private experiences he venerated could be explained in the public, behaviorist language of controlled stimulus and conditioned response.[118] (Bridgman's retort: "In the private mode [of human action] I feel my inviolable isolation from my fellows and may say, 'My thoughts are my own, and I will be damned if I let you know what I am thinking about'."[119]) The two men never had the meeting of minds their shared operational attitude seemed to portend.[120] "My efforts to convince you of the possibility of extending the operational method to human behavior," Skinner wrote regretfully to Bridgman in 1956, "have long since suffered extinction."[121]

Skinner was even keener to distance himself from what he considered to be the pseudo-operationism of Boring and Stevens. Convinced that operationism was the preserve of the behaviorist, Skinner was suspicious of the attempts of Stevens, a psychophysicist, and Boring, a methodological pluralist, to deploy operational analysis in the study of mental or introspective concepts. Skinner and Stevens would likely have encountered each other as early as 1932 while working in the laboratory of Hallowell Davis at the Harvard Medical School.[122] The young Skinner was not implacably hostile to Stevens's opening statements on operationism in psychology, insofar as he considered them "the best statement of the behavioristic attitude toward subjective terms now in print."[123] But he recognized as early as 1935 that Stevens was committed to the operational elucidation of what were for Skinner outmoded mentalistic notions of sensation and discriminatory capacity.[124] In his formal contribution to the 1945 *PR* symposium, Skinner sharply rejected what he saw as Stevens's attempt, in his 1939 paper "Psychology and the Science of Science," to tease apart behaviorism and operationism.[125] Evidently vexed by the temporizing of his fellow participants in the symposium, however, Skinner used his rejoinder to launch an astonishing attack upon the Boring-Stevens line on operationism and to push his argument about the identity of behaviorism and operationism yet further.

Asserting that the advent of behaviorism in psychology was "a revolution comparable in many respects with that which was taking place at the same time in physics [i.e., the quantum and relativity theories]," Skinner recalled that he had offered in 1930, as a substitute for his controversial thesis on the reflex, to prepare an "operational analysis of half-a-dozen key terms from subjective psychology." With the behaviorist revolution of Watson and Pavlov in place, such a task appeared to Skinner as "a *mere exercise in scientific method*" or "a bit of hack work," the results of which were as mechanically "pre-determined as that of a mathematical calculation." But his offer had been refused, and the patient behaviorist-operationist deconstruction of psychology's mentalistic vocabulary was abandoned by behaviorists in favor of the development of "a fresh set of concepts derived from a direct analysis of the newly emphasized data" emerging from behaviorist laboratories. Surveying the divided Harvard psychological scene in 1945, Skinner pointedly asserted that "the Harvard department would be happier today if my offer had been taken up." In lieu of the behaviorist elimination of mentalistic metaphysics, the "operationism of Boring and Stevens" endeavored to "acknowledge some of the more powerful claims of behaviorism (which could no longer be denied) but at the same time to preserve the old explanatory functions unharmed." To Skinner, "the position taken" by Boring, and implicitly by Stevens, "is merely that of 'methodological Behaviorism,'" according to which "the world is divided into public and private events, and psychology, in order to meet the requirements of a science, must confine itself to the former." This was a position "least objectionable to the subjectivist"—among whom Skinner numbered both Boring and Stevens—"because it permits him to retain 'experience' for purposes of self-enjoyment and 'non-physicalistic' self-knowledge." For the behaviorist, however, this stance was "not genuinely operational because it shows an unwillingness to abandon fictions. It is like saying that while the physicist must admittedly confine himself to Einsteinian time, it is *still true* that Newtonian absolute time flows 'equably without relation to anything external'."[126]

Skinner's rejoinder was well designed to exacerbate Harvard psychology's still-festering internal tensions. Polemic aside, however, he was right that his sense of operationism was very different from that of Boring and Stevens. Boring, by temperament a generalist in psychology, was only a lukewarm operationist and seems to have seized on operational definition as a means of tamping down controversy surrounding psychological concepts across the experimental and applied domains of the discipline.[127] His curiously errant remark, in the second edition of his *History of Experimental Psychology*, that "the Harvard experimentalist psychologists who took up with operationism were not behaviorists or animal psychologists"—what, glaringly, about Skinner?—"but men who were looking for rigor of definition in laboratory situations where introspection once ruled" indicated the low stakes for which Boring had played in the operationist debates of the 1930s and 1940s.[128]

Stevens was a different case. His vision of operationism was indeed distinct from that of Skinner and Bridgman.[129] The differences with Bridgman were clear. Throughout his 1930s essays on operationism, Stevens insisted that the fundamental criterion of a scientific fact was social agreement, "for science deals only with those aspects of nature which all normal men can observe alike. . . . Unless the psychologist can report his experience in such a way that others can verify it, we are left with no dependable operation for establishing it."[130] Stevens went on to pick out a basic, publicly accessible, "dependable operation" in terms of which all other operations could be understood: the act of "discriminatory response." If discrimination sounded like an introspective category, Stevens was at pains to insist that he meant "the concrete differential reactions of the living organism to environmental states, either internal or external." The basic human capacity for discrimination, on which rested the ability to make out more complex operations like measurement and calculation, was "a 'physical' process, or series of natural events."[131] What these comments make clear is that Stevens developed his brand of operationism independently of Bridgman, and indeed Stevens felt confident enough to describe the *Logic* as "rich in example but poor in percept" for talking of "'operations' without giving an explicit definition of the term."[132] Conversely, Stevens's search for a social criterion for an operation—which rendered it, in essence, a psychological capacity rather than an intuitive feature of one's expert practice—was anathema to Bridgman. "I simply cannot make him see," Bridgman wrote of Stevens in 1936, "that his 'public science' and 'other one' stuff is just plain twisted. I have also discussed with him his 'basic act of discrimination' without making much impression, and I have rather washed my hands of him."[133]

The Failure of Disciplinary Synthesis

Stevens's differences with Skinner were yet more significant. The source of this discord among Harvard's operationists can be traced back to the experimental practices they sought to defend. Stevens had followed in Bridgman's footsteps when he declared that "the development of operational principles is properly an empirical undertaking." The operationist would ask: "What do the science-makers do? What methodology has the maximum survival value? When do propositions have empirical validity?"[134] The problem for Stevens and his fellow operationists was that the "science-makers" in psychology did many things and engaged in a multiplicity of experimental practices, a good number of which were deemed illegitimate by rival factions in the profession. This reflexive turn, so typical of the Harvard complex, ensured that operationists in psychology divided along the lines of their favored forms of experimental practice.

The differences between Harvard's two most dedicated operationists, Skinner and Stevens, are certainly explicable in terms of distinctive modes of experimental practice. Both men, it must be noted, were in the early stages of their careers when they began to explore the possibilities of operational analysis. Skinner had arrived at Harvard as a novelist manqué in 1928.[135] From the beginning of his formal psychological studies, he was committed to elaborating what he called "a science of the description of behavior."[136] Resident in Cambridge until 1936, he wondered openly whether his science of behavior even "included" psychology, and he saw his role as a crusading behaviorist as much more important than his professional identification as a psychologist.[137] Part of the reason for Skinner's reticence was that he had hit upon psychology as a placeholder for prior interests in epistemology. At Hamilton College, the undergraduate Skinner had been introduced to the works of Jacques Loeb, the positivist biologist and devotee of Ernst Mach. During his self-described "lost year" working as a writer after graduation, Skinner was steered toward the Russian physiologist Ivan Pavlov's *Conditioned Reflexes* (1927). Such reading convinced Skinner that the behavior of organisms, sentient or otherwise, could best be understood as automatic responses to environmental stimuli. But it was the discovery of the work of Bertrand Russell and John Watson, through the review pages of the literary magazine *The Dial,* that converted Skinner to the behavioristic creed and drove him to graduate school at Harvard.[138] Of central importance was Russell's *Philosophy* (1925), which found the British philosopher in the throes of an intellectual affair with Watson's psychology.[139] Watson's elimination of internal states and private experience in favor of behavioral descriptions of epistemological concepts like perception, inference, testimony, and memory appeared, to Russell, exactly the sort of nonmetaphysical account of knowledge that philosophy required. Skinner was transfixed and soon devoured Watson's *Behaviorism* (1925).[140] What grabbed Skinner was the manner in which Russell's Watson showed how the Loeb-Pavlov schema of organisms and environmental variables rendered so refined an enterprise as epistemology an empirical, descriptive science free from talk of "ideas" and other private mental contents.[141] Skinner's first, abortive book, prepared in the early 1930s, was entitled "Sketch for An Epistemology," and he admitted later in life that he "came to Behaviorism . . . because of its bearing on epistemology, and I have not been disappointed."[142]

Skinner's predispositions were confirmed early in graduate school when he took George Sarton and L. J. Henderson's class on the History of Science. Henderson, as was his wont, pointed Skinner toward Mach's *Science of Mechanics*.[143] Given that Mach was the source of Loeb's positivism, it was not surprising that he appealed to Skinner. His treatment of physical theory as an evolutionary extension of the human animal's transactions with its environment fitted Skinner's behaviorist conception of epistemology.[144] Skinner found in Mach his paragon of the critical scientist and soon moved on to Mach's other works, including *Knowledge and Error* (1905) and the *Analysis of Sensations* (1914). Late in life, Skinner characterized his descriptive science of behavior as "following a strictly Machian line, in which behavior was analyzed as a subject matter in its own right as a function of environmental variables *without reference to either mind or the nervous system.*"[145]

Wedded to these epistemological positions was a heterodox experimental practice. Skinner remained largely aloof from the work of the Psychological Laboratory during his time at Harvard, even though his official dissertation advisor was Edwin Boring. When Skinner claimed that his PhD thesis "had only the vaguest of Harvard connections" this testified as much to his everyday laboratory experiments as to his fealty to his own brand of behaviorist epistemology.[146] Even as a child, Skinner had preferred to design and build his own toys and machines, and he remained a DIY experimenter throughout his professional life, as evidenced by the instruments with which he is identified: the "Skinner box," the air crib, the teaching machine, and the verbal summator.[147] The practices and experimental situations to which Skinner appealed in operational definitions were very peculiarly Skinner's own. More concretely, most of Skinner's laboratory research at Harvard took place not in the space made available for psychology at Emerson Hall, but instead, after 1930, in the basement of the biology department. Skinner also worked part time in the Medical School with Hallowell Davis on the physiology of the central nervous system. Skinner's last three years at Harvard, between 1933 and 1936, were spent not in the psychology department but as a Junior Fellow in Henderson's Society of Fellows.[148] Unsurprisingly, Skinner's sense of the "operations" involved in psychology was not the same as that of Boring or Stevens.

Stevens had different challenges with which to contend. When Stevens arrived at Harvard in 1930, Skinner was already well on the way to his PhD. While Stevens worked alongside Skinner, part time, in Davis's Medical School laboratory, his research followed a different path from Skinner's. Importantly, Stevens was Boring's protégé. Whereas Skinner had abandoned any research techniques that smacked of introspection, Stevens eschewed strict behaviorism for the more established experimental tradition of psychophysics. The difference between Wundt and James, on the one hand, and what Stevens conceived as his goal, on the other, was that Stevens believed he could substitute quantitative, mathematical correlations between measured stimulations and experienced, but objective, sensations, for the first-person, phenomenological "experience" of the new psychologists. Boring's aforementioned remark that "the Harvard psychologists who took up with operationism" were "men who were looking for rigor of definition in laboratory situations where introspection had once ruled" is intelligible principally as a characterization of Stevens's project. And it was in fact Stevens's desire to find a way to make the experimental procedures of psychophysical research fully objective that drew him to operationism. What the latter gave him was a means of making private experience public by translating the discriminatory capacities measured by the pyschophysicist into the language of laboratory procedure and publicly observable behavioral phenomena.

Another crucial feature of Stevens's embrace of operational definition was the uncertain future of his own career and of the experimental tradition he championed. Stevens wrote his first papers on operational definition at a time when his job prospects were, at best, unknown.[149] His suggestion that the measurement techniques of psychophysics were the best means of understanding the *ur*-operation of all science, discriminatory response, spoke to Stevens's need to make his research practices indispensible, or at least defensible, in the study of psychology. By 1936, Stevens had moved on to the project of demarcating his own notions of experimentation as the legitimate form of experimental practice in psychology; in pursuit of this goal, he founded, in collaboration with Edwin B. Newman, the Psychological Roundtable. Stevens's purpose in founding this group was to distinguish his work and that of his colleagues from the purportedly less-scientific activities of the Society of Experimental Psychologists, established by Titchener during the early years of the American Psychological Association.[150] Stevens was by then an assistant professor at Harvard. What was clear was that "operationism" was being invoked to defend practices that were diverse even within the confines of Harvard's own community of psychologists.

Such differences were to prove fatal for operationism as an agent of unity in psychology at Harvard. A split between the biotropes and the sociotropes in the Harvard department led rapidly to a breakdown in the community. Gordon Allport and Henry Murray, the sociotropes, never adopted the operationist vocabulary; Allport, indeed, used his 1940 presidential address to the American Psychological Association to deliver a stinging blow to the methodological and epistemological positivism inherent in the operationist rhetoric of his colleagues Boring and Stevens.[151] He presumed, rightly, that the operationist stance was intended to render the meaning of *his* concepts in the field of personality research and dynamic psychology "meaningless"— a view that Stevens was not reticent about voicing.[152] Push came to shove, in the early 1940s, over departmental matters of examination requirements for graduate students and the tenure decision for Stevens. Sensing that Lashley, Boring, and Stevens sought to marginalize abnormal and dynamic psychology by pedagogical fiat, Allport made a case for a twin-track system of examination in which his own vision of the field could be given its due. He also tried to push back against the attempt to equate "experimental psychology" with the biotropic enterprises of "behavioristics" and psychophysics. The tenuring of Stevens, after more than a decade of hand-to-hand combat in the department between biotropes and sociotropes, was the final straw for Allport.[153] He and Murray managed to arrange for a visiting committee to assess the department. Before this group could produce its report, however, Allport and Murray jumped ship in 1946 into the Department of Social Relations, the senior membership of which was composed of Harvard faculty in sociology and anthropology with whom Allport and Murray had been in sympathetic contact since the late 1930s.[154]

Despite all of the internal contractions and tensions in which the operational stance was enmeshed, it did not, in fact, die off as the Department of Psychology began to break apart in the years before World War II. Stevens's 1939 article on "Psychology and the Science of Science" served as the prelude to a second wave of the operationist debates, which culminated in the American Association for the Advancement of Science roundtable in 1953. How are we to account for the continuing potency of operationist reason? The answer can be found in the title of Stevens's survey article of 1939. Operationism, at Harvard, was bound up in an efflorescence of the tradition of scientific philosophy in Cambridge during the late 1930s and through the 1940s. That experimental psychology in the operationist vein, whether behaviorist or psychophysical in orientation, should have forged these cross-disciplinary connections was not a surprise. Both Skinner and Stevens sought an empiricist epistemology free from the metaphysical flights of fancy and a priori stipulations of traditional philosophical accounts of knowledge. In trying to get science and knowledge into a scientific, empirical perspective, they were joined by the logical empiricists of the Vienna Circle, with whom they teamed up in an uneasy and sometimes fraught alliance. Skinner declared himself a charter subscriber both to the Vienna Circle's organ *Erkenntnis* and its Anglophone counterpart *Philosophy of Science*.[155] Stevens would form, in 1940–1941, at the visiting Rudolf Carnap's instigation, a "Science of Science Discussion Group," in which a growing band of émigré scientific philosophers and practitioners of the human sciences at Harvard would congregate to explore common problems. To make sense of operationism's afterlife, we therefore need to examine the wider story of the transmission of logical empiricism into Harvard's discourse about the conditions of knowledge.

4

Radical Translation

W. V. Quine and the Reception of Logical Empiricism

Among the disciplines examined in this book, philosophy would seem to fit least well the profile of the Harvard complex. Throughout the early decades of the twentieth century, philosophy at Harvard was not in any obvious sense an interstitial enterprise—quite the opposite. Around 1900, the Department of Philosophy was "the undisputed philosophic center in the United States."[1] Its doyen, George Herbert Palmer, proclaimed that in these years the Harvard department boasted "the first well-rounded staff for teaching philosophy organized in this country . . . made up of extraordinary men, too eminent for praise."[2] In this "golden age," Harvard philosophy was identified with the luminaries William James, Josiah Royce, and George Santayana. Indeed, philosophy had a long and distinguished history in Cambridge. As the theological home of the Unitarianism, Harvard produced in the early 1800s a cadre of sophisticated moral philosophers, which was tasked with defending Unitarian orthodoxy.[3] The publication of Charles Darwin's *On the Origin of Species* sparked among Cambridge-based academics a further round of philosophical rescue operations for liberal Christianity.[4] Even after the golden generation of James, Royce, and Santayana had passed, the Department enjoyed a revival of professional influence after World War I. When a special committee of the American Philosophical Association compiled a survey of the profession in the late 1920s, they collected contributions from six present or former members of the Harvard faculty, easily the largest contingent from any university.[5]

Whatever else Harvard philosophy was by the 1930s, marginal it was not. But this is not quite the full story. This chapter describes how certain Harvard philosophers, under the influence of the tradition of scientific philosophy, took a self-referential turn in epistemology similar to that taken by

L. J. Henderson, Percy Bridgman, B. F. Skinner, and S. S. Stevens. In fact, the changes we shall track in philosophy were bound up with the march through Harvard's interstitial academy of Paretian social science and operationism. One of the principal attractions of scientific philosophy to Harvard's human scientists was that it made possible the pursuit of epistemology by other means: knowledge was to be understood in terms of the research practices of the sciences themselves. American philosophers, too, were influenced by the search for an epistemology more keenly attuned to the practices and empirical findings of the sciences. This had been the defining mark of the pragmatism of Charles Sanders Peirce and William James. During the interwar years, however, philosophers who moved within the interstitial academy were drawn to different epistemological currents in scientific philosophy. Of special concern in this connection was the philosophy of logical empiricism, which was carried to American shores by former members of the Vienna Circle during the 1930s. At Harvard, logical empiricism entered an academic milieu in which the insights of Mach, Poincaré, and Bertrand Russell had already been worked into new methodological shapes by Henderson, Bridgman, Skinner, and Stevens. Shunned for the most part by the Department of Philosophy, logical empiricism would be adapted for new philosophical purposes within the Harvard complex.

The Quinean Lens

In the work of the logical empiricists, the various strands of scientific philosophy were self-consciously united. In their 1929 manifesto, "The Scientific Conception of the World," the Vienna Circle signaled their allegiance to this distinguished if eclectic lineage. "The following," they wrote, "were the main strands from the history of science and philosophy that came together [in Vienna], marked by those of their representatives whose works were mainly read and discussed":

1. Positivism and empiricism: Hume, Enlightenment, Comte, J. S. Mill, Richard Avenarius, Mach.
2. Foundations, aims, and methods of empirical science (hypotheses in physics, geometry, etc.): Helmholtz, Riemann, Mach, Poincaré, Enriques, Duhem, Boltzmann, Einstein.

3. Logistic and its application to reality: Leibniz, Peano, Frege, Schröder, Russell, Whitehead, Wittgenstein.
4. Axiomatics: Pasch, Peano, Vailati, Pieri, Hilbert.
5. Hedonism and positivist sociology: Epicurus, Hume, Bentham, J. S. Mill, Comte, Feuerbach, Marx, Spencer, Muller-Lyer, Popper-Lynkeus, Carl Menger (the elder).[6]

Within this heterodox canon of scientific philosophers, two important elements would come to shape the reception of logical empiricism at Harvard. The scientific world-conception, wrote the Circle's representatives, was distinguished "essentially by *two features*": "*First* it is *empiricist and positivist:* there is knowledge only from experience, which rests on what is immediately given. This sets the limits for the content of legitimate science. *Second,* the scientific world-conception is marked by the application of a certain method, namely *logical analysis.* The aim of scientific effort is to reach the goal [of the scientific world-conception], unified science, by applying logical analysis to the empirical material."[7]

Recent scholarship has shown that the story of the reception of logical empiricism in the United States is complex and national in scope. But the basic orientation toward empiricism, on one hand, and the methods of logical analysis, on the other, defined the way in which the work of the Vienna Circle was taken up at Harvard.[8] The principal fruits of the encounter between the Harvard complex and logical empiricism were twofold: first, the rise of "analytic philosophy" as the main current in the American philosophical profession; and, second, the adoption of a position in epistemology known as "naturalism."[9] Whereas analytic philosophy was indebted to the methods and argumentative rigor embodied in modern logic, naturalism spoke to scientific philosophy's long-held commitment to identifying all positive knowledge with the empirical sciences. In making the Harvard sources of analytic philosophy and naturalism clear, this chapter will focus on a figure indelibly associated with both: Willard Van Orman Quine. Quine helped to "naturalize" logical empiricism in the United States, and he did so by naturalizing epistemology. For Quine, "naturalized epistemology" involved using "the very fruits of science" to explain "how we, physical denizens of the physical world, can have projected our scientific theory of that whole world from our meagre contacts with it."[10] Quine's method promised a scientific account of how science itself was possible, and it was conceived, explicitly, as a replacement for logical reconstructions of scientific theory such as those offered by the logical empiricists.[11]

It is crucial to note, however, that for Quine logical empiricism meant, above all, the philosophy of Rudolf Carnap. In his seminal 1928 treatise *Der Logische Aufbau der Welt,* Carnap had offered the most systematic formulation of the connection between empiricism and the "logistic" of Gottlob Frege, Bertrand Russell, and Alfred North Whitehead. Carnap framed the *Aufbau* as an application of logical analysis to the materials of immediately given experience, but he gave each term a special meaning.[12] With regard to "the given," Carnap proposed to treat basic phenomenal experiences, in their most elementary form, as given neither to a particular subject or self, nor as already divided into the real or unreal—these were all terms that would ultimately have to be constructed *from* the basic elements of experience. Carnap also refused to take specific sensations and feelings as primary: "elementary experiences" were "total impressions" consisting of a network or Gestalt system of experience.[13]

It was on these holistic total impressions that Carnap brought modern logic to bear. As Carnap saw it, logical analysis had been purified by Russell and Whitehead in their *Principia Mathematica* (1910–1913) so that logistic as a whole rested on the theory of relations. The logic of relations, Carnap maintained, opened the door to a new epistemology because it allowed one to create systems of propositions—or theories—from primitive terms by showing how such primitives could be ordered so as to produce more complex statements and operations. The *Aufbau* was thus "an attempt *to apply the theory of relations to the task of analysing reality.*" More specifically, Carnap presented a "constructional system" which involved "a step-by-step derivation or 'construction' of all concepts from certain fundamental concepts." The fundamental concepts were just the unanalyzable elementary experiences, from which, Carnap aimed to demonstrate, the objects of phenomenal experience, the natural sciences, and even the "cultural sciences" could be derived. Critically, Carnap did not view the work of rational reconstruction as the post-hoc regimentation of an already solid body of scientific knowledge. Rather, his application of logical analysis to primitive phenomenal experience would itself produce the objectivity of the sciences. What made the results of the sciences intersubjectively valid was not their grounding in phenomenal experience, for such experience was by defini-tion not intersubjective. What was comparable in all "streams of experience" were the structures and logical connections out of which the statements of the sciences were constructed, and these shared structures were precisely what would be vouchsafed by Carnap's constructional system. Objective knowledge, then, was not presupposed by logical analysis; it was achieved by the construction theory insofar as it revealed the intersubjective structures of experience.[14]

Carnap's technical epistemology, with its blend of psychology and logic, would become crucial to the young Quine's endeavor to fashion his own scientific philosophy. Often hailed as the greatest philosopher of the second half of the twentieth century, Quine is an obvious and important point of reference for any account of the Americanization of logical empiricism.[15] But what strikes the historian about Quine's response to logical empiricism is just how misleading it was, and it is this observation that will allow us see why the Harvard complex is central to understanding Quine's "naturaliza-tion" of logical empiricism. In a series of celebrated essays, Quine portrayed Carnap as a conventional, if logically sophisticated, empiricist.[16] Quine's Carnap was motivated by careworn empiricist concerns with the reduction of knowledge to sense experience. Moreover, Quine viewed Carnap's treat-ment of "analytic statements" as an attempt to ground the certainty of logi-cal and mathematical truths on meaning or "linguistic convention" alone. Quine's reputation as a philosopher rested on his searching criticisms of both of these "dogmas" of modern empiricism.

The problem was that none of this met Carnap on his own terms, as Quine himself sometimes admitted.[17] Certainly, Carnap's thought was evolving rapidly throughout the 1930s, as an early preoccupation with the syntax or logical grammar of scientific languages gave way to a more enduring "seman-tic" enterprise in which questions of confirmation and induction were pri-mary.[18] But all of this work took place under the rubric of Logic of Science or *Wissenschaftslogik,* which Carnap quite explicitly defined as a discipline that broke with the Kantian project of epistemology.[19] *Wissenschaftslogik* was that branch of mathematical logic that used systems of logic to produce for-mal descriptions of languages that would, in turn, enable the formalization of the language of science, in much the same way as Russell and Whitehead had formalized the language of mathematics.[20] Carnap thus treated discrete programs in logic and the foundations of mathematics as tools to be used in "language engineering."[21] Philosophy qua Logic of Science would aid scientific progress by studying the *"syntactic relations between the different languages that form part of the language of unified science,"* in this way providing a *"tool for the construction of a unified science."*[22] All of this was a long way from the old empiricist epistemological problem of reconciling abstract mathematical truths with the tribunal of sense experience.[23]

It was therefore tendentious of Quine to suggest that Carnap's project in the *Aufbau* and after was to "account for the external world as a logical construct of sense data."[24] As recent studies attest, in none of his set-piece critiques of Carnap as a traditional empiricist did Quine present "any *argument* against Carnap": his claims about the conventional nature of logical truth were "irrelevant to Carnap's conception of conventionality," while his vision of the epistemic situation of the scientific observer "at most begs the question against Carnap."[25] Moreover, Quine's claims against Carnap in his most famous essay, 1951's "Two Dogmas of Empiricism," "must," one commentator has written, "appear to Carnap to be unproductive burden shifting"—an outcome that makes for an "unsatisfying standoff" between the two philosophers.[26]

What were the motivations behind Quine's holding Carnap to account in ways the latter could not have found fully intelligible? Leading Carnap scholars indicate that much of Quine's critique rested on some strong, but largely unargued, assumptions about language, meaning, and the justification of belief.[27] Here we reach the nub of the issue that will frame the following discussion of Quine and the Harvard reception of logical empiricism. Quine's reading of Carnap revealed the legacy of an education that was both logical and empiricist, but not that of a logical empiricist. Quine was trained as a mathematical logician and was not fully socialized in philosophy. His bridges to philosophy were provided largely by logic and the study of the foundations of mathematics, on the one hand, and by an exposure to the empiricism of Russell, on the other. While logic provided Quine with resources for thinking philosophically, Russell's empiricism supplied Quine with his "fundamental paradigm for all other types of knowledge": all knowledge had to be reconciled with ordinary sense experience.[28] Upon arriving at Harvard in 1930, Quine's commitments on these scores were reinforced by his intellectual stimuli. Interactions with the epistemologist

C. I. Lewis, as well as with his friend B. F. Skinner and his sponsors Alfred North Whitehead and L. J. Henderson, confirmed Quine's empiricism and behaviorism. His general picture of the epistemic situation of the sciences was further informed by his participation in discussions about the "science of science" that took shape in the interstitial academy during the late 1930s. These cross-disciplinary, interfaculty meetings shaped his understanding of logical empiricism. Quine's naturalistic appeal to the actual findings of the sciences, especially physics and psychology, can be seen, at least in part, as a version of the scientific epistemology we have identified elsewhere in the Harvard complex. It is this peculiar education that explains the curious disconnection between Quine and Carnap.

An Unphilosophical Education

The most striking feature of Quine's induction into the philosophical profession is the absence of the usual commitments to existing canons and schools. We can get a sense of this weak socialization by looking at Quine's immediate predecessors in the field. Whatever their doctrinal differences, the distinguished philosophers who published their intellectual autobiographies in *Contemporary American Philosophy* in 1930 were as one in recalling moments of *Bildung* in which debates between rival philosophical camps were dramatically illuminated by a gifted teacher. John Dewey reminisced about his early exposure to Scottish common sense philosophy and, later, his philosophical awakening to the German Idealists; Clarence Lewis remembered adolescent "doubts and questions which went on until I faced the universe with something of the wonder of the first man"; Morris Cohen cited the abiding importance of youthful encounters with the writings of the American neo–Hegelians.[29] Interwar philosophers possessed a firmly established professional persona.[30] What stands out in the young Quine's curriculum vitae is the absence of many of these professional characteristics. Quine's "unphilosophical"[31] education in philosophy would help to define the form in which logical empiricism was received at Harvard.

America's bustling Midwest certainly gave the young Quine intellectual concerns more practical than cosmological. His early life in Ohio freed him from the usual educational obligations a neophyte philosopher might have been expected to incur. Born in 1908 and raised in Akron, Quine enjoyed a

childhood fired by an enthusiasm for science and technology that characterized much of American culture in the early decades of the twentieth century.
Looking back on those years, Quine remembered the industrial spirit of the
age: the proudly displayed Model-T Ford, which "gave our family wings"; the
corporate "rubber barons" of the tire industry in Akron, centered around
Goodyear and Firestone; and the scientific training in high school which
promised, vaguely, a career in engineering.[32] Quine enrolled at Oberlin
College in the autumn of 1926 evidently imbued with the technophilic lessons of his youth. In his earliest college essays he was happy to extol the
virtues of scientific inquiry in all fields of human learning. Knowledge was
a matter of good engineering, not philosophical dialectic. During his freshman year, he wrote in favor of "the universal coalescence of knowledge." "As
sciences progress," he claimed, "further evidences of internal similarity and
community of principle are continually brought to light . . . hitherto separate
fields coalesce, like dovetailing in the hands of a dexterous cabinet-maker;
specialization becomes conscious and deliberate where it was formerly blind
and accidental."[33]

   If already Quine took the engineer's perspective on questions of knowledge,
he was encouraged in this attitude by the circumstances of his undergraduate education. By the 1920s, Oberlin was fostering congenial conditions for
secular education. During the progressive era, the college had successfully
channeled much of the evangelical enthusiasm that placed it at the forefront
of the antislavery and social gospel movements into more secular causes. In
the words of college historian John Barnard, "much of the zeal that had once
gone into the advancement of evangelicalism was redirected to the pursuit
of learning. . . . Still officially contained within assumptions of Christian
thought and faith, the dynamics of learning and faith gave no guarantee
that they would always remain within that framework."[34] The undergraduate Quine was able to give full rein to a robustly secular attitude toward
questions of mind and language. Upon reading the work of John Watson
in a psychology course, Quine noted how well it "chimed in with my own
predilections."[35] Behaviorism, of a more complex sort than Watson's, would
become an important element in Quine's philosophical and professional
life: as a postdoctoral fellow at Harvard, he would become fast friends with
B. F. Skinner and a member of discussion circles that included operationists
like Stevens and Boring.[36] That he took Watson to be stating more-or-less

intuitive truths about the proper description of mental phenomena bore witness both to Quine's unschooled, resolutely materialist outlook and to his view of these problems as the concern of disciplines other than philosophy.

A related revelation accompanied Quine's discovery of the writings of Bertrand Russell. When Quine came across his work at Oberlin, Russell had very recently published *Philosophy* (1927), the text that had so captivated B. F. Skinner with its Watsonian redescription of epistemology. Russell's writings provided, almost by themselves, Quine's logical and philosophical training during his undergraduate years. At the end of his junior year, as he weighed his options for a major, Quine leaned toward either philosophy—in which he had taken no courses bar a mandated class in the philosophy of religion (Oberlin's Christian pedagogical framework proving tenacious in this case)—or mathematics, in which he excelled. Mathematics seemed the most obvious choice, but Quine worried it was a "dry subject" that "stopped short of what mattered most." It was at this point that a friend alerted him to "Russell, who had a 'mathematical philosophy'"; here was a way of doing mathematics that "promised wider possibilities." Quine decided to major in mathematics, "with honors reading in mathematical philosophy, mathematical logic."[37]

The aspect of Russell's philosophy that Quine admired most turned on an axiomatic conception of how the propositions of the sciences and mathematics were to be justified. In this view, typical of Russell's work around 1900, epistemology itself was to be considered a calculus in which scientific statements were grounded in more primitive terms. Part of Russell's inspiration for this view came from Gottlob Frege. In his *Begriffsschrift,* Frege described how a system of logical inference could be fully specified in the terms of a discrete set of axioms. Specifically, Frege provided an axiomatization of first-order logic: the calculus of propositions and the calculus of predicates. Frege went further with his system of logic and proved that first-order logic could itself account for all of the propositions and operations of arithmetic. And this was consequential indeed, for others had shown that powerful mathematical systems such as the calculus and Euclidean geometry were reducible to arithmetic: if arithmetic could provide the axioms of higher mathematics, and if logic could account for arithmetic, then, Frege claimed, all of mathematics itself could be grounded in a complete and consistent system of logic. Thus the program of "logicism" was born.

"To apply arithmetic in the physical sciences," Frege breezily asserted in *The Foundations of Arithmetic* (1884), "is to bring logic to bear on observed facts; calculation becomes deduction."[38]

Frege's system was soon beset by a paradox attendant upon his use of set theory to define cardinal numbers. Russell and Whitehead attempted to mend this tear in the logicist project by developing, in their *Principia Mathematica,* the theory of types. Russell's importance as a philosopher lay less in the fortunes of his own brand of logicism (itself to be dealt a fatal blow by Gödel's incompleteness theorem) than in the transfer of Frege's model of logical analysis—the reduction of arithmetic to first-order logic—from the study of the foundations of mathematics to matters of epistemology, ontology, and metaphysics. Russell's paradigm of this philosophical technique was his seminal 1905 paper in *Mind,* "On Denoting."[39] To be sure, a large part of Russell's purpose in this essay was to dispute Frege's treatment of denotation and meaning in propositions. But its enduring significance was to show how the techniques of logical reduction or explication—showcased in Frege's treatment of arithmetic—could be put to work in resolving puzzles in philosophy. Russell sought to demonstrate how statements containing denoting phrases ("everything," "nothing" "some," "the," etc.) could be paraphrased as logical statements in which it could be shown that, despite appearances, the denoting phrases in question did not refer to anything on their own, but only as part of the proposition in which they occurred. In essence, Russell used the predicate calculus to show that even apparently nonsensical yet intelligible statements, such as "the present King of France is bald" could be logically reduced to claims that could be straightforwardly treated as either true or false. Crucially, however, Russell presented his account of denotation as a "logical theory" to be "tested by its capacity for dealing with puzzles"—in this case, puzzles about the referents of terms in a proposition. For Russell, philosophy was in the business of logical analysis or explication: "every philosophical problem," he declared in *Our Knowledge of the External World* (1914), "when it is subjected to the necessary analysis and purification, is found either to be not really philosophical at all, or else to be, in the sense in which we are using the word, logical."[40] The sense in which Russell used the word was to characterize knowledge of the logical "forms" or structures that underpinned the obscure and often misleading expressions of natural language.[41] "Logical analysis" involved the retrieval of the logical forms that structured ordinary language, a task that was to be achieved through logical paraphrase, purification, and reduction.

Quine's early philosophical education would be that of a mathematician conducting extra reading in Russell's philosophy and in modern mathematical logic. Quine devoured Russell's writings on science, logic, mathematics, and philosophy. These texts—in addition to two survey courses that Quine did not remember fondly—constituted the bulk of his exposure to philosophy during his time at Oberlin. He read several of Russell's major works, including *Our Knowledge of the External World* (1914), *Introduction to Mathematical Philosophy* (1919), *A-B-C of Relativity* (1926), *An Outline of Philosophy* (1927), and *Sceptical Essays* (1928).[42] Quine's honors reading was principally in "logistics." His notes from this period show he read widely in the field, from Mill's *A System of Logic* to specialist treatises by European mathematicians such as Georg Cantor.[43] Certain strains of contemporary scientific philosophy, notably Wittgenstein's *Tractatus Logico-Philosophicus* (1922) and Hans Reichenbach's meditations on the epistemological implications of relativity theory, also featured on Quine's bibliographical menu.[44] Once more, however, Quine's lodestar was Russell, notably his *Principles of Mathematics* (1903) and the epochal *Principia Mathematica*. His honors thesis, which proved a general law in set theory in the notational system of *Principia*, was an exercise in Russell-style logicism.[45]

Russell's project suited the young Quine's interests very well. The Ohioan was a gifted mathematician and was attracted to philosophical speculation; Russell's use of logic and empiricism to gain insight into the perennial problems of philosophy provided Quine with a method of pursuing his twin interests in mathematics and philosophy while avoiding the narrower, specialist paths of research those two disciplines offered. Quine emerged from his undergraduate studies at Oberlin with two sets of concerns. First, he learned how to be a logician at Oberlin—not a philosopher. Off to one side of this newfound expertise, however, Quine also acquired Russell's commitment to empiricism and, especially, to the notion that logic could be used to reduce scientific statements to a basic sense-datum language—the enterprise that had occupied Russell in *Our Knowledge of the External World*. But whereas Russell had been, from the outset of his career, equally at home with Kant and Hegel as with Cantor and Hilbert,[46] Quine had none of that philosophical erudition; as we shall see, he would not acquire it in graduate school at Harvard. If Quine was going to get to matters of epistemology, it would have to be through logic and the philosophy of mathematics.

Shortly before graduating, Quine spoke before Oberlin's Mathematics Club on his findings in mathematical logic. Modern logic, he told his listeners, was the "most abstract form of mathematics," the "trunk supporting every branch." Its explicit formalization of "exact implication" promised increasing mathematization across the physical and even the social sciences. Strikingly, Quine was careful to insist that at the core of all such mathematical advances lay the bedrock of sense data: "Our raw material is always experience."[47] Both of Quine's doctrinal commitments, to logic and to empiricism, were therefore on display as his time at Oberlin drew to a close. Nevertheless, the relationship between the two was one that Quine had yet to work out: he could be a productive logician, it seemed, but could he entitle himself to epistemological claims in the empiricist vein?

As an ardent admirer of *Principia Mathematica*, Quine was drawn to graduate study at Harvard, where Alfred North Whitehead was a leading light. When Quine applied to Harvard in 1930, moreover, its Department of Philosophy was widely considered the preeminent center in the nation for study in mathematical logic. Alongside Whitehead, at least two other logicians of the first rank, H. M. Sheffer and C. I. Lewis, held faculty positions. Sheffer had become internationally famous among logicians for introducing a small but highly significant simplification into the logical notation of *Principia*. Lewis, on the other hand, was a distinguished modal logician and author of a key American textbook in mathematical logic.[48] A glance at the course offerings in Philosophy during Quine's first year of graduate study, 1930–31, would seem to underscore Harvard's reputation in logic. In addition to a general introduction to logic, designed for undergraduate freshmen, Sheffer offered two courses in his specialist area of "Relational Logic" and a "Seminar in Logic" intended principally for graduate students. Whitehead too taught a Seminar in Logic, as did a visiting John Dewey. The mathematician E. V. Huntington, whom Quine had encountered in his honours reading, offered to philosophers one of his courses on "The Fundamental Concepts of Mathematics." Finally, the connections between science and epistemology were covered in courses taught by Whitehead and Lewis. Whitehead's Philosophy 3, "Philosophy and the Sciences," complemented his "Cosmologies Ancient and Modern." Lewis explored similar territory in Philosophy 10, "Theory of Knowledge," and Philosophy 15, "The Kantian Philosophy."[49]

In spite of these auspicious signs, Harvard's graduate school would not be the scene of Quine's socialization into philosophy. Nor was it the place where the gifted young logician would draw together his instinctive but undertheorized empiricism with his expertise in mathematics and logic. Several factors militated against the crystallization of Quine's interests. For primarily financial reasons, Quine rushed through his PhD studies in two years. All told, the period from enrollment in the autumn of 1930 to the submission of a dissertation in April 1932 was just a little over eighteen months. Quine set himself this Herculean schedule, in part, because in these early years of the Depression he was unsure of financial support for a prolonged bout of graduate study.[50] So he rushed through a program of philosophical courses while devoting himself to the production of a dissertation that emerged from his ongoing engagement with *Principia Mathematica*. Quine took Whitehead's "Cosmologies Ancient and Modern," Lewis's "Theory of Knowledge" and reading course on Kant, one of James Haughton Woods's courses on Plato, and, according to his memoirs, a class by David Wright Prall on Russell's great paragon, Leibniz.[51] This took up most of Quine's first year; half of his second year was devoted entirely to his dissertation, and it would seem he also heard Sheffer's lectures on the logic of relations.[52] Quine was a talented student, receiving high marks, and demonstrating an aptitude for expository prose.[53]

Quine's hurried acquaintance with philosophy was compounded by the disappointment he felt with Harvard's provisions in mathematical logic. Another reason why Quine sped through his degree was that the study of logic at Harvard was not at the cutting edge of new philosophical discoveries in mathematics or the natural sciences. Whitehead was no longer working in logic; Lewis taught no courses in the field; Sheffer published little and was secretive about his research. Too often for Quine's liking, the "logic" invoked in the course catalogue was not the axiomatic systems of *Principia* but that of a Hegel or John Dewey: a mixture of philosophy of science, metaphysics, and symbolic logic. Graduates trained in "logic" at Harvard tended to follow their teachers in presupposing the unique philosophical status of logic as an object of study, rather than addressing technical problems of proof, formalization, and the simplification of procedure.[54] In the 1920s and early 1930s important advances in mathematical logic were being made in Europe, but very few of these were known at Harvard.[55]

Quine was therefore largely unmoved by the teaching he received. He had arrived at Harvard keen to address precisely the kind of technical problems in logic that had fallen out of favor in Emerson Hall. Whitehead gave him encouragement in this area, but little more.[56] While attending to the demands of his coursework, Quine continued his researches in logic largely unaided and was able to submit a dissertation based on his work in April 1932, during his second year of graduate study. His dissertation, rich in "technical detail and mathematical rigor," was unlike anything a PhD committee in philosophy at Harvard had seen since Norbert Wiener's 1913 dissertation—also a mathematical exploration of *Principia Mathematica*.[57] It was recognized as a major achievement and accepted for publication at the Harvard University Press. Yet there was very little in it of the work of the misleadingly named "Harvard school of logic."

In Search of a Theory of Knowledge

As a graduate student, Quine was in the Department of Philosophy at Harvard, but not of it. Finding Cambridge moribund in logic and rushing through his requirements in history of philosophy and the theory of knowledge, there was little in Quine's brief graduate career to give him a way of bringing logic and scientific empiricism together; he remained a logician with ongoing concerns in the philosophical and scientific commitments of mathematics. This is not to say, however, that Quine was entirely unmoved by his experiences at Harvard. Baffling though Quine found his lectures, Whitehead mentored the young Ohioan. The British metaphysician ran a Sunday-afternoon salon in his Cambridge apartment, at which Quine was a regular presence.[58] Something of the old-world flavor of these occasions is captured in Lucien Price's *Dialogues of Alfred North Whitehead*;[59] for Quine, their importance was less Socratic than social. Whitehead's salon was a gymnasium for educated speech that moved across topics, never falling into the specialist jargons of the professions. One can imagine that these sessions would have been especially welcome to someone like Quine, who already fit so uneasily within his chosen field. More saliently, however, the salon provided good practice for a later such forum into which Whitehead would induct Quine: the Society of Fellows. The Society was an interstitial space well suited to Quine's attempts to draw together his commitments to logic, empiricism, and philosophy. Finally, in coming to know Whitehead, Quine was put in touch with a figure whose remarks on the scientific attitude of the contemporary world resonated powerfully within the interstitial academy. Whitehead's Lowell Lectures, *Science and the Modern World* (1925), became the one of the most cited texts among those who moved in the interstitial academy, and especially among Henderson's followers.[60] These were connections that would develop further when Quine returned to Cambridge, full to the brim with Carnapian philosophy and European logic, in 1933.

The other figure who, in an even more subterranean fashion, influenced Quine's later work in philosophy was Clarence Lewis. Quine is widely known to have assailed the pillars of Lewis's philosophy: modal logic, intensions, and analyticity. While Lewis insisted that the truths of logic had nothing to do whatever with the "given" of ordinary sense experience, the mature Quine was to argue that the content of mathematical and logical truths could only be grasped in terms of checkpoints in experience. Nevertheless, Quine the graduate student found in Lewis's writings a way of framing a problematic about logic and mathematics, on the one hand, and the epistemic role of the sensory given, on the other—the two poles of the young Quine's nascent philosophical project.

Lewis had been trained in the Harvard philosophy department of William James, Josiah Royce, and Ralph Barton Perry and thereby in the competing yet overlapping doctrines of pragmatism, idealism, and realism represented by each.[61] Royce introduced him to logic, and Lewis developed from his reading of *Principia Mathematica* the foundations of modal logic. Lewis's logical studies eventually found philosophical expression. In *Mind and the World-Order* (1929), Lewis affirmed the "historic connection which exists between mathematics and exact science on the one hand and conceptions of knowledge on the other." The reason for this close connection, according to Lewis, was that "mathematics, of all human affairs, most clearly exhibits certitude and precision. If only one could come at the basis of this ideal character, the key-conceptions of epistemology might be disclosed."[62] Since Plato, this prospect had encouraged philosophers to frame their epistemological theories with reference to the best concepts that the mathematical sciences had to offer. Lewis felt that recent developments in physics such as quantum theory, alongside Russell and Whitehead's reduction of mathematics to logic, had revealed the analytic or purely definitional character of scientific knowledge. There was, as Lewis put it, a growing recognition of "the independence of the conceptual and the empirical."[63] Lewis was thus provoked to make a categorical distinction between "the given"—by which he meant the immediate sense data of experience—and a "pragmatic a priori" conceptual scheme that was applied to the given in order to make sense of it and to predict future experience.

Even as a graduate student, Quine seems to have been unconvinced by Lewis's definitions of the a priori, but he was not yet able to make his dissatisfaction an issue of philosophical importance. His essays from this period are thus marked by ambivalence. Quine read Kant's *Critique of Pure Reason* under Lewis and made use of Kant's formulation of the analytic/synthetic dichotomy in his coursework. In one essay, he advocated the doctrinaire Kantian understanding of judgments as being "either analytic or synthetic."[64] Yet in the same essay, Quine outlined a theory of "conceptual systems" which placed the analytic/synthetic dichotomy firmly in an empiricist and pragmatic framework. Describing the importance of empirically verifiable "working hypotheses" for the economy of conceptual systems, Quine noted that

> only the working hypothesis can stand which has endured without the emergence of any anomaly in the whole mass of experience since its inauguration. Analytic propositions are deduced on its basis . . . any violation of one of these by a subsequent experience would be a violation of the parent hypothesis. Failing any such violations, the system continues to grow; for other hypotheses have corresponding adventures, the successful ones remain and continue to beget analytic offspring, and groups of such hypotheses and their offspring unite in forming the basis for yet further propositions.[65]

Analytic propositions were seen as part of a fully empirical system of knowledge, capable of amendment not, as Lewis would have it, on the grounds merely of conceptual or syntactic conventions, but on the basis of empirical experience. Such propositions found themselves bound up in an interconnected system of concepts "exhibiting the maximum simplicity compatible with the accommodation of every item of experience falling within the field of that study." "If a recalcitrant item of experience, belonging to the field in question, should subsequently arise," Quine observed, "modification *somewhere in the system* must take place."[66] We are witnessing here the first statement of Quine's holistic theory of scientific knowledge, in which, as Quine would contend in "Two Dogmas of Empiricism" (1951), "a conflict with experience at the periphery occasions readjustments in the interior of the field."[67] These claims were far from orthodox in the early 1930s. In Quine's graduate writings, they lay cheek-by-jowl with received wisdom, with as yet no attempt to square the two.

Similar tensions emerged in Quine's other graduate essays. In papers touching on Lewis's *Mind and the World-Order*, Quine seemed happy to endorse Lewis's fundamental distinction between "the given" of sense data and the "pragmatic a priori" used to organize sense experience.[68] The former was immutable, the latter open to pragmatic adjustment as experience dictated. There are obvious affinities in Quine's thought to that of Lewis, not least in Lewis's advocacy of "conceptual pragmatism" and the importance of predictive capacity in conceptual schemes.[69] But reservations about Lewis's strict separation of the given and the conceptual were evident in Quine's coursework, particularly in papers he did not have to submit to Lewis. Although Lewis never went so far as to contend that the mind had unmediated access to the given, the latter remained an irreducible component of knowledge. For Quine, on the other hand, Lewis's categorical distinction between concepts and sense data made no sense. In one striking passage, he made his own vision of "conceptual pragmatism" clear.

> My experient career is not a simple matter of consciously taking odds and ends and amorphous bits of unidentified data and fitting them into a system; what I see before me is a *chair,* not an array of varicolored quadrilaterals which I consciously assemble and classify as a chair. My immediate experience, rather than consisting of raw material to be interpreted, is already seething with interpretation; in peeling off the interpretation I am peeling off a goodly portion of the immediate datum. . . . In a word, my thesis is that no analysis of a given experience can yield any other experience which is, in any full sense, the "bare datum" of the form of experience; any such analysis is, rather, merely

a further interpretation. It may well suit the purposes of the neurologist or the psychologist to take the presentational aspects of an experience as anterior, and to trace the remainder of the process through neural connections to arrive at conditioned reflexes and general habit responses; but it must be remembered that such treatment, exactly like the physicist's procedure in reducing the perceived object to its electronic constituency, depends upon the prior adoption of a whole system of concepts and hypotheses. Philosophy, if it would enquire into the nature of all such conceptual systems and hypotheses, must certainly endeavor to remain aloof from the initial adoption of any one such system. Let the physically prime be what it will, let the psychologically prime be what the psychologist finds most efficacious; for philosophy, no one item is initially certified as . . . more fundamental or ultimate than any other.[70]

Quine did not flinch from explicitly revising Lewis's conception of the given in the light of these remarks. The given was to be "regarded merely as the determination of *sign*—positive or negative—with respect to the applicability of every concept to the experience in question."[71] In other words, the given was defined as that item in experience which made a concept true or false—no more could be attributed to the given. There was no kernel of preconceptual or uninterpreted experience to which one might appeal in constructing a theory of the world.

As yet, Quine had not succeeded in overturning Lewis's arguments or in making a compelling contribution of his own. Rather, Lewis's account of the pragmatic a priori gave Quine the beginnings of a framework within which he could bring together his twin commitments to empiricism and logic. Further leads were opened up in Quine's dissertation, which addressed the ontological implications of *Principia Mathematica,* in particular those relating to meaning or "intensions." Russell and Whitehead had been vague regarding how the core logical notion of implication ("p implies q" or, in notation, "p ⊃ q") was to be defined. Lewis had spotted this some years earlier and made it the focus of his attempts to clarify the place of intensions in logical studies, a project that had provided the basis for his development of modal logic. Lewis's modal logic supplemented the predicate calculus with a notation that could be used to specify the mode of truth or falsity of statements, for example, to specify whether a given statement was either *necessarily* true or *possibly* (that is, contingently) true. To Lewis, this appeared to be a worthwhile distinction, which might elucidate the difference between mathematical and empirical truths. But Quine was wholly dissatisfied with this use of modal notions in logic, which he felt begged more questions than it answered. He sought to provide mathematical logic with an ontology committed solely to "extensions" (that is, classes of things or the values of bound variables in quantification theory) and one abstract notion, class membership, which could account for problematic pseudo-entities in mathematics such as numbers and functions. Ontology was thereby kept to a minimum.[72]

Carnap's Apostle

Quine's treatment of the foundations of logic and mathematics in his dissertation laid down lines of research he would follow out during the rest of his career. But his engagement with matters of ontology and meaning took place within the horizon of the logical system first limned in *Principia Mathematica*. The scattered remarks about analytic elements in scientific knowledge in Quine's papers for Lewis were paltry by comparison; the young Quine remained a talented logician with only a smattering of training in philosophical disciplines outside of mathematics. This may well have been sufficient for a career in philosophy much like Henry Sheffer's, as an expert in mathematical logic and the foundations of mathematics. The courses Quine offered during his early years as an instructor in Harvard's Department of Philosophy, beginning in 1936–37, provided a staple diet of classes on logic and the philosophy of mathematics: Philosophy 1, "Logic"; Mathematics 19a and 19b, "Mathematical Logic"; Philosophy 20m, "Seminary in the Philosophy of Mathematics" (covering the "derivation of mathematical concepts from logic" and the "problems and methods of metamathematics").[73] Like Sheffer, Quine never threatened to teach outside of his bailiwick in some of the Department's longstanding courses on Kant, the theory of knowledge, or the theory of value. Logic alone could have secured Quine's living as a philosopher.

In fact, however, the only course that Quine taught in these early years that was not squarely within mathematical logic revealed a more ambitious program in scientific philosophy. That course was Philosophy 16, "Logical

Positivism," first offered in the autumn of 1937.[74] Beneath this minor addition to the Harvard course catalogue lay an experience of conversion and discipleship that allowed Quine finally to see his interests in logic and the philosophy of mathematics as part of an attempt to draw lessons about ontology and epistemology from the natural sciences. Two equally important events catalyzed this transformation: Quine's encounter with Rudolf Carnap in Prague in the early spring of 1933 and his election to the Society of Fellows at around the same time. While the former made him see how he could make mathematics the foundation of a wider philosophical project, the latter gave him access to a web of interstitial disciplines in which he could develop his thought free from the constraints of orthodox professional philosophy.

Upon the submission of his dissertation in the spring of 1932, Quine was awarded a Sheldon Travelling Fellowship for the following academic year. The Sheldon Fellows selected at Harvard were expected to extend their horizons in their chosen field—a happy obligation that often meant a Grand Tour around the European universities. Quine was primed to go to Europe: his frustrating experiences at Harvard must have been tempered by his growing awareness that the cutting edge of logic was in Vienna, Berlin, and Warsaw, not in Cambridge, Massachusetts. Two friends at Harvard, John Cooley and Herbert Feigl, gave him an extra push toward Vienna and toward Carnap. Feigl had been one of the coauthors of the Vienna Circle's manifesto, and Cooley had recently discovered Carnap's *Aufbau*. Quine's course was set.

When Quine arrived in Vienna, however, the Circle was beginning to disintegrate. Although the membership at this time included the mathematician Kurt Gödel and the philosopher Freidrich Waismann, many of the most influential members had drifted away.[75] Nonetheless, in a move that signaled the growing ties between European scientific philosophy and American currents in logic and philosophy of science, Moritz Schlick invited Quine to attend the Circle's weekly meetings. The first paper the young American heard was a précis, by Waismann, of Bridgman's *The Logic of Modern Physics*. Soon after, Quine presented a summary of his dissertation research.[76] Absent from these meetings, however, were two figures who Quine was particularly keen to meet: Carnap, who held a chair in Prague, and Alfred Tarski, the leader of an important group of Polish logicians based in Warsaw. In late February 1933, Quine struck out for Prague.

Carnap was extraordinarily welcoming. Quine attended lectures in which Carnap outlined his view of philosophy as logical syntax; more importantly, he lent Quine portions of the manuscript of *Der Logische Syntax der Sprache* (1934) as they emerged from his typewriter and discussed them with the young American during long evenings at Carnap's apartment.[77] Quine was transformed by the six weeks he spent studying with Carnap. "It was my first experience of sustained intellectual engagement with anyone of an older generation, let alone a great man," he wrote toward the end of his life. The contrast with his education at Harvard was stark. His days with Carnap were his

> most notable experience of being intellectually fired by a living teacher rather than by a book. One goes on listening respectfully to one's elders, learning things, hearing things with varying degrees of approval, and expecting, as a matter of course, to have to fall back on one's own resources and those of the library for the main motive power. One recognizes that the professor has his own work to do, and that the problems and approaches that appeal to him need not coincide in any very fruitful way with those that are exercising oneself. I could see myself in the professor's place, and I sought nothing different. I suppose most of us go through life with no brighter view than this of the groves of Academe. So might I have done, but for the graciousness of Carnap.[78]

It is not hard to see why Quine would have found Carnap's way of thinking revelatory. The *Logische Syntax* was an intervention in what became known as the "foundational crisis" in mathematical logic and philosophy of mathematics during the late 1920s and early 1930s. In response to the paradoxes afflicting both Frege's attempt to ground arithmetic on logic and Cantor's set theory, three positions had emerged in the foundations of mathematics in the 1920s: the logicism of Frege and Russell, the formalism of David Hilbert's axiomatic proof theory, and the radical intuitionism of L. E. J. Brower.[79] The claims of these programs, and especially those of logicism and formalism, were thrown into question by Gödel's famous incompleteness theorems in arithmetic and by Tarski's metamathematical work on the semantic view of truth. These were developments that lay squarely within Quine's existing work in the logicism of *Principia Mathematica*. Carnap was searching, in the *Logische Syntax,* for ways of reconciling the contending positions on the foundations of mathematics so as to make the philosophical terrain they covered clearer to all concerned. That approach alone would have grabbed Quine's attention, but Carnap pushed this agenda further by insisting that the techniques opened up by Gödel's proofs, and by the systems of Russell, Hilbert, and Brower, made it possible to replace hitherto fuzzy philosophical notions of epistemology and metaphysics with the exact science of logical syntax. Any study of the logic of science, or logical methodology in science, was "nothing else than the *syntax of the language of science"*—a syntax that, framed in terms of this or that foundational system, was "nothing more than a *combinatorial analysis,* or, in other words, the *geometry* of finite, discrete, serial structures of a particular kind."[80] Quine was being told that his logical expertise was not merely confined to the philosophy of mathematics; it covered whatever was respectable in philosophy, namely, the study of the logic of science. Here was a route through mathematical logic to the core of philosophy since Descartes: epistemology.

It was during Quine's stay in Prague that he received a telegram informing him that he had been elected to a three-year term in the Society of Fellows. This second piece of serendipity had important consequences both for Quine's efforts to make Carnap's philosophy more widely known in the United States and for his endeavor to use Carnap's ideas to unite logic with empiricism. Quine would be a disciple first and a revisionist second. After spending a few weeks in Warsaw with a group of pioneering Polish logicians gathered around Alfred Tarski, Quine made his way back to Cambridge, Massachusetts.[81] His time with Carnap and Tarski was short, but he would later describe this period as "the intellectually most rewarding months I have known."[82] First and foremost, Quine returned to America an eager advocate of Carnap's logical empiricism. "In Prague," he informed a new crop of graduate students at Harvard upon his return, "there is but one figure so far as logic and related matters are concerned, namely Carnap; but because of Carnap alone a long stay in Prague would be amply rewarded."[83]

Unencumbered by teaching duties, Quine spent much of his fellowship pursuing two related agendas. In mathematical logic, he mapped out the terrain of post–*Principia* logical studies, in the form of both original contributions and survey lectures. In the broader realm of scientific philosophy, meanwhile, he began advocating Carnap's program of logical syntax. The first goal quickly changed from a pure research program into a pedagogical mission. One reason for this shift was that Quine's studies in advanced logic quickly became bogged down in Whitehead and Russell's Theory of Types. In *Principia Mathematica,* the Theory of Types had been used to stave off the paradoxes of class membership that had plagued Frege's earlier attempt to provide a logical basis for arithmetic. For Quine, the Theory relied too much upon a cumbersome and somewhat arbitrary "metamathematical grillwork."[84] Much to his frustration, however, Quine could not produce a formal theory of classes that made the prescriptions for class membership a simple matter of definition rather than metamathematical fiat. Nonetheless, Quine's revisions of *Principia* confirmed in print his earlier rejection of excessive ontological commitments in philosophical logic.[85]

More importantly, these studies laid down paths beyond *Principia* that he hoped others would follow. After his appointment as Instructor in the Department of Philosophy at Harvard in 1936, Quine spent the rest of the 1930s giving frequent lecture courses and invited papers at Harvard and elsewhere on the basics of modern logic and on the current state of the field.[86] Toward the end of his life, he emphasized that "the pedagogical motive has dominated my work in logic."[87] All of his major treatises on the foundations of logic and set theory were designed for use in classroom instruction.[88] In the medium term, Quine, in tandem with other logicians such as Princeton's Alonzo Church, succeeded in effecting within American philosophy a shift toward a formalistic, mathematical conception of logic. Before World War II, philosophers could still claim that the formal mathematical logic introduced by Frege, Russell, and Whitehead was but one version of logic, to be ranged alongside the logical theories of Aristotle, Kant, Hegel, F. H. Bradley, Royce, and Dewey; the efforts of Quine and others helped to ensure that, by the early 1950s, the word "logic" was more or less a synonym for mathematical logic.

If Quine was forceful in advocating the new logic, he was even more active in his promotion of Carnap's philosophy. He brought word of Carnap's work to L. J. Henderson in the Society of Fellows; this resulted in Quine's being invited to give a short lecture series at Harvard on Carnap's logical syntax program. The 1934 lectures on Carnap were largely expository—Quine later described them as "completely uncritical."[89] But it would be a mistake to suppose that Quine had adopted Carnap's epistemology in the *Logische Syntax* as his own at this point.[90] One entry in Carnap's professional log during Quine's stay in Prague shows that, even then, Quine had raised serious objections to the use of the analytic/synthetic distinction in the *Logische Syntax*.[91] Yet it would also be wrong to claim that anything like the mature Quine's naturalized epistemology is to be found in the 1934 lectures.[92] The lectures were not intended to display their author's philosophical progress, or to plunge him into the to-and-fro of philosophical disputation. Rather, they gave Quine an opportunity to engender within the Harvard community excitement about the scientific philosophy that he had discovered in Europe. In the peroration of his final lecture, Quine claimed that Carnap had "shown conclusively that the bulk of what we relegate to philosophy can be handled rigorously and clearly within syntax. . . . Whether or not he has really slain the metaphysical wolf, he has shown us how to keep him from our door."[93]

Quine's name soon became closely associated with that of his German mentor. Largely as a result of Quine's efforts, Carnap was awarded an honorary degree at Harvard's tercentenary celebrations in 1936. Speaking at the ceremony, Quine hailed Carnap's philosophy as "a program of universal breadth: a program not only of exploring the philosophical foundations of mathematics and natural science, but of attacking the great metaphysical problems themselves with all the rigor of scientists and mathematicians." Logical empiricism, he concluded, "must be recognized as one of the decisive philosophical movements of modern times."[94] In numerous presentations, Quine invoked Carnap.[95] Quine's efforts on behalf of Carnap and logical empiricism were rewarded in 1939 when the Fifth International Congress for the Unity of Science was held at Harvard. The Congress was, as Quine later observed, basically "the Vienna Circle, with some accretions, in international exile."[96] Here was a turning point in the naturalization of logical empiricism in the United States.

Translating Logical Empiricism

Convened at just the moment when Germany invaded Poland and Europe descended into war, the 1939 Congress is often viewed as the official beginning of the "Americanization" of logical empiricism. Several valuable accounts have been offered of this process of acculturation and adaptation.[97] Gerald Holton, a physicist and historian of science who was party to some attempts to revive the Vienna Circle in the United States, has elegantly traced the fortunes of the "scientific world-conception" during and after World War II. He notes that, in the wake of the Congress, some of the Harvard figures involved, Quine included, instigated a series of interfaculty discussion circles and seminars on the methodology of science. In 1940–41, S. S. Stevens organized a "Science of Science Discussion Group," which was reconstituted by the physicist and former Vienna Circle member Philipp Frank from 1943 onward. Frank's "Inter-Scientific Discussion Group" was soon formalized in the Rockefeller Foundation-supported Institute for the Unity of Science—a division of the American Academy of Arts and Sciences that explicitly sought to continue the scientific and educational activities of its eponymous European ancestor. For Holton, such enterprises could plausibly be seen as a continuation of a conversation about positivism, science, and philosophy begun by the likes of Ernst Mach and William James in the 1890s.[98] Other scholars, by contrast, have discerned a sharp reorientation in both logical empiricism and Unity of Science movement upon its forced emigration to America. According to Peter Galison, the new "interdisciplines" of World War II, which were based on collaborative, problem-based scientific teamwork and guided by technological novelties like radar, the general purpose computer, and cybernetics, fundamentally changed the meaning of "the unity of science." Even stalwarts of the Circle such as Frank recognized that hybrid fields like mathematical biophysics and operations research made older conceptions of unification through syntax and semantics outmoded.[99] But it has also been suggested that the Unity of Science program of the old Vienna Circle was shorn away in the era of McCarthy, when the frankly socialist sympathies of leading logical empiricists made the notion of unity a target for anticommunist crusaders in the academy, thereby pushing scientific philosophy onto "the icy slopes of logic."[100]

These accounts highlight the national dimensions of the story. Harvard was but one site for the communication of logical empiricism into the American academy. In the early stages of the migration, New York and Chicago were far more important conduits for the work of the Vienna Circle. New York was the spiritual home of the Unity of Science movement in the United States. When Otto Neurath, the devoted publicist of the scientific world-conception, visited New York in 1936 in order to spread the word about the Unity of Science program, he found a receptive audience among left-leaning philosophers and public intellectuals. Neurath's advocacy on behalf of an antimetaphysical, social democratic, and scientific blend of philosophy and politics registered with a cohort of academic liberals and socialists, many of them Jewish and indebted to the pragmatist teachings of James and Dewey.[101] New York-based philosophers and publicists such as Ernest Nagel, Sidney Hook, Horace Kallen, William Malisoff, Meyer Shapiro, William Gruen, and Dewey himself forged relations with Neurath during the second half of the 1930s.[102] While Dewey and Nagel wrote favorably, if by no means uncritically, about logical empiricism in professional forums such as the *Journal of Philosophy* and Neurath's *International Encyclopedia of Unified Science,* contributors to leading periodicals such as *Partisan Review* gave positive notices to the Unity of Science project that was reassembling itself in the United States.[103]

Meanwhile, Chicago became the principal organizational hub of movement. The driving forces behind this relocation from *Mitteleuropa* to the Midwest were Carnap, who secured a teaching position at the University of Chicago in 1936, and Charles Morris, a member of Chicago's Department of Philosophy who befriended Carnap and other members of the Circle in 1934.[104] Although Neurath kept nominal control of the Institute for the Unity of Science from his base in the Netherlands, during the later 1930s much of the publishing and publicity work for the Unity of Science movement was conducted in Chicago. Monographs for the *International Encyclopedia of Unified Science,* edited by Carnap, Morris, and Neurath, were, from 1938, published in English by the University of Chicago Press.[105] From 1937 onward, Morris also worked together with the émigré artist Laslo Maholy-Nagy on reconstructing in Chicago a "New Bauhaus" that would bring art, architecture, and scientific philosophy together in a modernist cultural politics.[106] However, the New York and Chicago networks proved brittle. Dewey, for example, held deep reservations about the strictly antimetaphysical line of Neurath, while Horace Kallen and Sidney Hook, early sympathizers with the program of the Viennese scientific philosophers, came in time to denounce what they saw as the totalitarian bent of the Unity of Science movement.[107] Charles Morris became increasingly assertive in his reservations over Neurath's politicized vision of scientific empiricism.[108]

The Harvard complex, and Quine in particular, would therefore serve a unique function in this broader reception of logical empiricism. In New

York and Chicago, the threads of concern that bound together Morris, Dewey, and Nagel with Neurath, Carnap, and Feigl were drawn from the fabric of the Unity of Science movement. This was the wing of the Vienna Circle that sought to apply the resources of logic and scientific empiricism to social politics–to urban reform, to adult education, or to the crusade against religious dogma. But logical empiricism was also a cluster of philosophical problematics and investigative tools. It was this more technical side of the movement that was addressed at the Cambridge site of the emigration. That logical empiricism should have shown up as one view of how knowledge in the sciences was made, rather than as a broad gauge cultural movement, should not be surprising: case-based epistemologies and varieties of operationism were already making their way through the Harvard complex. The Harvard complex was crucial to exactly this normalization of logical empiricism as another proxy for epistemology.

Logical Empiricism in the Harvard Complex

Interstitial networks facilitated Quine's encounter with logical empiricism from the outset. Herbert Feigl, one of the friends who had urged Quine to go to Vienna during his Sheldon Fellowship, was at once an emissary of the Circle and a member of Harvard's interstitial academy. His place in the annals of the Vienna Circle was secure even before he took up a Rockefeller Foundation fellowship at Harvard in 1930. It was Feigl and Freidrich Waismann, as students of Moritz Schlick at the University of Vienna, who in 1924 first broached with their mentor the idea of forming a discussion circle for those interested in practicing philosophy in the spirit of the modern mathematical and natural sciences. From that year until 1930, Feigl was present at the creation of logical empiricism: making sense of Wittgenstein's *Tractatus* and engaging its author in the cafes of Vienna; watching Carnap and Hans Reichenbach fight it out for the position of *Privatdozent* at the city's university; aiding in the composition of the *Wissenschaftliche Weltauffassung*.[109] Throughout, Feigl was active in both the technical and cultural wings of the Circle. On the technical side, he produced a PhD thesis on induction and probability in the epistemology of the natural sciences, and later published a book on the philosophy of physics, *Theorie und Erfahrung in der Physik* (1929). At the same time, he took his philosophy to the Dessau Bauhaus, where he encountered Paul Klee and Wassily Kandinsky.[110] After reading Bridgman's *The Logic of Modern Physics* and recognizing the parallels between Bridgman's thoroughly antimetaphysical theory of concepts and the philosophy of the Circle, he set sail for New York to work with Bridgman at Harvard. While in Cambridge in 1930–31, Feigl moved within the multidisciplinary spaces opened up by the interstitial academy; in addition to becoming acquainted with Bridgman in the Department of Physics, Feigl came to know C. I. Lewis and Henry Sheffer and also attended two important Harvard salons: Whitehead's Sunday "soirées" and a discussion group organized by the philosopher and Radcliffe tutor Susanne K. Langer.[111] Meanwhile, Feigl was able to forge connections between the Circle and his American acquaintances, notably through informal contact with younger figures like Quine but also through the publication of a survey article on "logical positivism" in the *Journal of Philosophy*.[112] Feigl would continue to play an important role in evangelizing the natives through a series of articles and professional leadership roles during the 1930s and 1940s.[113]

When Quine returned from Europe in 1933, Harvard's interstitial networks continued to provide the main source of support for logical empiricism in Cambridge. Largely stymied in the Department of Philosophy, Quine found himself pushing on open doors in his advocacy of Carnap in the interstitial academy. In the mid-to-late 1930s Quine was especially active in cross-disciplinary discussions circles dedicated to the promotion of scientific philosophy. There were three phases in this Viennization of the Harvard complex. From 1933 until around 1939, Quine faced an uphill battle in gaining recognition for scientific philosophy. Between 1939 and American intervention in World War II, however, the ideas and ideals of the Circle briefly flourished owing to a combination of academic serendipity, forced migration, and a critical mass of activism. The third and final phase began with the formation of an interdisciplinary research culture at Harvard and elsewhere during World War II, which in turn produced a reconceived vision of the "unity" project of the original circle.

Whatever success Quine was able to achieve in spreading the Carnapian word after 1933 was due, in large part, to his association with the Society of Fellows and the resources it made available. Quine's 1934 lectures on Carnap were delivered at the invitation of Henderson, as chairman of the Society.[114] Henderson heard enough about the Vienna Circle from Quine to nod toward the *Logische Syntax* in *Pareto's General Sociology* (1937).[115] Quine's lectures, in turn, prepared the way for Henderson's Executive Committee of the Tercentenary Conference of Arts and Sciences to award Carnap an honorary degree.[116] During his three-year tenure in the Society, Quine did not lose contact with Harvard's philosophers: David Wright Prall, Henry Sheffer, and Clarence Lewis were sufficiently stirred by the 1934 lectures to engage Quine on the topic of "logical positivism," and both Prall and Lewis pursued their interests in Carnap in their professional work. Nevertheless, Quine came to see his philosophical project as belonging to the cross-disciplinary enterprises promoted by the Society and embodied by the methodological and epistemological discourses organized around Henderson, Bridgman, Stevens, and Skinner.[117] Unsurprisingly, Quine was treated to Henderson's Paretian sermons, which had few obvious effects bar one: for the length of his professional career, Quine would speak in Henderson's vernacular of the "conceptual schemes" in terms of which agents constructed their theory of the world.[118] Yet more important was the reinforcement of Quine's Russellian conviction that modern empiricism was the provenance of *all* natural and mathematical sciences and not a provenance of professional philosophy alone. Vital to this developing sensibility was Quine's quickly established bond with B. F. Skinner, who confirmed Quine's behavioristic view of meaning and who directed him toward linguistic theorists such as Otto Jespersen and Leonard Bloomfield.[119] Contacts of this sort helped Quine, who was already only weakly affiliated with the mainstream of American philosophy, to embrace a naturalistic, cross-disciplinary view of philosophy, which he identified with "the kind of thing the Society of Fellows was bent on doing, encouraging—rubbing out boundaries between fields."[120] Skinner remembered conversations with Quine being just as valuable for his own work.[121]

At the end of his fellowship, in 1936, Quine was appointed an instructor in the Department of Philosophy, with special responsibility for teaching in mathematical logic and the philosophy of mathematics. Excluding a half-year course on "Logical Positivism" that ran from 1937–38 onward, these highly technical classes were Quine's pedagogical staple.[122] The attitude of the Department toward Carnap and logical empiricism was skeptical at best, hostile at worst; until 1947, Quine would remain unhappy with the philosophical scene in Emerson Hall.[123] However, between 1938 and the attack on Pearl Harbor, Harvard's interstitial discourse on science began to focus on the Vienna Circle-in-exile then taking shape in Cambridge. As more and more scholars fled fascism after 1933, Cambridge, Massachusetts, was an obvious port of call. Not only did distinguished former members of the Vienna Circle like the physicist and philosopher Philipp Frank find it natural to reach out to those sympathetic to their worldview such as Percy Bridgman; under the astronomer Harlow Shapley, Harvard became a center for displaced scholars seeking academic asylum in the United States.[124] A number of these exiles had ties to the Unity of Science movement, and several found teaching positions at Harvard, including Frank and the mathematician Richard von Mises (each in 1939).[125] Around the same time, as war loomed in Europe, American supporters of logical empiricism, such as Quine, Morris, and Bridgman, began to consider how they might bring parts of the movement to North America. The upshot of these maneuvers was that a full debut of the Vienna Circle was arranged for the Fifth International Congress for the Unity of Science, of which Bridgman was chairman and Quine local secretary. In an enthusiastic plug for the Congress in the November 1938 number of the journal *Synthese,* Morris described the Harvard meeting as "the first of the congresses to be held in the United States, and it is wished to acquaint American thinkers with the whole range of activities embraced in the Unity of Science movement." The theme of the conference, in broad Carnapian terms, was to be *"Logic of Science,"* but Morris was at pains to insist that the Congress would combine the best of Viennese exactitude in logic and mathematical science with the strongly "naturalistic" approach of American scholars to psychology and the social sciences. From a "sociological viewpoint," wrote Morris, the United States appeared to be "singularly prepared to play an important place [sic] in the incorporation of the socio-humanistic studies into the unity of science movement."[126]

When we consider the permeation of "socio-humanistic studies" at Harvard by the practice-oriented concepts of the Harvard complex–Paretian systems thinking, operationism, and now logical empiricism–Morris's remarks seem prescient. The program of the Congress bears out this impression. To be sure, the American Organizing Committee for the Congress had a certain austere philosophical pedigree: professional philosophers like C. I. Lewis, Morris Cohen, William Pepperell Montague, and Ernest Nagel were among its members. But the full roster tells a different story. In addition to

Bridgman and Quine, the committee counted among its membership L. J. Henderson, Susanne Langer, Karl Lashley, and the Harvard-based historian of science George Sarton. Several émigré logical empiricists were also on the Committee, including Carnap, Feigl, and Reichenbach. More striking still was the presence of heterodox practitioners of the human sciences such as Clark Hull, Edward Tolman, Charles Morris, Wolfgang Köhler, and Edward Sapir.[127] Those who were struggling to make sense of the nature of knowledge-making in the human sciences, especially in Harvard's interstitial academy, saw the reconstituted Circle as one further extradisciplinary venue in which to pursue their various projects. At the conference itself, which took place as the invasion of Poland began, the logicians and scientific philosophers of Europe rubbed shoulders with central players in the interstitial academy. Bridgman, Sarton, Quine, Pratt, Stevens, Henderson, and Talcott Parsons all gave papers. Speakers from the now-vanished world of middle-European scientific philosophy included Tarski, Neurath, Reichenbach, Carnap, Hempel, and Frank. The themes covered in the papers were strikingly eclectic: Henderson spoke on "A Relation of Physiology to the Social Sciences" while other discussed logic, probability theory, contemporary physics, and the sociology and history of science.[128]

The 1939 Congress turned out to be the prelude to a more sustained interdisciplinary enterprise, which also marked a further extension and elaboration of Harvard's interstitial networks. Remarkably, the academic year 1940–41 saw the thronging of the leaders of scientific empiricism and logic in Cambridge.[129] By far the most notable visitor was Bertrand Russell, who arrived in the winter to deliver the William James lectures, which would be published in 1941 as *An Inquiry into Meaning and Truth*.[130] Also visiting Harvard that year were Rudolf Carnap, on an exchange with Sheffer from the University of Chicago; Herbert Feigl, on another Rockefeller Foundation fellowship; Alfred Tarski, who was still in search of a permanent post; and the historian of science and soon-to-be MIT faculty member Giorgio de Santilliana. And then there were the local scholars with interests in the foundations of science: Bridgman, Quine, Henderson, Frank, von Mises, Langer, Stevens, Sarton, and Boring. Soon Quine, Carnap, and Tarski formed their own discussion circle, in which Quine and Tarski were quick to take issue with Carnap's formulation of the analytic/synthetic distinction.[131] A more formal group was to emerge by the end of October 1940, when Carnap approached Stevens with the idea that the latter might convene a "monthly discussion group of those whose interests could be said to overflow the fences surrounding their narrow professional specialties." The core topic was to be the "science of science."[132]

Carnap had already tried a similar venture at Chicago with his Logic of Science Discussion Group.[133] But the Harvard network was much more explicitly an attempt to recreate something of the coherence and diversity of the old Vienna Circle. The Science of Science Discussion Group meetings were held on the same day and time as the Schlick discussion circle, 6:30 PM on Thursdays.[134] There were also enough veterans from Vienna days on hand to lend an air of authenticity to the simulacrum; in practice, the SSDG took on as much of the complexion of the Harvard complex as it did the Vienna Circle or the Unity of Science movement. The epistemic import of the methods of scientific investigation had been a salient issue among Harvard's human scientists as early as Henderson's Pareto seminar and had been taken up in fields as diverse as sociology, psychology, business studies, and philosophy. The invitation to the first session of the SSDG spoke directly to this decade-long engagement. "As an effort in the direction of debabelization," the invitation read, "the undersigned committee is organizing a supper-and-discussion-group to consider topics in the *Science of Science*. Scientists from various fields, together with some logicians and methodologists present this year at Harvard, will debate issues concerning the meaning and methods of science."[135] The signatories were Carnap, Stevens, Boring, Frank, Quine, Feigl, and Pareto circle member Joseph Schumpeter. Among the target audience for this invitation were, in addition to Bridgman and his cosponsors, L. J. Henderson, Talcott Parsons, Clyde Kluckhohn, George Sarton, and the logician Nelson Goodman. Henderson presented a topic to the Discussion Group, as did Stevens, Frank, and von Mises. Quine, Goodman, Bridgman, and Stevens, among others, attended the first meeting on October 31, 1940.[136]

What needs to be underscored in this account of the SSDG is that it must be viewed as a continuation of the interwar Harvard concern with understanding science in scientific terms—explaining the efficacy and legitimacy of science's beliefs in terms of the actual practices of the sciences themselves, whether pedagogical, psychological, or experimental. We have seen how such a move made sense for Paretians and operationists among Harvard's marginal human scientists. But in the case of Quine, we can observe the reabsorption of these scientific proxies for epistemology back into philosophy itself. For it was in the Society of Fellows, the Tercentenary Conference, the Fifth International Congress for the Unity of Science, and the Science of Science Discussion Group that Quine was able to press an agenda in logical empiricism that was not encouraged in the Department of Philosophy. It is telling that Quine launched his attacks on analyticity in the SSDG and in his discussion circle with Tarski and Carnap.[137]

After his encounter with Carnap in Prague, Quine sought to promote an ambitious philosophical program for logic and the foundations of mathematics, but was still wedded to a free-floating empiricism-cum-behaviorism that was the legacy of his formal education. Quine's experiences in the Harvard complex taught the young logician that his empiricist picture of human knowledge ought to be the one to which all knowledge, logical or otherwise, should be brought to account. His philosophical education had never given him the sophisticated epistemological qualms that thinkers such as Lewis possessed; further maturation in Harvard's interstitial academy emboldened Quine to suppose that all philosophers, too, even logical empiricists, were in the business of framing their claims in terms of the ongoing findings and practices of the sciences. It was small wonder, then, that Quine should have felt that the technical difficulties encountered by Carnap in his syntactical and semantical researches were best answered by beginning philosophical inquiry with the picture of the nature of human knowledge provided by physics and psychology. For Carnap, this was a question-begging move because questions of valid knowledge could come on the philosophical agenda only after the logic of scientific discourse had been formally stipulated. In contrast, Quine's experiences in the Harvard complex had taught him that matters of logic and linguistics were best considered as part of the ongoing enterprise of knowledge-making in the empirical sciences. Quine's naturalism thus possessed a distinctly Harvard pedigree.

5

The Levellers

Harvard's Social Scientists from World War to Cold War

W. V. Quine's response to logical empiricism was notable for its neglect of the Unity of Science movement. Neither Rudolf Carnap's program of the Logic of Science nor Otto Neurath's grandiose attempts to orchestrate a working alliance of the sciences registered strongly at Depression-era Harvard. Nevertheless, as S. S. Stevens's organization of the Science of Science Discussion Group made clear, members of Harvard's interstitial academy were deeply concerned with matters of cross-disciplinary communication and coordination. This was evident from Quine's conviction that logic, natural science, and philosophy had important things to say to one another. The problem of interdisciplinary understanding had from the outset been bound up with the formation of the Harvard complex. Henderson's model of case-based scientific reasoning was attractive to proponents of the human sciences because it appeared to offer an opportunity for those disciplines to make common cause with the natural sciences. In a similar fashion, one of the motivations driving Harvard's psychologists to embrace Bridgman's operationism was that it promised to render unambiguous the hitherto hazy concepts involved in the study of mind and behavior. The Harvard complex was a nursery for disciplines and subfields seeking to arrange a place for themselves in the order of human knowledge—and thereby on the administrative chart of Harvard University.

As the torrid intellectual climate of the Great Depression gave way to the busy and anxious years of the 1940s, the meaning of interdisciplinary cooperation was gradually transformed. The mobilization of science and scholarship during World War II permanently altered the funding, scale, and organization of academic research in the United States. This unprecedented militarization of the academy amounted to a new scientific revolution.[1] Harvard was centrally involved in these developments. Famously, President James

Bryant Conant chaired the National Defense Research Council, a role that included oversight of the Manhattan Project.[2] Harvard's wartime mobilization extended far beyond Conant's government duties: much of the student body of the College was drafted into the armed services; professors headed to Washington for government jobs or worked in campus laboratories on classified military projects; and thousands of trainee officers enrolled for special programs of instruction provided at Harvard by the Army, Navy, and what remained of the faculty. Harvard was no different in these respects from its Cambridge neighbor MIT, or from Chicago, Berkeley, and the California Institute of Technology.[3]

Within months of the attack on Pearl Harbor, new patterns of research emerged in the academic disciplines. Across the board, the war and its aftermath opened new possibilities for the human sciences. Not only did sociologists, political scientists, anthropologists, economists, and psychologists find gainful employment applying their expertise—assaying the social structures of the Axis powers, measuring the morale of Army personnel, modeling supply chains for the armed services, and so forth—but those who engaged in war research were also inducted into a distinctive vision of scientific inquiry. They worked across disciplinary lines and did so with tools and models drawn from emergent, hybrid fields connected with electrical engineering, computer science, radar, and nuclear technology. Even those human scientists who were not involved with this technological avant-garde found themselves immersed in team-based projects on civilian morale, propaganda, and enemy intelligence. Experiences of this kind engendered ambitious new conceptions of interdisciplinarity and scientific unity. In particular, they gave rise to the belief that a new "behavioral science" had taken shape, which, like the physical and biological sciences, could be a source of both fundamental laws and technological control.

Harvard laid down a marker for the nascent behavioral science movement in 1946 when it established the Department of Social Relations, the first interdisciplinary department of its kind in the United States. Like "behavioral science," the term "interdisciplinary" was itself a word with which to conjure in the postwar academy.[4] Historians, with good reason, have emphasized the sense of novelty and optimism that drove the leaders of the DSR. However, the Department owed its existence to more than wartime cultural capital and Cold War scientific imperatives. The Social Relations program would have been inconceivable without the emergence, during the 1930s, of the Harvard complex. This longer history, when combined with the wartime experiences of several of the senior faculty of the DSR, made Harvard an especially hospitable venue for the sorts of challenges that interdisciplinary behavioral scientists had to face. Foremost among these was the need for theoretical foundations— principles that would allow social scientists to synthesize the data, which they had proven so adept at collecting during the war, into the components of a full-fledged science of human behavior. The blurring of epistemology, pedagogy, and investigative practice that had been so crucial to the formation of the human sciences at Harvard continued in new ways after 1945, this time under the banners of interdisciplinarity and behavioral science.

Beyond the Pareto Circle

Grand visions of a unified science of human behavior did not emerge de novo in the 1940s. During World War I, psychologists involved in Army personnel selection considered mental testing an applied social technology and later sought to convert their knowledge into profit through the establishment of consultancies like the Psychological Corporation and the Scott Company. After the war, one of the founders of the Scott Company, Beardsley Ruml, used his stewardship of the Laura Spelman Rockefeller Memorial, one of a family of Rockefeller philanthropic foundations, to funnel nearly $50 million into interdisciplinary social-scientific research.[5] Ruml was convinced that the social sciences, when fully committed to the provision of objective social expertise, could provide tools for the reform and control of human behavior.[6] Spreading LSRM money liberally across American and European universities, Ruml's largest grant was awarded to the leading interdisciplinary institute of the interwar decades, the Institute of Human Relations at Yale. Guided principally by the "psy" sciences—social psychology, experimental psychology (of a neo-behaviorist stripe), and elements of Freudian psychoanalysis, with anthropology and sociology in important but ancillary roles—the IHR remained for much of its existence an eclectic enterprise.[7] Only the so-called Dollard-Miller hypothesis, which linked the frustration of psychic drives to aggression and related sociopathic character traits, would enjoy a significant afterlife in the post–World War II behavioral sciences.[8]

In the case of Harvard, thicker lines of continuity can be drawn between interwar social science and the post–1945 behavioral sciences. Another of

Ruml's pet projects for an integrated social technology was the Elton Mayo-led Hawthorne experiments. Influenced by Henderson's Paretian model for social science, as well as by Mayo's suggestive combination of social anthropology, psychiatry, and social psychology, coworkers on the Hawthorne project like Fritz Roethlisberger and T. North Whitehead would place "human relations" research at the center of the Harvard Business School's agenda up to the 1950s. The connections between the Hawthorne studies and the Pareto network point to a yet more significant feature of interwar Harvard social science, a feature that would encourage the ambitious interdisciplinary program of the Department of Social Relations. If L. J. Henderson offered any lessons to Harvard's marginal human scientists it was that social data were best treated as "cases" of general social systems. Among the fruits of this perspective on scientific reasoning were the "model societies" described in studies by George Homans, Clyde Kluckhohn, and Fritz Roethlisberger. This epistemological disposition toward finding a general theory at work in individual cases became particularly valuable as Depression gave way to World War. After World War II, sound theoretical principles seemed to many aspiring behavioral scientists the most vital foundation for a unified account of human behavior.

But this concern did not involve a straightforward importation of the Hendersonian model of social science into Cold War intellectual culture. Henderson's principles would first be transformed into a more general account of the role of abstraction in the practices of the social sciences. No one connected with the Pareto circle was more adept at revising Henderson's social theory along these lines than the sociologist Talcott Parsons. Appointed to an instructorship in Harvard's Department of Economics in 1927 after graduate studies at the London School of Economics and the University of Heidelberg, Parsons devoted his earliest writings to the relationship between economic and sociological theory. Having written his doctoral thesis on the concept of capitalism in the works of Werner Sombart and Max Weber, Parsons attached great significance to the sociological dimensions of economic rationality.[9] Not only was it historically true, Parsons argued, that economic behavior in capitalist markets rested upon a broad system of cultural values; it was also increasingly the case that "individualistic" and "rationalistic" Anglo-American economics implied a range of "residual" sociological categories not easily absorbed into the narrow assumptions of calculative rationality upon which much orthodox economic theory was based.[10] During the decade before he wrote his first book, *The Structure of Social Action* (published in 1937), Parsons moved from the former problem—the origins and sociological structure of capitalist societies—to the latter question of the residual sociological categories of neoclassical economic theory. This attempt to demonstrate the theoretical autonomy of sociology mirrored Parsons's own move from Harvard's Department of Economics into what was, at Harvard, the fragile and suspect field of sociology.[11] Parsons's double-aspect argument—at once theoretical and intramural—expressed the young sociologist's "cosmopolitan localism": his tendency to give "general . . . significance to questions and issues growing out of one's immediate intellectual context."[12]

Parsons's turn, in the early 1930s, toward Henderson and Pareto displayed a different side of his cosmopolitan localism. In his engagement with Pareto's *Trattato,* Parsons presented the Italian as a sophisticated theoretician who shared the young Parsons's belief that there were specifically sociological dimensions of economic life that demanded their own discrete theory. Soon after the first flush of the Pareto enthusiasm at Harvard, Parsons published a bravura reading of "Pareto's central analytical scheme," which simultaneously echoed and turned to his own purposes Henderson's ideas. With respect to the question of the epistemic status of Pareto's investigations, Parsons followed Henderson's lead in insisting that Pareto did not divorce the conceptual from the empirical. As in Henderson's account of the diagnostic use of conceptual schemes, Parsons argued that Pareto treated abstraction and facticity as continuous:

> most "facts" of science fall short of the empirically possible degree of concrete completeness—they state only certain aspects or elements of the concrete situation in hand. . . . Pareto does not set concept over against fact—the one abstract, the other concrete. This position, short of the radical empiricism which repudiates abstract concepts altogether, issues in the "fiction" theory that concepts are useful fictions, but somehow not "true." Contrary to this view, the element of abstraction is included in his concept of fact as such.[13]

Strikingly, in his gloss on Pareto's theory of inquiry Parsons refrained from depicting Pareto as Henderson did, as the Hippocrates of social

science. Not for Parsons Henderson's folksy picture of the physician's familiarity with the real, a familiarity social scientists might one day hope to
achieve. Parsons agreed with Henderson's emphasis upon the role of conceptual schemes in scientific reasoning, but the latter's deliberately informal
account of diagnostic skill in social science was revised to encompass a more
formal description of self-conscious and rigorous abstraction. Parsons conceived of sociology as an intellectually demanding theoretical occupation.

Parsons's most daring appropriation of Henderson's Pareto involved
a revision of the central notion of non-logical behavior. He reframed the
Trattato's distinction between logical and non-logical actions around what
was, in his view, the central problem of sociological reasoning: the place
of values as both ends and norms in social action. Parsons suggested that
Pareto had put himself in a bind by defining the non-logical "residually," that
is, as those actions in which means were not "intrinsically related to their
end in a scientifically verifiable way." Being thus negatively defined, Pareto's
criterion of non-logical behavior—actions that the canons of scientific reason would not count as rational—dumped into the same category elements
in non-logical action that Parsons considered distinct. Specifically, Parsons
argued that Pareto left himself unable to distinguish "between the elements
of heredity and environment on the one hand and what may be called 'value'
elements on the other." A proper understanding of the spectrum of possible
means-ends relationships in social action, Parsons suggested, revealed that
the logical/non-logical distinction was best made in terms of a spectrum
of possible means-ends relationships, arranged according to the relative
importance of conditions (heredity and environment) and ends (values).[14]
Hence, for Parsons, the "great importance" of Pareto's logical/non-logical
distinction was indicated not by Pareto's own view of the matter (viz., that
humans often behave in a way scientists would view as irrational), but rather
by the "fact" of Parsons's sketch of a system of possible means-ends relationships in social action. What looked like a pious affirmation of Pareto's—and
Henderson's—analytical scheme was in fact a defense of Parsons's own.

The Levellers

Driving Parsons's reformulations of Henderson's social theory was a set
of increasingly grand disciplinary ambitions. From Henderson, Harvard's

fugitive sociologists, anthropologists, industrial psychologists, and management scientists learned that the articulation of systematic general theories could serve as a potent means of advancing the cause of the new social sciences. That conviction would propel the establishment of the Department of Social Relations in 1946. But the payoffs would not be immediate. Until the end of World War II, the prospects of the human sciences at Harvard were bleak. As Parsons pointed out to the Dean of the Faculty of Arts and Sciences Paul H. Buck in 1944, Harvard's "big three" social science departments—history, government, and economics—had "dominated the social sciences situation at Harvard for a generation."[15] The new social sciences, meanwhile, lived on the margins of the organizational chart. We have already surveyed psychology's unhappy separation from philosophy. Meanwhile the Department of Sociology, established in 1931, had only a handful of tenured staff and was governed by an interdepartmental faculty committee.[16] Up to the end of World War II, Harvard anthropology was equally lacking in autonomy and status.

Faced with continuing indifference from the Harvard administration, in the mid-1930s a group of young, nontenured faculty calling themselves "the Levellers" began an informal "shop club" in which they discussed matters of shared scientific concern.[17] Although the nod to the dissidents of the English Civil War was no doubt intended to signify their insurgent status within the Harvard academic community, Parsons later observed that the name was chosen "in consideration of the many levels on which behavioral phenomena required consideration." The Levellers had first gathered in Henderson's seminar on Pareto and Methods of Scientific Investigation. Among the core members who went on to form the shop club were Parsons, Clyde Kluckhohn, O. H. Mowrer, and Henry Murray. While closely associated with the Pareto network, the Levellers also sought alternative forums in which to advance the cause of the "new social sciences."

A breakthrough occurred in 1939 when Parsons was appointed chairman of a Harvard teaching committee, the Committee on Concentration in the Area of Social Science. The CCASS soon became a vehicle for the ambitions of the Levellers, and especially for Parsons, who found himself stymied by Pitirim Sorokin, the senior professor in the Department of Sociology. As chairman of the CCASS, Parsons placed at the top of the Committee's agenda the question of common ground among the social sciences—which, for the purposes of the "concentration" (or major) included history, government, and economics as well as sociology, anthropology, and psychology. Parsons and Kluckhohn collaborated with other members of the Committee on a 1941 report entitled "Toward a Common Language for the Area of Social Science." Parsons later referred to this document as "a kind of theoretical charter of basic social science."[18]

A charter it most certainly was. The primary thesis of the report was that new abstractions were needed for dealing with the antecedent analytical abstractions of individual social sciences. As the authors put it:

> Psychology, Government, Economics, Anthropology, and Sociology can all be viewed as constituting essentially different abstractions of human society. Each has been primarily concerned with eliminating some features of human behavior from consideration and concentrating on explaining the rest. The Area Experiment constitutes an attempt to build up a single model of human behavior in societies in addition to several separate and specialized abstractions.

Such an attempt at unification was not, for Parsons and his collaborators, a matter of convenience: they worried about the lack of "uniformity" evident in "the types of 'things' or 'relationships'" discussed in recent studies by the likes of George Homans and the anthropologist Ruth Benedict. "Still more disconcerting," the report went on,

> is the absence of any explicit common language in either reading or discussion. Not only has it been shown that words such as 'marriage' and 'property' mean different things in each of [the societies examined by Homans, Benedict, and others] but in talking about the various communities as teachers or students there is certain to be confusion unless some rigor in the use of terms can be established. . . . We urgently require a conceptual scheme whereby "behavior," "culture," "society," "property," "authority" (and many other well-known abstractions of the existent social sciences) can be reduced to or articulated with certain elementary categories which can then be integrated in a single coherent framework.[19]

Theory, it seemed, was part of the problem, insofar as the profusion of theoretical languages in the social sciences hampered communication among the disciplines. Yet the authors of "Toward a Common Language" insisted that the appeal to abstract concepts was unavoidable given the complexity and specificity of human behavior. Echoing Henderson, the report rejected the "naïve" suggestion that "if one could get away from 'theories' to 'facts' all would be well." The facts "*never* 'speak for themselves'. They must be cross-examined in terms of a set of abstractions or concepts." In a declaration that bore the mark of Parsons's involvement in the Pareto circle, the report further called into question a central tenet of atheoretical empiricism. It was wrong to suppose that "'facts' themselves rest entirely on the authority of the senses. Rather, as Professor L. J. Henderson says, 'a fact is a receptor experience ('sense datum') *in terms of a conceptual scheme.*' A conceptual scheme is a set of abstractions which provides a convenient way of thinking about observed events or things."[20] Hence abstraction and the creation of theoretical frameworks were hardly to be discouraged; what mattered was finding a conceptual scheme in which the abstractions of the various social sciences could be consolidated and rendered a convenient "tool" for thinking about human behavior in the round.[21]

The authors of the 1941 report had little doubt that the social sciences under their purview *ought* to share a "common language." Pushed together by their shared marginality, the Levellers were keen to find points of connection where the theoretical precepts of their disciplines could be brought into alignment. Parsons's committee outlined an avowedly provisional "Conceptual Scheme for the Area Experiment," which rested on the master concept of behavioral "patterns" and their mediation by individual "status" and "role" functions. These specifications were murky at best, a state of affairs not eased by the repeated splitting of concepts.[22] (There were "ideal patterns" and "behavioral patterns"—essentially, systems of social norms and observed behavior—but there were also "situational patterns," "instrumental patterns," and "integrative patterns"; the mapping of one set of definitions onto another was far from straightforward.[23]) More revealing was the summary of where each discipline was supposed to slot into the common conceptual scheme. It read like the carving up of conceptual territory among the Levellers:

> it is clear that, although the "social scientists" differ among themselves in the abstractions they select, their distinct abstractions nevertheless have important properties in common as opposed to those selected by

physical scientists. These common properties rest upon these facts: all "social sciences" deal with living organisms who have certain physiological propensities ("psychology") but who learn (also "psychology") to surrender, to some degree, their physiological autonomy to cultural ("anthropology") control. The traditional modes of behaving ("anthropology") concretely take form, however, as interactions of human beings living in organized groups and institutionalized relationships (sociology). This culturally patterned ("anthropology") behavior ("Psychology" [*sic*]) in organized groups and institutions ("sociology") tends to become structured into types of activity of peculiar importance ("economics," "government"). Finally, the precise nature of these types (and the general nature of the culture and society as well) has a perspective in time ("history"). The processes that determine events are imbedded not only in the innate nature of human beings ("psychology") and in the structured interrelationships which prevail between human beings at a given time level ("anthropology," "economics," "government," "sociology") but also in the sequential development of these forms in time ("history").[24]

There was a strong element of wish fulfillment in this recipe. Harvard's big three social science departments had been reduced to special instances (of admittedly "peculiar importance") of groups and institutions that were best grasped with the purportedly more basic social sciences of psychology, anthropology, and sociology. This would indeed become the organizational gambit of the Levellers in the early 1940s. In the meantime, however, rapid changes in the intellectual weather during the war made the theoretical hard line of Parsons and his coconspirators more appealing to the Harvard administration.

Harvard's War and the Making of a Behavioral Science

World War II gave the institutional claims of the Levellers extra purchase. It also dramatically transformed their meaning. Two factors drove these changes forward. First, the comprehensive mobilization of resources for the war effort, from manpower and manufacturing to science and technology, weakened the intellectual and administrative boundaries of American universities, Harvard included. This jolt of energy delivered into the academy opened up the scientific playing field, especially for the human sciences. Second, the very success of human scientists in attracting government research contracts and generating empirical results rapidly produced its own problems. Skilled as they were at making data, project leaders for war morale studies and personnel selection programs quickly discovered that they had no shared theory or conceptual framework with which to interpret their results. They were thus compelled to carry out the piecemeal work demanded by their military or civilian masters, without a guiding scientific rationale for research design. By the end of the war, the combined result of these two driving forces of academic change was the exciting prospect of a "behavioral science." It was in this context that the theoretical and interdisciplinary concerns of the Levellers made possible the foundation of the Department of Social Relations.

Because World War II is so often taken to mark a sharp break in the history of the American university, it is worth underscoring the persistence, throughout the war, of Harvard's organizational and pedagogical traditions. In his presidential report for the academic year 1942–43, Conant was keen to point out that while Harvard was "contributing a large share of its scholarly resources to the war effort" the "essential core of the University" had been preserved. What he meant was that not one "faculty has even temporarily suspended its teaching activities; there has been no breach in the continuity of any essential phase of our academic life." No departments or schools had been closed; the Faculty of Arts and Sciences and all of the graduate schools remained in business so that "when the period of demobilization begins we shall be able to pick up our normal peace-time functions in all sections of the University without undue loss of time."[25] No doubt Conant, himself frequently absent from University Hall on government business, wished to reassure the alumni that their alma mater had not become a military academy or (even worse) a technical college. But the glimpses of academic life on campus provided by the *Official Register of Harvard University* told a different story. As Paul Buck, in his Dean's report for the Faculty of Arts and Sciences, put it with respect to the same academic year of 1942–43, "the appearance of uniforms, bugle calls, and formations in the Yard made a marked change in the atmosphere of the College."[26]

So it did. Army and Navy special schools came onto campus months before Pearl Harbor. First to arrive, the Naval Supply School set up shop in the grounds of the Business School and graduated its inaugural class in September 1941. In the summer of the same year, Harvard's Cruft Laboratory laid on a new course in electronics for the Army's Signal Corps and taught Navy officers the rudiments of radar technology.[27] With the shift to a war footing after December 1941, a wide variety of special schools and intensive training programs came to Harvard—some involving instruction from faculty, particularly in the mathematical sciences, others run exclusively by the armed services themselves with Harvard providing only physical and administrative resources. Components of the University were drawn into these programs to a greater or lesser degree: while the Medical School offered a more or less regular slate of courses during the early years of the war—both the Army and Navy needed as many doctors as the Universities could produce—the Business School was given over entirely to the teaching of war-related curricula, including special programs in statistics and the management of war production.[28] By the end of the summer of 1943, some 2,300 men in uniform were receiving instruction in the College and in the medical and dental schools under the auspices of two major education initiatives: the Army Specialized Training Program and the V-12 Navy College Training Program. These programs were designed in part to train officers for the armed services; in part to swell the precipitously declining enrollments of America's colleges and universities; and in part to equip the Army and Navy with the expertise in foreign languages, medicine, and engineering technology needed to fight a modern, global war.[29] In the same period, however, the number of nonenlisted students in the College and professional schools plummeted. By January 1944, there were more than 5,000 Army and Navy enrollments in the University, spread across the special schools, the ASTP, and the V-12 program. The total of civilian students stood at 1,826, less than a quarter of the prewar level of 8,076. As early as 1942, moreover, four hundred faculty had departed Cambridge to take up war-related roles elsewhere.[30]

The Harvard campus itself was a hub of scientific activity. A symptom of the intellectual changes wrought by the war could be found in the names of the special service schools. As of January 1, 1945, the Army had created at Harvard such entities as an Air Force Statistical School (located at the Harvard Business School), an Electronic Training School, and a Civilian Affairs Training School for officers taking on administrative roles overseas.

Meanwhile, the Navy boasted a Radar School, a Communications School, and its own Civilian Affairs Training School.[31] Radar, electronics, communications, civilian affairs—these were novel but vital ventures for America's armed services, and they pointed toward the crucial place technologies such as radar and radio, as well as administration and management science, held in the war effort. Only after the war, however, could Conant and Dean Buck begin to disclose "the contributions of the Faculty in developing the new weapons of war." The press was quick to pick up on Conant's guiding hand in the Manhattan Project; for his part, Buck hailed the work of Conant's student (and Harvard faculty member) Louis Fieser in developing napalm—the substance used to ignite the houses of Tokyo and Yokohama.[32]

Around Harvard Square, several wartime laboratories brought together physicists, psychologists, physiologists, electrical engineers, statisticians, chemists, and many others to address the technological problems of war. Professor of Applied Physics Frederick V. Hunt directed an Underwater Sound Laboratory, which was tasked with finding ways of using radar to track and destroy German submarines. A narrow disciplinary enterprise this most certainly was not: Hunt had to call on people with expertise in his own field, but they had to work together with specialists in electronics, oceanography, and naval strategy.[33] This was an interdisciplinary pattern replicated elsewhere in the name of deeply technical, yet necessarily "applied," real-world challenges of defense, detection, and battle. S. S. Stevens, in the Psycho-Acoustical Laboratory, and the acoustician Leo Baranek, manager of the Electro-Acoustical Laboratory, engaged psychologists, radio engineers, and others in the problem of mitigating noise pollution in aircraft cockpits.[34] An enormous Radio Research Laboratory sprawled over an entire wing of the Biology Building and employed as many as six hundred researchers to work on circumventing radar detection by German air defense.[35] Down the road at MIT, some of Harvard's faculty belonged to other radar research teams based at one the of biggest wartime labs, the Radiation Laboratory.[36] During the war, Howard Aiken's Computation Lab produced an electromechanical calculating machine, the Mark I, which was used in data processing projects central both to military operations and the design of the atomic bomb.[37] All told, Harvard collected $31 million in research contracts from the Office of Scientific Research and Development, putting it behind only MIT and Cal Tech in the receipt of government grants.[38]

As many historians have observed, World War II produced a culture of scientific research in which a premium was placed on getting one's hands dirty and mucking in with specialists from other fields.[39] Physicists and mathematicians could hardly stand aloof from engineers and manufacturers when faced with the challenge of installing radar equipment light enough to be taken on board military aircraft; nor could the theoretical physicists who laid the groundwork for the atom bomb ignore practical matters like the design of proximity fuses. This was not simply a matter of temporarily setting aside disciplinary protocols for the sake "getting the numbers out" in a time of national emergency. For many scientists commissioned by the wartime laboratories, cross-disciplinary collaboration exposed them new hybrid sciences, or what the historian of science Peter Galison has called "interdisciplines": nucleonics, informatics, ballistics, cryptography, cybernetics, game theory, operations research, and computer science.[40] These were fields that straddled, or triangulated, the natural, mathematical, and human sciences, much as research in the Psycho-Acoustic or Underwater Sound laboratories smudged the lines between individual disciplines and between pure science and engineering. Here were sciences of the artificial, of "complex systems" and "cyborg" man-machine networks.

The human sciences were drawn inexorably into this research culture. Even as seemingly tradition-bound an activity as the training of "overseas" (i.e. colonial) administrators in academic military programs encouraged thoughts of cross-disciplinary synthesis. The School for Overseas Administration, created by Conant and the Harvard faculty in 1943, was a case in point. Coopted by the armed forces as American troops captured ground in the Pacific and southern European theatres, the School was soon subject to the same task-oriented demands that kept the likes of Hunt and Stevens busy in their laboratories. The military requirements of the day "often developed so rapidly that the staff of the School was hardly able to plan and put into effect a program to meet one situation before another still more pressing required its attention." Led by the Government professor Carl Friedrich, the multidisciplinary faculty came to "consider itself a team, and . . . there developed naturally a desire on the part of the staff to handle a number of topics by panels of several members whose specialized knowledge would be of value." Friedrich reported to Buck the revelation that "information and types of competence which have in the past been separated into different academic disciplines can be effectively integrated," and the Dean himself noted that "the use of the combined resources of the Faculty on certain areas deserve, and are now obtaining, serious consideration by both this Faculty and other bodies."[41] What had been a promissory note in Parsons's 1941 report on the Harvard social science concentration was becoming, in the research culture of World War II, an ethos of interdisciplinarity. Team-based, problem-focused, applied research was helping to bring down divisions between disciplines, thereby creating a new institutional landscape for the human sciences.

However, it was an open question exactly what kinds of integration were best suited to the human sciences. Many practicing economists, psychologists, anthropologists, and sociologists were attracted by the war-induced "borderland" sciences of cybernetics, computer science, game theory, and operations research. Harvard psychologist E. G. Boring, along with the neurophysiologist and psychiatrist Walter McCulloch, were drawn to the theory of intelligent machines developed during the war by Alan Turing and MIT's Norbert Weiner; psychology's dalliance with cybernetics would be consummated in the famous 1948 Hixon Symposium at Cal Tech, where the field of cognitive science crystallized.[42] Mathematically sophisticated neoclassical economists, meanwhile, threw their lot in with operations research and information theory, a shift that transformed the practices and problems of their discipline in the process.[43] Even representatives of the "softer" social sciences such as Margaret Mead, Gregory Bateson, Clyde Kluckhohn, and Talcott Parsons flirted with Wiener's cybernetics immediately after the war in a series of conferences sponsored by the Josiah Macy, Jr., Foundation.[44] But the future of sociology, psychology, and anthropology of the sort championed by Parsons and Kluckhohn lay not with the "cyborg sciences," but with a constellation of war-engendered research problems. Thousands of social scientists were conscripted into the war effort to tackle issues surrounding human management—from the selection of personnel in the armed forces to the administration of populations under military rule— and to handle the acquisition of intelligence on the social structure, belief systems, and the "national character" of enemy nations.[45] So it was that a range of military and civilian agencies became the proving ground of the new human sciences: agencies like the Office of Strategic Services (OSS), the Office of War Information (OWI), the Psychological Warfare Division of the

Supreme Headquarters of the Allied Expeditionary Force, the Sociological Research Project of the Poston Relocation Center for Japanese-Americans in the Colorado River Valley, and the Research Branch of the Information and Education Division of the War Department.

Within this nexus of wartime research, a new "behavioral science" was conceived. But it was to have a complicated birth. In the first instance, most of those involved in these wartime agencies, from military patrons and civilian administrators to the social scientists themselves, thought they would serve a role somewhat similar to that of the research bureaus and fact-finding advisory commissions through which academic economists, sociologists, and management specialists had furnished their expertise during the New Deal era. For the most part, this commission model of government-sponsored social science placed the experts in the position of providing data, advice, and, occasionally, administrative aid in implementing policy determined by the officers of the state.[46] Hence the survey researchers and statisticians who went to work for the Research Branch in the War Department initially saw it as their job to gather information on the adjustment of conscripted men to life in the United States Army and to gauge morale among the men, with a view to improving the efficiency of America's armed services. In practice, however, this paradigm was rapidly replaced by another, in which the prospect of the prediction and control of human behavior was to the fore. The sources of this shift were complex, but they stemmed from the peculiar social forces unleashed by the war. Very quickly, human scientists convinced themselves that the men and women they studied were plastic creatures, whose beliefs and patterns of behavior were programmed by the culture in which they had been formed, but which, in conditions of social dislocation produced by mass mobilization and total war, could be decoded and manipulated by those with training in the psychology and sociology of "human relations." Researchers on the Army morale study found men unable to articulate a common rationale for America's involvement in World War II; psychologists involved in the selection of candidates for the OSS believed they could strip away the trappings of social identity to reveal the bare human personality beneath; anthropologists assessing the nature of the Japanese or German enemies were convinced that they could "crack the codes" of alien cultures simply by limning the dynamics of personality formation. What began as a data-gathering operation ended as a nascent behavioral science.[47]

Or so it seemed. Figures like Mead and Benedict, who worked in the Foreign Morale Analysis Division of the OSS, could believe that their dialectical vision of "culture and personality" was the key to the scientific understanding of human behavior. But sociologists engaged in the Army morale studies looked to the theory of small groups for techniques of social engineering. More importantly, all of these scholars had taken their research plans from nonacademic military or civilian overlords; unlike their colleagues on the Manhattan Project, they had no clear antecedent hypotheses to test or mathematical laws to fall back on. Indeed, they often disagreed among themselves about what they saw in their data. And data it was—mountains of it—that defined wartime social science. The Army wanted information on its "human resources"; their experts duly gathered it, much like their forebears in New Deal commissions. But if a Newtonian behavioral science was to come into existence, these studies would have to be treated as mere instances of universal laws or general theories. A city, for example, would, in the words of the leader of the Research Branch morale studies Samuel Stouffer, need to be assessed as a case study for "the exploration of general sociological ideas, just as government is such a locus, the factory, or the school, or China, or the consuming public. In other words, I would argue that it is a process which should be the main object of study, and the process should be studied in whatever setting it is most easily available for examination."[48] To gaze through "the school" or "China" at the process beneath was to draw on an overarching set of theoretical principles. The development of a fundamental theory was the vital next step in the creation of a behavioral science.

Founding the DSR

It is not hard to see why Harvard should have become central to this agenda. Process was exactly what the Paretian model societies constructed by Henderson's followers were meant to extract from each "case" of a social system. Likewise, the call for more and better abstraction in the social sciences in "Toward a Common Language" evinced a desire to yoke discrete empirical studies to an overarching conceptual scheme. Predating American intervention in World War II, these commitments made Harvard's social scientists well equipped to move the behavioral science project forward after 1945. Moreover, among those who would go on to found the Department of

Social Relations, several had undertaken research and intelligence work for the wartime agencies. Stouffer, as we have seen, led the Army morale studies, later published as *The American Solider* book series, and would soon be tapped to manage the DSR's Laboratory of Social Relations.[49] Kluckhohn was co-chief of the Foreign Morale Analysis Division of the OWI. The clinical psychologist Henry Murray, meanwhile, led the evaluation of candidates for the OSS, a role that allowed him to deploy his theories of personality formation in a selection "bootcamp" held on a secret country estate known as "S."[50] The OSS became a magnet for Harvard scholars after William Langer, a senior historian on the faculty, was put in charge of its Research and Analysis Branch: among its Harvard alumni were Alex Inkeles, Barrington Moore, Jr., and H. Stuart Hughes.[51] Even Parsons, although he did not hold any formal role in a government agency, gave instruction on the Japanese and Chinese social systems for Harvard's ASTP program and wrote sociological surveys of the Nazi regime.[52] For reasons of institutional structure as well as wartime experience, then, Harvard would become a focal point for the attempt to establish a behavioral science.

It was against this wider background of wartime social science that the Department of Social Relations took shape. The Levellers had bided their time; now, in the mid-1940s, they were ready to catch the wave. In the summer of 1943, a number of dissatisfied faculty members approached the administration to demand a fundamental reorganization of the social sciences. Led by Gordon Allport, this group included Parsons, Henry Murray, and Clyde Kluckhohn.[53] Existing departmental divisions, they asserted, were impeding the development of "basic social science"—a claim that, in the context of the war, would not have sounded far-fetched. Asked to expand on their views by Dean Buck, the group formed a committee that included Allport, Parsons, Murray, and Kluckhohn, with the psychologist (and former Leveller) O. H. Mowrer added to the team. Putting their position to the Dean, the Allport Committee maintained that there existed "an independent field of human social development, interaction and social integration. This study of individuals and their adjustment to each other and to the impersonal environment may be said to be 'basic' to the more specialized studies of economics, government, and the like."[54] The lines of basic social science, the Allport Committee contended, "do not follow present departmental disciplines with any exactitude, but the nucleus of knowledge involved is at present dealt with mainly in the fields of social and clinical psychology, social and cultural anthropology, and institutional sociology."[55] Parsons, watching the formation of the behavioral sciences in government agencies from afar, was particularly bold in his support of this view: "I will stake my whole professional reputation," he wrote Buck in 1944, "on the statement that [the emergence of basic social science] is one of the really great movements of modern scientific thought, comparable, for instance, to the development of Biology in the last third of the 19th century."[56]

By the end of the war, Buck and Conant were ready to accede to the wishes of their long-marginalized social psychologists, sociologists, and cultural anthropologists. Buck's reasons for meeting the Allport Committee's demands were eminently pragmatic: he wished to keep a promising crop of junior social scientists at Harvard, and the creation of an umbrella social science department to house them all seemed the most practical solution.[57] In addition to addressing this long-festering institutional sore point, however, Buck was willing to believe that the new department "represented a logical development of scientific thought during the last generation." The rise of problem-based "interdisciplines" during the war was at the forefront of his mind in this regard: noting in his report for the 1945–46 academic year that, along with the DSR, a new Department of Engineering and Applied Physics had been formed in acknowledgment of the fact that the "most useful man in our laboratories in solving engineering problems under war pressure was the man who was thoroughly grounded in all the basic sciences," he went on to observe that many saw a parallel with the synthesis of specialisms in the foundation of the DSR. "In the eyes of many members of the Faculty this step was in its own sphere quite as radical as that taken in the field of Engineering Sciences and Applied Physics."[58] As Parsons put the matter in one of his many attempts to persuade the Harvard authorities of the need for administrative reform in the social sciences, "the essence of organized research is relating the activities of many research workers to a common empirical subject matter. . . . The conceptual schemes, techniques, and specialized knowledge of *all* [relevant disciplines] will prove essential to adequate treatment of any major problems in the field."[59] The DSR promised to produce scholars as well grounded in all of the relevant sciences as the best laboratory engineers had been during the war.

On the basis of such promises, the Department of Social Relations was established in 1946, formed around the nucleus of the Allport Committee. The principal challenge for the nascent behavioral science was to find some basis for integration of its extremely heterogeneous parts; now the Department of Social Relations stood ready to carry out this synthesis. The financial weather was set very fair for such a venture. The Carnegie Corporation provided grants to the fledgling Department that totaled $335,000 in the first decade of its existence. Of the Corporation's money, $275,000 went to Stouffer's Laboratory of Social Relations to support informal, interdisciplinary pilot studies.[60] Despite this generous helping of manna, the Allport Committee found itself a victim of its own success, for the Harvard administration had acceded to the foundation of the new Department before a consensus had been reached either on the nature of "basic social science" or the means by which that prospective science should be taught. Almost as soon as it opened its doors, the DSR came up against the problem of interdisciplinary communication. An early faculty seminar on methodology was designed to produce the fundamental concepts of basic social science, but, according to George Homans, it "did nothing to integrate the Department methodologically."[61]

Blessed with money but not with scientific consensus, the founders of the DSR relied on tried and tested Harvard modes of organizing research and pedagogy.[62] Graduate education, especially, followed established lines. After the first-year introduction to the field, students had to pass a series of written examinations in order proceed to doctoral studies. In addition to the production of a dissertation, graduate students were expected to meet a series of requirements in statistics, foreign languages, and fieldwork. These characteristic benchmarks of graduate studies in the arts and sciences masked a deeper tension between the interdisciplinary ethos of the Department and the professional imperatives of graduate training. The first-year proseminar in social relations, designed to transmit the interdisciplinary orientation of the DSR, was in practice a pedagogical circus: "exciting, sometimes confusing, and for many a little frightening" was how two early graduate students described it.[63] The coherence of the graduate program as a whole was also uncertain. Although they were granted considerable freedom to roam across disciplines, students had to take their degree in just one of the DSR's constituent fields. One could not take a PhD in "social relations." No other department of that kind existed outside of Harvard.

Several other factors, wound into the DNA of the DSR, also worked against synthesis. Perversely, the Carnegie money, provided with so few strings, was itself part of the problem. Because Stouffer was able freely to dole out

Laboratory funds, and for avowedly speculative projects, the Laboratory of Social Relations was never forced—unlike, for example, Columbia University's Bureau of Applied Social Research—to agree on an official program of research or a single methodological training regime for graduate students.[64] An undoubted intellectual success, the Laboratory nevertheless did little to advance or clarify the case for a fundamental behavioral science.[65] The DSR's two major fieldwork enterprises of the late 1940s—Kluckhohn's Ramah Project in Mexico and the Mobility Project in Greater Boston—were, like the diverse research ventures of the Laboratory, contingent accretions to the Department rather than carefully worked out testing grounds for basic social science. While they provided valuable fieldwork experience for graduate students, these projects did not further the cause of integration.

The DSR's fragmented research and teaching profile was further exacerbated by a lack of unity among the faculty. This fragmentation had a physical dimension: the Department had no single building of its own and consequently its staff was scattered across the Harvard campus.[66] Ideologically, moreover, fissures were opened both among the founding members of the Department and between senior and junior faculty. George Mandler recalled of his days on the faculty that "there was little collaborative research" among its members.[67] His views were echoed by the anthropologist David Schneider, one of the first graduate students in Social Relations and later a lecturer in the Department. "There was a sense," he recalled in an interview given toward the end of his career, "that [Clyde] Kluckhohn was competing with Parsons." Battle was joined over students and the conceptual primacy of "culture" (i.e. anthropology) versus "social systems" (sociology). Stouffer, meanwhile, "tended to want statistics to explain everything." Schneider found these conceptual struggles one of the Department's strengths; among its foremost "attractions" to graduate students was its unintentional pluralism.[68] Clifford Geertz was even more forgiving in his assessment of the disorganization of the Department, which he found as a graduate student to be a "maze of grand possibilities, only loosely related, and some even in serious tension with one another."[69] But such tensions undermined attempts to forge departmental unity. The founders failed utterly to win over the next generation of DSR faculty to their way of thinking. Early junior members of the Department such as George Homans and Jerome Bruner were skeptics from the outset. Little effort was made to integrate subsequent cohorts of nontenured staff. During the 1950s, when the youngest tranche of Social Relations faculty should have been rising through the ranks, a major proportion of junior staff resigned their posts for positions elsewhere.[70]

These early struggles testified to the subtle institutional legacies in play in the DSR. While in many respects on the cutting edge of the postwar behavioral science revolution, the DSR was as much a product of the interwar Harvard complex as it was of the avant-garde interdisciplines of World War II. Indeed, the enthusiasm for interdisciplinary behavioral science after 1945 can be seen as the trigger for the Levellers to institutionalize the networks in the human sciences they had pieced together during the 1930s. Only when we understand the Department of Social Relations as an institutional extension of the interstitial academy can we make sense of the problems it rapidly faced after its foundation. Nowhere were these problems more evident than in Parsons's attempt to develop a unified theory for the behavioral sciences—the Holy Grail of the postwar human sciences. The lessons Parsons had learned in the Harvard complex about the conditions of knowledge in the human sciences would prove extraordinarily unhelpful when put to work in grounding the activities of a formal department of the university.

The Carnegie Project on Theory

In the autumn of 1948, when Parsons began discussing a prospective theoretical research program with Carnegie Corporation officials, it was already evident that the two-year-old Department was not living up to its founding principles. Under the circumstances, it was not surprising that Parsons and his senior colleagues should have turned more energetically than before to the creation of a unified theory of human behavior. Not only was this the goal on which those involved in wartime research had set their sights; it was also the enterprise to which Henderson's Pareto network had devoted itself. Parsons soon won a grant from the Carnegie Corporation for a dedicated program of research on the theoretical foundations of the new department. But there was an irony inherent in such work from the outset: conceived as a service to "the social sciences generally," the Carnegie Project on Theory was a response to an eminently local institutional failure.[71]

At stake for those invested in the fortunes of the Project on Theory was the very meaning of "theory" in the social sciences: its modes, proper domain, and utility as a tool for connecting disciplines. During the late 1940s and early 1950s, there were other routes to disciplinary synthesis in the social sciences. Survey research institutes and area studies programs offered a model of integration based on empirical studies. The Center for Advanced Study in the Behavioral Sciences, on the other hand, exemplified the belief that social scientists would make common cause when given the time and space for the free exchange of ideas, methods, and research findings.[72] Parsons's insistence that theoretical research should be the motor of disciplinary integration was a view typical of the Harvard complex. It is not hard to see why Parsons would have felt that theory could provide the scientific glue that fieldwork, lab research, and methods seminars had failed to furnish. Both the Levellers and the Allport Committee had bonded over concepts: they concerned themselves with the basic features of human action and they agreed that each of their chosen fields captured an important dimension of the "action frame of reference." These had been the animating convictions of the shop club, of the "Common Language" mission statement, and of the Allport Committee's 1943 report on basic social science. It was natural for Parsons to suppose, as he put it once the Project was underway, "that serious concerted effort directed to theoretical work *as such* pays off."[73] Attending to the foundations of knowledge in this way had been common among human scientists at Harvard since the early 1930s.

In the postwar context, however, the situation had changed. Whereas the Levellers had thrived as an insurgent group in informal settings—clubs, dinner parties, interdisciplinary seminars, and workshops—the DSR Project on Theory had to formalize this theoretical interchange in order to translate the organizational combination of the Department into an interdisciplinary scientific program. In his grant application to the Carnegie Corporation, Parsons explicitly connected the theoretical project with the disciplinary matrix represented in the DSR. The funds, he told his sponsors, were intended "for the purpose of working with assistance on the best possible general formulation of the theoretical fundamentals of the field of social relations, that is, of Sociology, Social Anthropology, and Psychology, insofar as they converge in terms of a common conceptual scheme."[74]

If Parsons was acutely aware of the daunting task faced by the Project on Theory, he showed a fatal lack of foresight in his plans for its conduct. His first tactical error was the failure to include any member of the DSR beside himself on the full-time staff of the Project. Initially, Parsons hoped to include

M. Brewster Smith, but Smith decamped Cambridge for Vassar in 1949 and removed himself from the running.[75] Ultimately, Parsons recruited three men to form the core staff: the behavioral psychologist Edward C. Tolman from Berkeley, the sociologist Edward Shils from the University of Chicago, and the social anthropologist Richard Sheldon, a graduate student and fellow of Harvard's Russian Research Center. Tolman and Shils were scholars of significant reputation; Parsons could certainly defend their selection on intellectual grounds.[76] But this choice of collaborators expressed preexisting fissures within the DSR. The inclusion of Shils gave Parsons what he would otherwise have lacked: another social theorist on the Project. He might have chosen the only other social theorist on staff, George Homans, but differences of temperament and theoretical outlook—visible as early as the 1930s, when both men were involved in Henderson's Pareto seminar—blocked collaboration.[77] The recruitment of Shils therefore expressed one limit on the synthesis then taking shape. Tolman's presence was likewise a tacit admission of a constitutive exclusion in the Project. Harvard was home to three of the leading experimental psychologists in the United States: Edwin G. Boring, B. F. Skinner, and S. S. Stevens. In the early 1940s, however, Allport had bridled at Boring's dominance of the Department of Psychology and led the secession of the social psychologists in 1946.[78] The DSR had been born in revolt against Harvard's "biotropic" psychologists and could scarcely incorporate them into the Project on Theory.

Parsons's next move was equally maladroit. He formed two working groups that met weekly to discuss themes and problems raised by the Project staff. The first was composed of the Project leaders plus Allport, Murray, Kluckhohn, Stouffer, and Robert Sears—and thus encompassed the six tenured professors in the DSR. A second seminar, again including Parsons, Tolman, Shils, and Sheldon, was designed to solicit the opinions of interested parties among what Parsons euphemistically called the "younger" members of the Department.[79] The symbolism of this distinction was not lost on the junior membership: there was a "senior" seminar, composed of Levellers and founding members of the DSR,[80] and a general seminar for the junior faculty. This view was confirmed by the reading protocol followed by both seminars. In almost every case, the texts provided for discussion were written by the senior staff: Kluckhohn, Murray, Parsons, Tolman, Shils.[81] Junior faculty played no role in setting the theoretical agenda.

What could have led Parsons to exacerbate the Departmental fragmentation that the Project was supposed to mitigate? The composition of the senior seminar suggests that Parsons saw the Project on Theory as a continuation, albeit in a more formal setting, of the work of the Levellers, the Area of Social Science group, and the Allport Committee. Parsons's solution to the fragmentation of the DSR was a heavy dose of the same conceptual kibitzing that had defined his own most valued intellectual experiences at Harvard. He seems to have thought that the procedures of the Harvard complex could be carried over into the organization of an interdisciplinary department. This was not as solipsistic as it sounds. Despite the ongoing bureaucratization of American academic life and scientific research during the 1940s, research and funding in the social sciences during the early Cold War years relied upon strikingly informal but tightly overlapping networks of scholars, foundation officers, and government contractors.[82] The academic commons, with its seminars and freewheeling interdisciplinary exchange cultures, was viewed by many of its members as a model for democratic society.[83] By no means would it have been aberrant for Parsons to believe that the route to consensus in the DSR lay in the extension and formalization of the template of interdisciplinary cooperation established by the interstitial academy.

In practice, however, Parsons seems to have been less interested in extending the Leveller template than in replicating it outright. Expected to fill the role of underlaborers for the Project on Theory, several members of the junior faculty revolted. One after the other, David McClelland, Alex Inkeles, Richard Soloman, Leo Postman, Jerome Bruner, and George Homans stepped forward to dispute with Parsons about the conceptual premises of the Project. Parsons held that the question of the functional significance of a given pattern of behavior was central to the understanding of social systems qua systems. The notion of function, relative to a complex social structure, was necessary "in order to have a generalized and systematic frame of reference."[84] The dissidents refused to acknowledge that the concept of function did any explanatory work at all and cast doubt on the validity of the systems perspective that underpinned Parsons's vision of theory. A report on the meeting of "Group 2" (the general seminar) held on October 3, 1949 noted "a feeling of uneasiness about the term 'functional': Ordinarily by the term 'functional' is meant that actions produce consequences which will maintain structure. An empirical question immediately presents itself: Are all actions functional? Do they maintain the system?"[85]

In the same meeting, McClelland and Bruner "pointed out that in their recent joint seminar they tried to think of examples of behavior that were not adjustive or adaptive in some sense and could not do so." The problem, as McClelland saw it, was that any sort of behavior could be viewed as functional from one or another perspective: "the results of functional analysis all depend on what you are looking at. Take as an example the training of Army officers. If it is good training it is functional to the army, disfunctional [*sic*] to the officers' families (the officers get killed), and what would be the function to personality [*sic*]."[86]

Homans pressed the point further. Far from being necessary to theoretical explanation, he argued, the concept of function was redundant if it could not characterize exclusively a given form of action.[87] The most it could do was pick up on independent variables in social behavior. In the general seminar of October 10, Homans "offered a challenge to the group: 'Give me a statement in the form "A has a functional significance for B," and I will state it in the form "if not A then not B"'."[88] He repeated the challenge on October 24 by translating Parsons's functional analysis of the medical profession into conditional statements of causal relationships.[89]

At the same meeting, questions about the explanatory utility of the concept of function were extended to the idea of "system" itself. Parsons had parried criticisms of functional explanation by suggesting that only the concept of function could relate isolated instances of behavior to the wider aggregation of interlocking institutions that formed a total social system. This claim rested on Parsons's belief that the structure of social systems could be stated independently and, as it were, pretheoretically, before any functional hypotheses about this or that pattern of behavior had been proffered. Bruner, Inkeles, and Soloman voiced skepticism that structure could be stated in a theoretically neutral fashion or that talk of functions did not presuppose a substantive theoretical understanding of the operation of a system. Responding to Parsons's attempt to give a structural description of two different social systems, Bruner noted that Parsons had invoked notions of "disfunction, threat, and injury." Such talk presupposed "some kind of internal economy of the structure, some kind of equilibrium; and it follows that you have to state, very precisely, what its characteristics are or else decide that you don't want to play that way." Given a further example of structure by Parsons, "Dr. Soloman said that these categories did not derive directly out of functional theory, and that other students, approaching the same problem, would arrive at different categories, depending upon their past experiences in the field."[90]

Having made little effort to bring junior members of the DSR staff into the Project on Theory, Parsons found little encouragement for his theoretical enterprise in the general seminar. Mired in debates on fundamental issues that Parsons hoped would be taken for granted, the seminar petered out before the Project came to a conclusion.[91] Parsons and Shils concentrated their efforts in the senior seminar, among the ex–Levellers. Even here, however, a certain narrowing of scope was visible. While the group did agree upon the "General Statement" on theoretical foundations, published in the collective volume *Toward a General Theory of Action* (1951), the practical dimensions of collaboration and team research in the Project fell away. It was Parsons who drafted the almost-two-hundred-page centerpiece in the *General Theory*, "Values, Motives, and Systems of Action." Accounting for nearly half of the book, this "coauthored" Parsons-Shils monograph was eloquent testimony to the design flaws visible in the Project while it was still on the drawing board.

Theoretical Practices

Organizational problems notwithstanding, a key challenge remained. How to make theorizing seem a legitimate and discrete professional activity, on a par with the small group studies and macroscopic data collection projects also sponsored by the philanthropic foundations? How, in other words, could theory unite the practices of the postwar human sciences? At times, Parsons's attempts to provide some bona fides for his precariously balanced enterprise were almost endearingly ingenuous. Asked by the Carnegie Corporation to describe the procedures used to derive the "results" of the Project on Theory, he replied that they were "essentially those of cogitation in relation to many types and fields of empirical knowledge. This cogitation was partly individual, in the solitude of the study, partly *a deux* in prolonged and intimate discussions, partly in completely informal groups of two, three, or four, and partly in formal meetings of eight to fifteen persons.

. . . The relevant individuals 'racked their brains' as best they could, and attempted to mobilize their empirical knowledge in relation to the theoretical problems."[92]

Perhaps aware that such accounts made it sound as if the procedural norms of the theorist got no more scientific than kicking around campus and attending the odd seminar—racking one's brain was hardly an observable scientific practice—Parsons used other occasions to stress the practical significance of the Project's activities. The theoretical propositions advanced in the Project on Theory, he insisted, should admit of "operational testing out"—that is, they would stand or fall on their consequences for empirical analysis.[93] But Parsons and his staff placed more scientific weight than usual on the theoretical side of the epistemological equation. Theory did not just go cap-in-hand to the data, asking for its measure of legitimacy; it also acted as tutor or director to the unruly play of bare empirics. According to the *General Theory,* a system of scientific theory was uniquely able to codify "our existing concrete knowledge," and thereby "help to promote the process of cumulative growth of our knowledge." As such, it could act as a "guide to research" and as a source of "hypotheses to be applied and tested" in social inquiry. On this reckoning, the theorist was the master synthesizer and orchestrator of scientific research.[94]

The theorist's importance in principle still had to be demonstrated in practice. The Project on Theory deployed a variety of mechanisms to make the elusive practice of theory into an "event" similar to those captured on fieldwork assignments and in the laboratory. Here we find a late version of the identification made in the tradition of scientific philosophy between empirical knowledge and the practices of scientific research. One technique especially favored by Parsons, Shils, and Tolman was the construction of tables and diagrams. These abstract representations and cross-categorizations were intended to reify the occurrence of theory. A prominent example was the Parsons-Shils diagram of the "Components of the Action Frame of Reference."[95] This was designed to display the symmetries and transformations inherent in the system of categories that underpinned the general theory of action as a whole. Capturing in a single spatial organization of statements the key elements of the action frame of reference, the figure was intended to demonstrate how the same two subsystems of an action system—namely, the personality system and the social system—could show up both among the features of actor-subjects in a given action system *and* among the objects toward which actor-subjects might need to orient themselves. The figure also showed how a third kind of system, the cultural system, cut orthogonally across the subject-object distinction in all action systems and was thus to be deployed in the analysis of systems that composed actor-subjects and the objects to which they oriented themselves.

Such instruments, moreover, wore their materiality on their sleeve, because they were also, by virtue of their very existence, proof of something: they were evidence that theorizing in the field of social relations could produce fruitful instruments and mechanisms for organizing the data of social science. In the Carnegie seminars, theoretical tools had to be produced for the first time as evidence of the scientific status of the general theory of action. That bid to embody preexisting theoretical and institutional commitments in a set of tools and practices of thinking had been a central feature of the Harvard complex since its inception. In the Project on Theory, this reflexive move came once more to the fore.

Nowhere was this clearer than with respect to the pattern variables. The pattern-variable scheme was developed by Parsons and Shils in response to the challenge of finding patterns in the orientations of actors.[96] Given the complexity of the situations in which actors found themselves, how methodically to determine their orientations? Drawing on his earlier analysis of professional roles in modern societies, Parsons elaborated a series of five dichotomies that defined the possible value-orientations an actor could have in performing any kind of action.[97] These "pattern variables" attempted "to formulate the way each and every social action, long- or short-term, proposed or concrete, prescribed or carried out, can be analyzed into five choices ([which may be] conscious or unconscious, implicit or explici) formulated by these five dichotomies."[98] This was tantamount to the claim that all actions were oriented toward a system of values, a position that Parsons had sought to vindicate in his earlier work in economic sociology and the nature of social action.

On the face of it, the pattern variables were themselves uncontroversial. It seemed reasonable to ask of an action whether or not it was carried out with regard to its consequences beyond the personal gratification of the actor concerned; whether it was undertaken only in terms of its personal significance for the actor or with regard to its consequences for a collective;

whether the actor assessed the object of orientation in a disinterested manner or from the perspective of its personal importance for the actor as a private individual; whether, if the object of action was social, it was viewed as a composite of particular sorts of contingent action or of fixed, ascribed qualities; and, finally, whether the social object could make only a specific claim on one's own activities, and vice versa, or whether its diffuse demands could be made and received with respect to that object. Modern medical doctors, for example, would be expected *in their professional duties* to assess the consequences of their actions, consider their implications for the collective good and accepted moral codes, treat their patients in a disinterested manner, and ask and receive only specific claims related to their role as medical professionals. Outside of their medical practice, however, the same doctors' actions towards family members in the home would likely be marked by personal and diffuse orientations. Here was one way of marking the distinctions between professional and kinship roles.

What made these categorical distinctions important in the context of the *General Theory*, however, was that Parsons attempted to make them into one of his key theoretical tools. He and Shils reported that the "pattern variables have proved to form . . . a peculiarly strategic focus of the whole theory of action."[99] They couched their five sets of categories in a jargon, in which the pattern variables were listed as follows:

1. Affectivity—Affective neutrality.
2. Self-orientation—Collectivity orientation.
3. Universalism—Particularism.
4. Ascription—Achievement.
5. Specificity—Diffuseness.

These dichotomies were presented as transcendental conditions of action, "one side of which must be chosen by an actor before the meaning of a situation is determinate for him, and thus before he can act with respect to the situation." Parsons and Shils also asserted the logical completeness of the variables with respect to the action frame of reference; as such, "in the sense that they are *all* of the pattern variables that so derive, they constitute a system."[100] As a systematic set of variables that defined all possible forms of social action, Parsons and Shils conceived the pattern variables as an analytic machine. A series of boxes defining the value components of both actors (the domain of psychology) and social roles (the object of sociology) were designed as tools for deriving the value patterns of any general system of action taken as an object of empirical analysis. Insofar as they appeared to demonstrate the systematic interrelation of the pattern variables, these categorizing grids bore witness to the technical sophistication of the general theory of action as a whole.

In addition to building an instrumentarium for social theory, Parsons also deployed distinctive rhetorical strategies to give theorizing a temporal and scientific specificity. Time and again, in grant reports, intragroup memos, and the published outputs of the Project, Parsons used the classic scientific rhetoric of the "breakthrough" or unexpected result—the "Eureka" moment so often invoked in the history of scientific discovery. While still in the midst of the Project meetings, Parsons sought to tantalize (and no doubt also to mollify) his contact at the Carnegie Corporation with the claim that "within the last two weeks a very important theoretical break-through has occured [sic] which will allow us to attain a level of order and clarity of the whole field which I don't think any of us thought possible when we began."[101] Soon after, Parsons prepared for the Project members "A Narrative Account of Theoretical Developments: Dec. 3–18, 1949," in which the sheer sequence and timing of breakthrough events in the theoretical debates were recounted with the precision of date-and-time lab results.[102] This concern for the "happening" of theory was carried through into the meetings that were held in December, during which Parsons dramatized the lunchtime discussions and postseminar encounters in which this or that revision of a problem was hammered out.[103] In both *Toward a General Theory of Action* and *The Social System,* Parsons was emphatic that theorizing was subject to patterns of discovery and advance similar to those in experimental-technical research. He excused any inconsistencies in his theory of social systems with the boast that "the development of theoretical ideas has been proceeding so rapidly that a difference of a few months or even weeks in time may lead to important changes."[104]

This obsession with making theory into a scientific-empirical event marked the Project from the outset. Parsons arranged to have the discussions of the two weekly seminars taped, summarized (later, simply transcribed), and then analyzed as two studies in "small group" dynamics—one "an interaction analysis of the actual processes that have gone on in the group," and the other an account of "the intellectual processes in reaching the final agreement" on the theoretical fundamentals of social relations.[105] Neither study seems to have been completed; the collapse of the junior seminar no doubt made such self-scrutiny less enticing. But the existence of the transcripts testifies to the desire to give the nascent subculture of social relations theory a spatiotemporal embodiment: the transcripts were designed as evidence of *work* of a certain sort. Research practice and theory were one. The belief that such records could be evidence of science-in-action underscored the faith that Parsons and some of his colleagues placed in the significance of academic conversation and tool-sharing as substantive scientific practices in their own right.

In some respects, the seminar records betray the grand claims made on behalf of theory-as-event by exposing them to the harsh light of word-by-word transcription, with its mangled syntax, trivial asides, and bathetic absence of authentic "break-through" moments (Allport: "Don't tell me you're embracing functional autonomy!" Tolman: "I've had a little too much to drink; I don't know what I'm embracing"). But they also bear witness to the self-consciousness of those who sought to practice theory in the DSR: it was not enough that they should have a theory of society; the creation of that theory had itself to be understood in objective, instrumental terms.

The fate of the Project on Theory was decided in 1951, when a routine DSR staff meeting became the scene of a high-stakes academic standoff.[106] Parsons opened proceedings by placing before his colleagues the newly published *Toward a General Theory of Action*. The message of the book, as finessed by Parsons and Shils, was clear: with a system of categories for the analysis of action in place, social scientists were on the brink of formulating the general laws of behavioral science.[107] The chairman exhorted his staff to read the volume and suggested that the tenets of action theory might guide their future teaching and research. Parsons was not, however, preaching to the choir. As soon as he concluded his recommendation, George Homans, a leader of the restive junior faculty, leapt in with a rebuttal. There should, he stated, "be no implication that this document is to be taken as the official doctrine of the department, and no member shall be put under any pressure to read it." A pregnant silence followed. Homans had touched a raw nerve: was the Department to coalesce around the *General Theory* as a theoretical resource and pedagogical guide, thereby realizing the promise of "basic social science" adumbrated in the organizational fusion of the DSR itself, or would dissenting voices deny Parsons's programmatic ambitions and confirm a looser interdisciplinary principle as the guiding ethos of the Department? Homans himself had just completed a theoretical treatise profoundly at odds with Parsons's action theory.[108] At length, one of the senior faculty contributors to the Project on Theory, the statistician Samuel Stouffer, intervened. The *General Theory*, Stouffer conceded, was not to be taken as official doctrine. The question of a general theory was quickly dropped, at least at the Departmental level.[109]

Judged by its consequences, the Homans–Parsons confrontation seems like a minor academic spat. In its wake, Parsons moved quickly onto new theoretical pastures, and his colleagues returned to the ad hoc collaborations and eclectic research projects that had characterized the DSR from its inception.[110] "In no sense did [the Carnegie Project on Theory] result in an 'official' theoretical line for the Department," Parsons wrote in his decennial departmental review of 1956. The Carnegie seminars, he now breezily recorded, had been an act of intellectual housekeeping of the sort that might be carried out "at intervals of every ten or fifteen years."[111] As the DSR limped toward dissolution at the beginning of the 1970s, Parsons became even more low-key about his early ambitions for the Department: Social Relations, he observed, had been nothing more than an "experiment."[112] In fact it had been much more than that. For Parsons, his closest collaborators, and his sponsors, the Project on Theory was designed to produce an enduring synthesis in the study of human behavior. But the DSR, it seemed, was in no position to create such a synthesis. The loose weave of the Harvard complex could not, in the end, service the interdisciplinary ambitions of Parsons, Kluckhohn, and Stouffer.

6

Lessons of the Revolution

History, Sociology, and Philosophy of Science

The failure of the Carnegie Project on Theory revealed an important limitation of the Harvard complex. Parsons and his fellow Levellers had assumed that the protocols guiding seminar discussion and informal interchange in the interstitial academy would offer a suitable foundation for a general theory that could unite the Department of Social Relations. The blending of epistemology, pedagogy, and research practice seemed natural to the alumni of Henderson's Pareto seminar. But the ambitions of the DSR's leaders proved too grand. Harvard's interstitial networks were too diverse and amorphous to provide the basis for a coherent independent department or an integrated general theory. Much as the Harvard complex might seed novel fields and "interdisciplines," it could not, ultimately, create templates from which stable academic departments could be manufactured.

If the Harvard complex was not conducive to interdisciplinary synthesis on the Parsonian model, however, it did give individuals an education that escaped the confines of professional socialization. L. J. Henderson was among the first products of this interstitial intellectual culture, his tenure in the Royce Club serving as the prelude to a career teaching and working in fields outside of his scientific home in biological chemistry. In preceding chapters we have explored the forums that supported such careers: the Pareto Circle, Sociology 23, the Society of Fellows, the Business School, the Science of Science Discussion Group, the Inter-Science Discussion Group, and the Department of Social Relations. While attaching themselves to these enterprises, the scholars and scientists whom we have examined nonetheless came to acquire strong disciplinary and departmental commitments: Homans and Parsons in sociology, Quine in philosophy, Skinner in psychology. This final chapter examines the intellectual consequences of an education drawn from the resources of the interstitial academy alone. What would be the implications for one's understanding of the nature of knowledge to have been formed by a culture of scientific philosophy in which epistemology, pedagogy, and research practice were interchangeable?

One figure embodied the interstitial mind in its purest form: Thomas Kuhn. Kuhn's famous book of 1962, *The Structure of Scientific Revolutions*, displayed the heritage of the Harvard complex in condensed form. Elements of this Harvard background to Kuhn's theory of science have been highlighted before. In a survey of the history of the sociology of science, Robert K. Merton took care to record Kuhn's formative experiences in "serendipity-prone intellectual microenvironments" such as the Society of Fellows.[1] Kuhn himself was quick to point out the importance of his time in interior or extra-disciplinary associations for his methodologically eclectic view of science.[2] The sociologist Steve Fuller, meanwhile, has argued that the character and influence of *The Structure of Scientific Revolutions* cannot be understood apart from Kuhn's association with Harvard University, both in regard to its social prestige and to the elitist social theory that animated James Bryant Conant's vision of general education in science.[3] None of these accounts, however, have situated Kuhn where he belongs, in the dense interstitial "microenvironments" of the Harvard complex, where his serendipitous encounters with a diverse array of epistemological texts took place.

Singular though Kuhn's education in the interstitial academy was, his struggle to produce a theory of knowledge from the study of the history of science was not unprecedented. Kuhn's theory of science was one of a series of attempts at mid-twentieth-century Harvard to grasp the nature of the scientific understanding through historical analysis. L. J. Henderson had instigated this tradition in 1911 with the establishment of the first general history of science course at Harvard, History of Science 1. This early attempt to gauge the historical and social character of scientific knowledge was soon augmented by the scholarship of the Harvard scholars George Sarton, Robert Merton, Bernard Barber, Philipp Frank, and President James Bryant Conant. At stake in the work of these figures was relationship between scientific inquiry and society: in what sense could science be considered a social activity, and what were the benefits of the scientific attitude for a society? Although these questions had been posed in the United States for more than a century, they acquired special significance in the age of the atomic bomb and in the context of the practice-oriented epistemology promoted in the Harvard complex. Kuhn's ultimate embrace of a pedagogical theory of science in *The Structure of Scientific Revolutions* represented a distinctive response to these pressing problems of epistemology and society.

The Social Meaning of Science from Sarton to Bernal

The questions of science and history that Kuhn sought to answer had been posed inside the Harvard complex since the time of World War I. L. J. Henderson never considered himself a historian, but he saw enough value in the history of science to make it one of the focal points of his work in the interstitial academy. In addition to establishing History of Science 1, Henderson's gift to the history of science at Harvard was to bring George Sarton onto campus. This was the beginning of what would become a half-century relationship between the University and the Belgian historian of science.[4] Sarton's difficult and ultimately unfulfilled career in Cambridge exemplified the possibilities and limits of the history of science at Harvard during the interwar years.[5] Henderson and Sarton met through the Carnegie Institution in Washington, D.C., where Henderson was an associate in 1916. Three years earlier, Sarton founded the journal *Isis* as a forum devoted to the study of the history of science. Sarton was "above all a man of the nineteenth century," and nowhere was his attraction to the tropes and commitments of nineteenth-century thought more evident than in his plans for *Isis*.[6] As a disciple of the positivists Auguste Comte and Herbert Spencer, Sarton held a faith in the progression of mankind, which, he argued, was expressed in the record of scientific advancement. A corollary of this progressive faith was a committed internationalism, for the historical record showed that contributions to scientific knowledge had come from all the civilizations of mankind, whether medieval or modern, Arabic or Anglo-Saxon: the history of science was to provide the basis for a "new humanism" in the modern world. Finally, Sarton was convinced that the history of science itself had to be a quintessentially modernist enterprise, involving the cataloguing and recording of all available knowledge and artifacts and the use of the methods of advanced bibliography and librarianship.[7]

Because, for Sarton, the history of science had to be so much more than an academic discipline, it remained, during the Belgian's tenure at Harvard, a good deal less. After 1916, Sarton kept an arrangement with Harvard whereby he was granted the use of a suite of offices in the Widener Library in return for providing a course of lectures; the bulk of his salary was paid by the Carnegie Institution.[8] Although Sarton chafed at his marginal status at Harvard, he consistently missed the opportunity to build a discipline from the ground up. He let it be known that he considered his editorial work for *Isis,* along with the production of his monumental multivolume *Introduction to the History of Science,* much more important than his teaching commitments.[9] Although he delivered a large number of new lectures at Harvard after his arrival, he seemed keenest to skip ahead to the foundation of an "Institute for the History of Science and Civilization." The purpose of such an Institute, Sarton explained, would be to organize a team of scholars whose task would be "to interpret the ethical and social implications of science in all ages, and especially in our own, to integrate science into general education, in a word to 'humanize' science."[10] When this visionary project was met with a lukewarm response by the administrations of Lowell and Conant, Sarton found himself adrift in his small workshop in Widener, without a pedagogical or disciplinary foothold in the College or graduate schools. Only in 1934 was History of Science 2 added to History of Science 1—and this was an expanded version of Sarton's second half of Henderson's original course.[11] Up to the eve of Pearl Harbor, in fact, the listings for History of Science in the *Official Register* contained just a handful of courses.[12] In 1938, a Committee on Higher Degrees in the History of Science and Learning was established. But this was a Hendersonian operation, with the biochemist as chairman and allies like Crane Brinton and Arthur Darby Nock as members. Tucked away in his small institutional enclave, Sarton was nonetheless able, by 1935, to establish a seminar in the history of science that counted among its members such future leaders of the field as Henry Guerlac, Robert Merton, and Bernard Cohen. Still, his vaulting ambitions and unrealistic expectations left him constitutionally unable to give birth to the history of science as a discipline at Harvard.[13]

American intervention in World War II substantially changed the picture, yet the ultimate effect of the war was to preserve the history of science in its peculiar ideological position. Most obviously, the war placed the social relations of science at the center of both public and academic consciousness. The mushroom clouds over Hiroshima and Nagasaki in the summer of 1945 provided a minatory illustration of the apocalyptic power that concerted scientific research could unleash. Subsequent negotiations at the United Nations over the international control of atomic energy made clear to all the high stakes involved in the public administration of science and technology.[14] Yet the diminutive atom was in many respects just the tip of the iceberg, beneath which bulked large the body of a new phenomenon: Big Science.[15] Our survey of the transformation of Harvard in the preceding chapter gave us a glimpse of this structure. In place of the lone wolf investigator, who tinkered with his instruments at the laboratory workbench, the scaled-up, capital-intensive laboratories of World War II substituted large, interdisciplinary research teams and ever-more complex and extensive machines.[16] Harvard had its Radio Research Laboratory, MIT its Radiation Laboratory. Meanwhile, Ernest Lawrence built giant calutrons at Berkeley for the purpose of enriching uranium to be used in the creation of the nuclear bomb. The Manhattan Project itself spanned a national network of laboratories, manufacturing plants, and test sites, the nodes of which included Chicago; Oak Ridge, Tennessee; and Los Alamos, New Mexico.[17] This was science on an industrial scale, the products of which were wonders of planned, generously funded collective endeavor among academic scientists, engineers, and the military. Among those products were the atomic bomb itself, along with radar, rocket technology, Lawrence's cyclotrons, myriad vacuum tubes, early electrical computers, and much else besides.[18] Although state-funded public works projects in science and technology were not unknown during the New Deal era—the Tennessee Valley Authority's hydroelectric projects and the Golden Gate Bridge were early avatars of wartime Big Science and Technology—the scale of financing, coordination, and personnel involved in the laboratories of World War II was unprecedented.

That science was now a significant, if not a crucial, social estate was by the middle of the 1940s no longer in doubt. While it went without saying that the future of atomic energy would be near the top of the postwar agenda of the political class, the role that scientists could or ought to play in the decision-making process was much less clear and would soon bring figures like J. Robert Oppenheimer to an unhappy pass.[19] But the ripples in the social fabric created by the sheer standing fact of Big Science were immense. What to do with the artificially pumped-up scientific infrastructure created by the military-industrial complex during wartime? Debates on this issue were drawn inexorably into the struggle over the establishment of a National Science Foundation.[20] Some, like the president of the National Academy of Sciences, Frank Jewett, opposed the NSF bill in Congress because they believed it would create a virtual state monopoly in scientific research, thus forestalling the open competition for resources and recognition in which (so Jewett believed) the most creative scientists—the Edisons and Bells—prospered. Even those who supported the bill differed sharply in their views of how such a foundation should be organized. Democratic Senator Harley Kilgore envisioned the NSF as a civilian-run body modeled on New Deal agencies like the National Recovery Administration. Whereas Kilgore wanted to put science to work solving the problems of social welfare, scientific leaders such as Vannevar Bush wanted government funds without oversight or policy control by those outside of the scientific community.[21]

Beyond these knotty legislative battles, moreover, the brute reality of the new scientific establishment was reshaping American education, culture, and geography. The Manhattan Project, for example, had altered the very landscape of towns like Oak Ridge, which now housed thousands of professional scientists and engineers and their families.[22] At the same time, the universities were training many thousands of men to staff the industrial infrastructure needed for weapons research, machine maintenance, and the manning of air defense stations; the nation had begun to mass-produce physicists, engineers, and technicians.[23] Finally, as a result of the rapid upscaling of science, it began to dawn on American intellectuals and opinion makers that a new kind of figure, the research scientist or technician, had emerged. The working lives and beliefs of the laboratory natives were largely obscure to the average citizen. How was this new world to be integrated in existing political traditions and social norms?

Questions about the social relations of science had loomed large in the years leading up to the commencement of hostilities. Before the fraught years of the Great Depression, the history of science, especially in the United States, kept to its mission of inculcating appreciation of and respect for the scientific investigator's ability to divine the truths of nature by means of disciplined experimentation and rigorous thinking; like many teachers of the history of science in the United States, Henderson kept his instruction to dynamic accounts of how the objects of scientific knowledge emerged through the cognitive processes of inquiry. However, this "internalist" orientation in the field was spectacularly challenged in 1931 during an infamous meeting of the International Congress of the History of Science and Technology at the Science Museum in London.[24] A delegation of Soviet scientists and philosophers, led by Nicolai Bukharin, arrived unannounced shortly before the Congress began, and harangued their audience on the decadence of Western bourgeois science and the superiority of Soviet state-directed research. One of their number, Boris Hessen, delivered a paper that was to have a galvanizing effect on a generation of British scientists and historians of science. In "The Social and Economic Roots of Newton's *Principia*," Hessen argued that Newton's cosmology was the product and reflection of the practical needs of English capitalism of the period, especially the technical requirements of a proto-industrial empire. Hessen's crude if revelatory paper inspired a circle of intellectuals gathered around the Marxist molecular biologist J. D. Bernal to explore the social conditions and relations of science. Bernal and sympathetic colleagues such as the historian of Chinese science Joseph Needham and the journalist J. G. Crowther soon had enough traction to found within the British Association for the Advancement of Science a Division for the Social and International Relations of Science.

The purpose of the Social Relations of Science movement was not so much to explain scientific theories in terms of prevalent modes of production as to press the case for socializing science much like, under a socialist system of government, industry and the means of production would be nationalized. This argument was pressed most forcefully in Bernal's widely read 1939 treatise *The Social Function of Science,* which argued for the planning of research with a view to meeting the material needs of society.[25] Bernal struck a chord with an increasingly activist community of scientists and engineers in the United States, who gathered on the left of the New Deal coalition to press for the democratic organization of science and technology.[26] The Bernalists looked upon the burgeoning military-industrial research culture of World War II as an opportunity to bring scientific planning to bear on the conduct of science. During the opening stages of the war, the "operations research" of Bernal's friend P. M. S. Blackett had seemed to bode well for this program, but, perhaps because British Operations Research was associated with the leftist penchant for planning, state-directed research in Britain remained more of a gleam in the Bernal's eye than an administrative reality.[27]

All of this fervor for bringing science into the realm of social politics was not long in creating a backlash in the polarized intellectual culture of the late 1930s. Michael Polanyi, a professor of chemistry at the University of Manchester who had emigrated from Germany when the Nazis seized power in 1933, was especially vocal in his objections to the Bernalists. He worried that statist planning in science, much like statist planning of the economy, would stifle innovation and individual creativity. In fact, as Polanyi explained in a series of books and articles, science was progressive and efficacious because it rested on an implicit craft skill and intuitive feel for the right investigative techniques, which only those inducted into the guild could understand—without being able ever fully to state, and hence automate, their procedures.[28] Such was Polanyi's revulsion toward the Bernal circle's attempt to organize scientific research for utilitarian ends that he founded in 1941 an opposition group to the Social Relations of Science movement, the Society for Freedom in Science.[29] It is against this background of polemic in the history and sociology of science, in which usually Marxist "externalists" did battle with anti-statist "internalists" like Polanyi, that we must grasp the character of the historical and social study of science at Harvard after the cessation of hostilities.

One might have expected the undeniable results of the Manhattan Project to have settled the Bernal-Polanyi dispute decisively in favor of the former. The building of the atom bomb had been an intricately planned research enterprise: designed with a clear result in mind, bankrolled by the state, and executed by the military. Perhaps even Bukharin would have been proud. Not only did nuclear knowledge seem a direct product of careful social organization; Polanyi's convictions regarding the imperviousness of the scientific community's norms to rational planning or military discipline had to all intents and purposes been exploded. Counterintuitively, however, in the postwar United States Bernal's position was easily defanged and repurposed, while Polanyi's conception of science became more, rather than less, salient. Conant's Harvard would play a central role in this unexpected outcome.

Internal and External: Science, Society, and Pedagogy at Midcentury

Well before the outbreak of World War II, the historical study of science at Harvard had struck what was an increasingly comfortable modus vivendi between "internal" and "external" accounts of scientific inquiry. Internalism was common in historical explanations of scientific discovery. Three distinct motivations for this approach converged on one another in support of this position among the Harvard faculty: the classic pedagogical mission of American history of science to integrate scientific research into the humanist tradition; Sarton's prophetic mission to vindicate science as the source of cultural progress; and Henderson's commitment to understanding the logic of science in terms of the psychological or cognitive processes of the investigator. Henderson and Sarton's long-running course, History of Science 1, displayed the scientific mind in action, but it did not treat scientific knowledge as a reflection of underlying social conditions. After the war, the spirit of Sarton's internalist histories of science was adopted by his student I. Bernard Cohen in his apologia for modern science—and rebuttal of Bernal—*Science, Servant of Man* (1948).[30] By its very nature, Cohen suggested, science could not be subject to planning because scientists were fuelled not by social goals but by "innate curiosity" and the dogged desire to discover the truth.

Externalism, meanwhile, came onto the agenda at 1930s Harvard among those associated with Henderson's Pareto network. A young Robert K. Merton, in tandem with Talcott Parsons and Bernard Barber, defined this emerging field of sociological investigation. Merton had taken History of Science 1 and attended the discussions of Pareto's social theory in Henderson's seminar.[31] His evolving interest in the place of sentiments or value systems in the functioning of a social system, and especially in Weber's description of the vocation of the scientist, led Merton, with Sarton's encouragement, onto a prosopography of seventeenth-century English natural philosophers and inventors, from which he drew lessons about the kind of ethos that encouraged the efflorescence of scientific inquiry with a society.[32] It was already plain in Merton's doctoral dissertation—published in 1938 in Sarton's *Osiris* series of research monographs as *Science, Technology, and Society in Seventeenth Century England*—that he would treat scientific truths as given and seek to explain only the social conditions that proved especially congenial, or uncongenial, to their production.[33] Although Merton cited sympathetically Hessen's incendiary essay and the socialist writings of Bernal, his emerging sociology of science would forestall the brewing conflict between internal and external accounts of scientific development.[34] As the specter of Nazi and then Soviet totalitarianism began to affect American social and political thought from the *Kristallnacht* onward, the Mertonian focus on the social and cultural structures that most effectively supported the pursuit of scientific knowledge was easily transmuted into a defense of American liberal democracy as the regime most conducive to the promotion of free scientific inquiry.[35] Merton himself, in "Science and the Social Order" (1938), made the invidious contrast between science in a democracy and its perversion in Germany under Hitler.[36] Next, in an essay published in the midst of the war, Merton defined the "ethos of science" in terms that made explicit its reflection of democratic values.[37] Talcott Parsons would make similar arguments about science and (American, liberal) democracy in writings from the late 1940s.[38] So too would Merton's friend Bernard Barber, whose *Science and the Social Order* (1952) rang the changes on Merton's contrast between democratic and totalitarian states and the relative health of the estate of science in each.[39]

In this Harvard context, external studies of the social and historical development of science were to be welcomed. For to speak of the "social relations of science" implied neither the harnessing of science to ends set by the state, nor the determination of scientific knowledge by the means and relations of production, but rather the affirmation of democratic values that, by their very existence, created a climate for the serendipitous discovery of the truths of nature. From World War to Cold War, the ideological imperative of anti-totalitarianism had severed, at least in the United States, the links Bernal had traced between the results of scientific inquiry and its social organization. Harvard's scholars of scientific culture—Sarton, Henderson, Parsons, Merton, Cohen, and Barber—had been instrumental in effecting this separation. Nonetheless, the precise relations between the social support of science and the truths scientists could be expected to adduce were not unambiguously stated in the years after 1945. Only slowly would a synthesis emerge in the writings of Conant and Kuhn.

An illustration of the open-endedness of this transitional moment after the war is provided by Philipp Frank and his attempts to revive the Unity of Science movement in Cambridge. Surveying the rapidly changing landscape of American science during World War II, Frank seized the opportunity to reconstruct the program of the Vienna Circle for the age of nuclear energy and electrical engineering.[40] His Inter-Scientific Discussion Group, formed in September 1944, was devoted to exploring the dense latticework of "rapidly increasing cross-connections" between disciplines then being forged in the interdisciplinary wartime laboratories.[41] By 1947, with the support of the

Rockefeller Foundation and the American Academy of Arts and Sciences, Frank was able to formalize the ISDG into a reconstructed Institute for the Unity of Science.[42] In addition to the revised meaning of "unity" that animated Frank's activities in these years, one feature of the new movement stood out: it would not limit itself to the logical and empirical dimensions of science alone, but would also address the psychological and sociological conditions of the conduct of science. As the letter announcing the founding of the ISDG recorded, the group was "interested in considering science as a whole in terms of the scientific temper itself, and in the study of logic, psychology, and sociology of science."[43] In his writings of the 1940s and early 1950s, Frank repeatedly insisted that the scientific study of science would have to include "socio-psychological" as well as "logical-empirical" analysis.[44] Internal reconstructions of scientific theory thus needed to be supplemented by sociological, psychological, and historical assessments of how those theories were validated and accepted.

Frank's new recipe for the unity of science came with a pedagogical twist. This was revealed in his rationale for the new Institute. "It has been commonplace for over a century," Frank wrote in 1947, "that science is drifting more and more towards specialization," with the result that "nobody . . . can comprehend more than a very small fraction of the domain of science and every attempt towards integration is doomed to wind up in superficiality and dilettantism [*sic*]." Here the problem of education raised its head. According to Frank, "extreme departmentalization and specialization have become a serious threat to our culture and to the very goal of liberal education." It was not just that students would have to be "satisfied with a scholarly and competent understanding of a narrow fraction of a narrow department"; the danger was that students would be tempted by "unifying sources" that ranged from "books of popular science, movies and science columns in newspapers" to "deliberate quacks who exploit the longing for integration of knowledge in the service of some political, social, or intellectual creed."[45] But all was not lost. First, there were the "cross-connections" between fields that promised a practical, piecewise coordination of scientific fields, perhaps under the banner of physics. Hence the problem was more linguistic than institutional: "several field[s] of study use the same words in totally different meanings and, what is still worse, in slightly different meanings."[46] Second, this terminological and conceptual confusion could in turn be addressed by using the tools of the Logic

of Science, by which Frank meant the tradition of scientific philosophy beginning with Mach, Poincaré, James, and Peirce, and culminating with logical
empiricism and the operationism of Bridgman.[47] Logical empiricism, in particular, provided tools for analyzing the *formal* validity of scientific statements.
However, although logical empiricism offered a means of unifying the language of science, Carnap's Principle of Tolerance helped to show that "there
are many different kinds" of symbolic systems that could be used to "derive
the observational statements of our experience." Understanding the reasons
for which one system rather than another had been accepted raised the wider
scientific problem of the "effect of symbols on human behavior"—the domain
of the human sciences.[48] In the combination of logico-empirical and sociopsychological investigation, proponents of general education in American universities would find in the Logic of Science "an approach which works toward
integration, but at the same time works against superficiality by applying the
highest standards of logical thinking and psychological judgment."[49]

Frank soon became more explicit about his pedagogic concerns. His ideal
constituency was science majors who needed some way of placing the sciences in the humanistic traditions of Western culture. This could be accomplished only by revealing *"the human values which are intrinsic in science
itself."*[50] Once more, logico-empirical analysis would clear the philosophical
ground, but there would come a point where the "pragmatic" element in the
choice between the linguistic systems of scientific theory would become crucial.[51] The sociopsychological component of this education in the human values of science would be drawn from the historical record. In Frank's words,
the "history of science is the workshop for the philosophy of science. We
have to teach the student all the relevant principles which have been set up
in the course of history."[52] For Frank, this was the natural way to fill the gap
between science and the humanities, or "between physics and philosophy":[53]
the objective study of how values had contributed to the selection among different systems of theory over time would provide a way of seeing science as
part of the history of the human race. Just as importantly, through this kind of
education the *"student of science will get the habit of looking at social and religious problems from the interior of his own field"*—i.e. from within the domain
of scientific investigation, not from popular or mystical sources.[54] Frank was
able to put his designs into at least partial practice. Beginning in the spring of
1949, the physicist began to offer two courses for upperclassmen in Harvard's
General Education program, each running in alternate years. The first,

Natural Sciences 112, "Introduction to the Philosophy of Science," explored "the relation between science and philosophy" and "the role of experience, logical thinking, and free imagination in physics."[55] Meanwhile, Natural Sciences 113, "Contemporary Physics and its Philosophical Interpretations" examined the contribution physics could make "towards the solution of philosophical questions."[56]

Although Frank's ideas about science pedagogy and the relations between science and the humanities were geared toward the postwar world of Big Science, they also harked back to the social program of Neurath's left-leaning wing of the Vienna Circle.[57] The notion that sociology and psychology would form a necessary part of the attempt to unify the sciences had been central to the 1929 manifesto of the Circle, which had gone out of its way to flag up the unhappy possibility of a retreat to the "icy slopes of logic" among scientific philosophers.[58] Frank's postwar position can thus be understood as a rehabilitation of the cultural ambitions of the Circle, which had encompassed adult education, urban and economic planning, and architectural modernism. With a foothold in the physics department at Harvard, Frank pushed forward his ecumenical movement, the published results of which found a place in a regular series of special issues of the *Proceedings of the American Academy of Arts and Sciences* and *The Scientific Monthly*.[59] For the most part, Frank was able to fit these activities into the internal/external divide that characterized this history and philosophy of science at Harvard. Almost single-handedly, he kept alive the conversation about Bridgman's by now rather threadbare operationism, while attracting contributions from Harvard-trained historians of science like Henry Guerlac.[60] But shades of Neurath's socialist faith in the potential for scientific control of science showed through in some of Frank's remarks about the "behavioral" aspect of the validation of scientific theories. This view called into question the Harvard division of "internal" and "external" studies of science. Frank and his Institute would fall foul of the politics of anti-communism during the 1950s, when the call for a unity grounded in a social as well as a natural science could be portrayed as wanton fellow-travelling.[61]

Conant's Formula

A more enduring solution to the problem of science and social order, one rooted firmly in the Harvard tradition of history and sociology of science, was fashioned by James Bryant Conant. The stakes involved for President Conant were highlighted by two seminal postwar reports that the chemist had a hand in shaping. Conant helped Vannevar Bush prepare his 1945 report on the need for federal support for scientific research, *Science—The Endless Frontier*.[62] Both men had overseen the growth of Big Science during the war, Bush as Director of the Office of Scientific Research and Development, Conant as chairman of the National Defense Research Committee; each recognized that a new era of large-scale, personnel- and capital-intensive research was at hand, for which only the fiscal reserves of the state could provide adequate funds. Yet Bush and Conant were wary of state control of science, and drew a sharp distinction between "basic research," as the life-blood of scientific inquiry, and "applied science" or "technology," which could emerge only downstream from the results of "pure" or "basic" research.[63] As Conant put the matter shortly after the war, technological advances, intertwined though they were with pure research, "do not form a part of science."[64] Bush urged America's civilian leadership to invest in basic research and let the technological benefits trickle down into public health, national defense research, and other vital sectors. A second report, published a few months earlier than Bush's pamphlet, showcased Conant's concerns as a university president. *General Education in a Free Society* (1945) had been written by a Harvard faculty committee, with little direct input from Conant.[65] But, along with Dean Paul Buck, Conant had established and generously funded the group that wrote it and drove forward the agenda of the "Red Book" after its publication.[66]

Conant's interest in general education stretched back as far as his oration at the tercentenary celebrations in 1936, where he voiced concern with the ongoing specialization of knowledge and highlighted the need for a liberal arts curriculum that could impart a common culture to all.[67] His attempts to augment Lowell's system of concentration and distribution requirements in the College with a core curriculum in American history was an avowed failure, but Conant was able to use the peculiar conditions of the war to put a core group of faculty humanists together who, like their colleagues in the natural and social sciences, would devote a large portion of their research time to a national cause: in this case, education reform. While the Committee on the Objectives of General Education in a Free Society was busy producing its final report, Conant had become fixated on a specific pedagogical problem within liberal education: the transmission of an appreciation of the place of science in the humanist tradition. Conant's interest in this problem had a contemporary aspect. As he observed in the autumn of 1945, his experiences as a scientific administrator during the war had made him acutely aware of "the bewilderment of lawyers, business men, writers, public servants (and not a few Army and Navy officers) when confronted with matters of policy involving scientific matters."[68] In very similar terms, Conant's acknowledgement of the publication of *General Education in a Free Society* in his 1945 President's Report contained the admission that "those of us who have taught the sciences in college have failed for the most part in accomplishing for the future lawyer, public servant, writer, and business man what we set out to do." The future men of affairs were given general courses in the sciences that assumed they were physicists, chemists, or geologists in training—whereas what those being readied for public life needed was a feel for "the tactics and strategy of science, with the minimum emphasis on factual knowledge and the maximum attention to scientific methods as illustrated by historical examples."[69] Only this induction into the thoughtways of scientists as embodied in historical cases could allow the appreciation of science to enter into a common general education of the rising generation.

In the months and years after V-J Day, Conant hit upon a pedagogical model that united his two postwar preoccupations with state sponsorship and administration of basic research (and its products such as nuclear energy) and the teaching of the "tactics and strategy of science" to those future men of affairs who would not become scientists. Conant seems to have recognized that these were related problems, which were associated with the rise of large-scale science: the case for support for basic research would be easier to make if the public servants who held the federal purse strings could appreciate the habit of mind of the scientific investigator. At all events, the stance that allowed Conant to blend these issues was Henderson's psychology of inquiry, with its emphasis on the case method and the piecemeal construction of conceptual schemes.[70] From one perspective, this was a natural move. Henderson's teaching in History of Science 1 was anchored in the study of exemplary texts. Moreover, it was Henderson who had carried on the tradition of the case method at Harvard in his course Sociology 23 and who tied cases to the formation of conceptual schemes. To top all of this off, Conant was married to the Harvard chemist T. W. Richards's only daughter; Henderson was Richards's brother-in-law and thus Conant's wife's uncle. Conant was a protégé of Richards's and gained an appreciation of the value of the history of chemistry through his courses. Henderson, meanwhile, had served as an emissary for the Harvard Corporation during the selection process for the Harvard presidency in 1932–33.[71] During the 1930s, while holding Sarton's grander schemes for the history of science at bay, Conant allowed for concentration in "History and Science" in the College and also established the Committee for Higher Degrees in Science and Learning, under the chairmanship of Henderson.

Despite these very clear lines of continuity, Conant's adoption of the rhetoric of cases and conceptual schemes placed Henderson's ideas in the service of novel ends. In its migration from the Law to the Medical to the Business School and then into Sociology 23, the case method had been used to stabilize and legitimate the research practices and scientific claims of its advocates. Cases were used to inculcate in students of a given profession or discipline the conceptual schemes and habits of judgment they would need to practice the vocation in question. We saw in the case of the Pareto circle that Harvard's marginal human scientists used this model of knowledge creation and discipline building to produce their own model social systems, whether these were drawn from a Navajo reservation in New Mexico or the records of English medieval administrators. By contrast, in Conant's hands the case method in the scientific education of scientists would not create practitioners of this or that discipline; those who went through the case studies Conant offered in his general education courses on the natural sciences would not acquire the conceptual scheme of a specialist. Time and again, Conant stressed the peculiar attitudes and practices of modern scientists and argued that a "visitor to a laboratory, unless he is himself a scientist, will find it almost impossible to understand the work in progress."[72] Unlike all previous deployments of case-based pedagogy at Harvard, students would not learn *how* a particular conceptual scheme in a particular field worked; they would learn *that* the advancement of science relied on the development, application, and revision of concepts and conceptual schemes. The scheme itself would remain under glass, so to speak.

After Conant's initial exploration of these themes in his President's Report of 1945, he set about fleshing out his vision of science education for non-scientists. His 1946 Terry Lectures at Yale—later published as *On Understanding Science: An Historical Approach* (1947)—served as a dry run for the general education courses on the development of the natural sciences he would begin to teach at Harvard in the late 1940s. Although the book was not intended as a "manual for instructors," it did walk the reader through the purpose and pedagogical methods of a course on the "tactics and strategy of science."[73] Conant baptized this second-order perspective—which centered not on the conceptual scheme itself, but knowledge of the centrality of such schemes to the conduct of research—a "special point of view" known as "understanding science." "Note carefully," Conant told his readers, that this point of view "is independent of a knowledge of the scientific facts or techniques in the new area to which [the non-expert in a field] comes." Practicing scientists had this general understanding already, in addition to their own special expertise. The challenge was to "bridge the gap to some degree between those who understand science because science is their profession, and those who have only studied the results of scientific inquiry—in short, the laymen." To do this, non-scientists would need to "retrace the steps by which certain end results have been produced" in the history of science; in particular, Conant wanted "to show the difficulties that attend each new push forward in the advance of science."[74]

Here was the nub of Conant's argument. While he indicated a wish to "illustrate the interconnection between science and society about which so much has been said in recent years by our Marxist friends," it was evident that the Harvard president's sympathies lay with Polanyi's side of debate, as was made clear in his emphasis upon "the difficulties which historically have attended the development of new concepts."[75] What Conant liked most about the case-based exploration of the development of conceptual schemes was that they showed the "evolution of new conceptual schemes as a result of experimentation." It was in this "intricate interplay between experiment, or observation, and the development of new concepts and new generalizations" that the uniquely "accumulative," "progressive," or "dynamic" quality of science consisted.[76] Very much in line with Henderson's position, this was a description of a craft of scientific thinking that marked it out from common sense or speculative flights of fancy. Conant summed up this view in one especially vivid passage:

> The texture of modern science is the result of the interweaving of the fruitful concepts. The test of a new idea is therefore not only its success in correlating the then-known facts but much more its success or

failure in stimulating further experimentation or observation which is
in turn fruitful. This dynamic quality of science viewed not as a practical undertaking but as development of conceptual schemes seems to
me to be close to the heart of the best definition [of science]. It is this
quality which can be demonstrated only by the historical approach, or
else learned by direct professional experience.[77]

So much, then, for the social planning of scientific research if the very
efficacy of science—the complex evolution of conceptual schemes in relation
to experiment—could only be "demonstrated" to non-scientists via the case
method or "experienced" by the professional scientist in their day-to-day
work of investigation. Conant was hostile to positions that failed to mark
off science from non-science in this manner. He reserved special disdain for
the positivist position that science was an extension of the commonsense
interactions of human agents with their environment and that the creation
of scientific knowledge consisted in the more-or-less mechanical processing of common facts or raw sense experience. "If science were as simple as
this very readable account would have us believe," wrote Conant of Karl
Pearson's positivist tract *The Grammar of Science* (1892) "why did it take so
long a period of fumbling before scientists were clear on some very familiar
matters?"[78] Such fumblings, illustrative of the dynamic interplay of experiments and conceptual schemes, were exactly what Conant's cherry-picked
case histories were crafted to illustrate. His goal was to distinguish scientists as a people apart, whose ways could be appreciated but not straightforwardly accessed by those outside the profession. The congeniality of such a
view to the call of Bush and other leading scientists for a National Science
Foundation—funded by the state, run by scientists, and devoted to supporting basic research—was striking.

Conant put these principles into practice in the wake of his Terry Lectures,
which he treated as a "ground plan" for a new general education course at
Harvard on "The Growth of the Experimental Sciences."[79] Natural Sciences
4 (Natural Sciences 11a, in its first iteration) was "designed for those who
are planning to concentrate in the social sciences or humanities and intend
to take no further work in the natural sciences." A secondary school class
in physics, chemistry, or "general science" was all that was required of those
who enrolled. "The objective," ran the course description in the course

listings for 1948–49, "is to give a critical understanding of the procedures of modern science by an examination of cases drawn from the history of the experimental sciences."[80] Those cases would not be taken from contemporary discoveries like atomic energy or inventions like Henry Aiken's calculating machine. "Modern science has become so complicated," declared Conant, "that today methods of research cannot be studied by looking over the shoulder of the scientist at work." Instead, with the veil discreetly drawn over the Manhattan Project and the Radiation Laboratory, the student would be given cases focusing on moments when "a science is in its infancy, and a new field is opened by a great pioneer," for in these instances "the relevant information . . . can be summed up in a brief compass."[81]

A number of the cases utilized in Natural Sciences 4, some compiled by Conant, others by the young teaching staff he hired to assist him, were published from 1948 onward and in 1957 were gathered together as the two-volume *Harvard Case Histories in Experimental Science*. These studies were the final product of a wider project Conant had devised, with the support of the Carnegie Corporation, to provide the materials for college-level courses for lay students of science.[82] The case histories themselves were not so much singular "crucial experiments" as moments when received conceptual schemes were in flux and alternative concepts and working hypotheses were taking shape that would soon "revolutionize" the field in question. Thus Conant's cases included Robert Boyle's experiments with air pumps as part of the rise of "experimental science" and a shift in the conceptual scheme of the physical sciences; the "chemical revolution" of the late eighteenth century in which the traditional phlogiston theory was abandoned when the presence of oxygen was discovered by Joseph Priestley and Antoine Lavoisier; and Pasteur's wide-ranging investigation into the biochemical processes of fermentation and the operation of microorganisms.[83] The summary of the significance of the Priestley-Lavoisier case study illustrated Conant's determination to foreground the craft skill involved in experimental research: "By studying the evolution of a new conceptual scheme we can see the difficulties with which pioneers in science almost always have to contend, and the false steps that usually accompany even the most successful forward marches."[84]

In Conant's scheme for general education in natural science, the Harvard complex had arrived at a critical pass. The artisanal conception of epistemology that the merging of pedagogy, research practice, and knowledge-making had created in the era of Henderson had changed its meaning in the early years of the Cold War. Now the case study was being used not to make or validate practitioners of a craft, but as a way of encouraging those who did not possess a scientific conceptual scheme to appreciate the activities of those who did. Instead of seeding disciplines, the pedagogy-cum-epistemology of the Harvard complex was attached to a general education program the purpose of which was to reach beyond disciplinary boundaries. Conant's synthesis of the imperatives of science patronage and general education was a brilliant balancing act, combining as it did an acknowledgement of the new social role of science while keeping a strict separation between the internal logic of science and its external supports. He was able to lay claim to an august Harvard pedagogical tradition while validating the Harvard formula for keeping internalism and externalism in their separate spheres. Thus Conant could cite approvingly Merton's work on the conduciveness of open democratic regimes to science while tasking Nat Sci 4 assistants like Leonard Nash with the elaboration of the strictly internal norms of scientific inquiry.[85] No sooner had Conant struck this accord, however, than one of his protégés, Thomas Kuhn, reworked his approach.

The Conversion of Thomas S. Kuhn

The delicate balance Conant had arranged between internal and external accounts of science contained one crucial equivocation. To repeat one of the Harvard president's remarks quoted above, for Conant the dynamic nature of experimental science could "be demonstrated only by the historical approach, or else learned by direct professional experience." Conant's writings on science pedagogy were trained almost exclusively on "demonstration," because, as we have seen, only the case-based historical method could allow those who would never have "direct professional experience" of science to understand the "tactics and strategies" involved in experimental inquiry. About the acquisition of the intellectual skill of "understanding science" by research scientists themselves, Conant had less to say. His remarks on the "stumbling" evolution of conceptual schemes within the process of theoretical speculation, hypothesis, experiment, and analysis was little more than a gloss on Henderson's artisanal psychology of research. And, like Henderson, Conant limited his historical examples of conceptual change to the revolutionary discoveries of the Great Men of the tradition: Boyle, Priestly, Lavoisier, Pasteur. Conant therefore possessed a detailed pedagogy of how non-scientists should learn about science, but held no substantive theory about how scientists themselves, beyond the Great Men, conducted their investigations and moved toward revolutionary discoveries.

Thomas Kuhn's innovation in the study of the history of science was to collapse Conant's pedagogical distinction between "demonstration" for non-scientists and intuitive "experience" for practitioners. In *The Structure of Scientific Revolutions,* his famous book of 1962, Kuhn pushed Henderson's and Conant's pedagogy into the analysis of science itself and thereby combined it with an account of research practice and the creation of empirical knowledge. Like Conant, Kuhn insisted that science textbooks obscured the fundamental importance of the research process—its strategies and tactics—for understanding the nature of empirical knowledge. But Kuhn went further by suggesting that the textbook image of science had invidious consequences for professional philosophers of science, as well as for laymen. What we find in *Structure,* then, is a novel synthesis of the theories of pedagogy, epistemology, and research that had developed in the Harvard complex since World War I.

How did Kuhn arrive at this understanding of science? We can begin to answer this question by returning to what Kuhn often cited as the source of his engagement with the history of science. In the summer of 1947, Kuhn was a graduate student in physics at Harvard. He worked under the direction of John Van Vleck, an expert in quantum mechanics and the theory of magnetism. Kuhn was known to be a young scientist of wide interests, having served as both editor of the *Harvard Crimson* and president of the literary Signet Society. These twin affiliations with literature and science had led to an invitation in 1945 to give the "student's view" on *General Education in a Free Society* for the *Harvard Alumni Bulletin.*[86] Kuhn's extracurricular profile would seem to have encouraged Conant to invite Kuhn to prepare some lectures on the origins of seventeenth-century mechanics for the trial run of Natural Sciences 4.[87] This assignment led Kuhn to a reading of Aristotle's *Physics.* Kuhn wanted to find out, in classic Whig fashion, "how much mechanics Aristotle had known, [and] how much he had left for people like Galileo and Newton to discover."[88] Yet, not only did the young Kuhn find Aristotle lacking knowledge of even the rudiments of modern mechanics, but very little in the ancient philosopher's physics was intelligible as a theory of any sort of mechanics. It seemed, in fact, that the man who had formalized the theory of logic and otherwise proven himself an acute observer of natural phenomena had precipitated himself into all manner of illogical and error-strewn remarks about space and motion.[89] Puzzling at how Aristotle could have been so badly misled, Kuhn turned over in his mind the Aristotelian claims about motion and matter, until one day they fell into a new configuration:

> I was sitting at my desk with the text of Aristotle's *Physics* open in front of me and with a four-colored pencil in my hand. Looking up, I gazed abstractedly out of the window of my room—the visual image is one I still retain. Suddenly the fragments in my head sorted themselves out in a new way, and fell into place together. My jaw dropped, for all at once Aristotle seemed a very good physicist indeed, but of a sort I'd never dreamed possible.[90]

Kuhn had come to grasp that "motion" in the Aristotelian lexicon connoted an entire genus of movement involving the change from one quality to another; changes in the position of a physical body was but a single species of this sort of motion. The familiar concept of position, moreover, was to be understood as a quality of matter, such that a change of position involved a change in the qualities of a formless, omnipresent matter. For Kuhn, the crucial revelation was that these peculiar Aristotelian theses about motion, position, and matter were "deeply interdependent, almost equivalent notions." Claims "that appear arbitrary in isolation," like Aristotle's odd-sounding remarks about motion, could thus, when conceived as a comprehensive theory, "lend each other mutual authority and support."[91] Aristotle's theory could not be learned piecemeal, thesis-by-thesis: it could only be grasped as an all-or-nothing whole. Evidently, Galileo and Newton's conceptual scheme did not add a detail or two to Aristotle's picture: it replaced it altogether with a different set of "interdependent notions" that could be understood only as a conceptual whole. Such, Kuhn concluded, was the nature of "revolutionary" scientific discoveries like Copernicus's recognition that the earth revolved around the sun. In the shift from the Aristotelian to the Newtonian world picture, Kuhn had "discovered my first scientific revolution."[92]

Despite its formative significance, however, Kuhn's "Aristotle moment" did not mark a new departure in his thinking. When we trace the longer-term sources of his reading of Aristotle, the significance of the Harvard complex in shaping Kuhn's thought in the 1940s becomes clear. First of all, Kuhn had nurtured heterodox concerns with the epistemology of the natural sciences long before the summer of 1947. It is a striking fact about the theorist of "dogmatic initiation" in scientific training that Kuhn's own intellectual formation is notable precisely for its lack of disciplinary or professional boundaries.[93] Before entering Harvard in 1940, Kuhn was educated in a string of progressive and elite preparatory schools in and around New York State. The most important of these was the Hessian Hills School in Croton-on-Hudson. Hessian Hills belonged in a small cohort of Associated Experimental Schools in the area and was imbued with the principles of progressive education laid down by educational reformers such as the philosopher John Dewey. There were no grades, no rote learning, and "little in the way of set subject matter." As a pupil from sixth to ninth grade, Kuhn credited Hessian Hills with "teaching me to think for myself" and making "a major contribution to my independence of mind." He learned to pursue his interests outside of the teaching drills and rigid curricula of institutionalized disciplines.[94]

A compressed undergraduate career at Harvard did little to domesticate Kuhn. Impressed by the career possibilities open to someone with training in mathematics and the natural sciences, Kuhn declared his concentration in physics in 1940, his freshman year. But he was deeply ambivalent about his choice. The coming of the war locked Kuhn into his path in physics: the nation needed young men with expertise in science and engineering. Yet American intervention in the war also dramatically accelerated Kuhn's education, plunging him into special courses in electronics and preparing him for work in Harvard's Radio Research Laboratory.[95] This did not amount to a comprehensive induction into the physics profession, and in many respects it left Kuhn's other undergraduate enthusiasms undimmed. Of particular note in this regard is an essay for a sophomore English composition class on "The Metaphysical Possibilities of Physics," in which Kuhn endeavored "to outline the method of the physicist and to consider the possibility that he may ever gain complete knowledge of the structure and mechanism of the universe."[96] This was a frank attempt to assay the philosophical potential of physics and was connected to a reading of Kant's *Critique of Pure Reason* in another class. Kant's notion of the a priori conditions of knowledge, Kuhn later recalled, "just knocked me over."[97] We can understand, then, why Kuhn confessed to a relative shortly after graduation that, with uncertain prospects as a physicist in war and peace, "I lean . . . increasingly toward the academic, and to philosophy of & through science rather than to pure physics. But I am to [*sic*] far from knowing where it will end to waste your time here."[98] Although Kuhn elected to begin studies at Harvard for a PhD in physics in the autumn of 1945, he received special permission to pursue his interests in philosophy.[99] Quickly, however, Kuhn abandoned his education in the field after a brief, and not entirely successful, foray into metaphysics and mathematical logic.[100]

It was against the background of this attempt to practice philosophy "of and through science" that Conant's invitation to prepare a case history in seventeenth-century mechanics arrived in the spring of 1947. Almost immediately, and certainly before the Aristotle experience, Kuhn seized on the philosophical dimensions of his pedagogical challenge. In a memorandum written in May 1947 as part of his preparation for teaching on what would become Natural Sciences 4, Kuhn outlined a "list of topics which one might hope such a course would enable its students to discuss." Already, Kuhn's thinking about general education in the physical sciences was saturated with conceptions of science as a "process" of "stages." He was also interested in the formation of "conceptual structures" by the dialectic of theory and experiment, and in the place of logic as a "language" of science. Like Conant, Kuhn emphasized the "predominant role of experiment in science," but he went further than his mentor in specifying the encompassing, extralogical nature of theoretical systems.[101]

Most strikingly, Kuhn's outline included a major section devoted to "Symbols, Logic, and Mathematics as the 'language' of science." Although language, and especially the notation of mathematics, proved "essential . . . as a tool of thought," Kuhn, even at this early stage of the journey to *Structure,* was keen to point out the "limited role of pure logic," and indeed of all attempts to provide a formula for the connections between theory and experiment. What the philosophers called "induction" was for Kuhn "a name for 'the experimental method' rather than for a logically, or even operationally, defined route from the particular to the general (experiment to theory)." "Scientific generalization," Kuhn continued, was "a creative imaginative process."[102] With these convictions in place, Kuhn was primed to identify the gestalt switch he experienced in reading Aristotle with the pervasive reach of theory as an imaginative structure for thought. By September, Kuhn was taking notes on Conant's lectures for Natural Sciences 11a, in which the logic of conceptual schemes in experimental science was a central concern.[103]

Conant's course provided the bridge from science into philosophy that Kuhn had been looking for since the early 1940s. In Kuhn's *annus mirabilis* of 1947, the young physicist had come to see how studying the history of the process of research as a means of teaching laymen how to "understand science" could open the way to the epistemological and metaphysical problems that had attracted him as an undergraduate. Such an enterprise was neither science nor philosophy, nor yet history understood as the recovery of the past for its own sake. Recognizing that he needed to flesh out his holistic conception of scientific theories, Kuhn asked Conant to sponsor his application to the Society of Fellows.[104] As someone not fully professionalized as a physicist, and with a heterodox set of interests in general education, philosophy, history, and psychology, this move into one of the nodes of the Harvard complex made a good deal of sense; Kuhn was duly elected and took up a fellowship in 1948 with the plan to convert himself into a historian of science. In practice, however, Kuhn spent much of his reading time in the Society exploring a linguistic and psychological conception of theory change and scientific thinking. A set of Kuhn's reading notes from 1949 gives a flavor of the neophyte historian's interests in this period. According to Kuhn's reading log, between March and June he digested, in whole or in part, a number of central texts in the canon of scientific philosophy.[105] Alongside Kant's *Critique*, Kuhn examined John Stuart Mill's *System of Logic* (1843), Bertrand Russell's *The Scientific Outlook* (1931), A. J. Ayer's *Language, Truth, and Logic* (1936), Susanne Langer's *Philosophy in a New Key* (1942), and Rudolf Carnap's *Introduction to Semantics* (1943). Philosophical statements by practicing scientists were also present in the form of the theoretical biologist J. H. Woodger's *The Technique of Theory Construction* (1939), Norbert Wiener's *Cybernetics* (1948), and Erwin Schrödinger's *Science and the Human Temperament* (1935). Continuing his concern with the linguistic element in theory formation, Kuhn paid special attention to recent work in mathematical logic and linguistics: he consulted Quine's *Mathematical Logic* (1940), Alfred Tarski's *Introduction to Logic and to the Methodology of the Deductive Sciences* (1941), and Leonard Bloomfield's

Linguistic Aspects of Science (1939). Finally, and crucially, Kuhn read some seminal works in the sociology of science and the psychology of mental development, notably Merton's *Science, Technology, and Society in Seventeenth Century England,* an English translation of Max Weber's methodological writings, Jean Piaget's *Judgment and Reasoning in the Child* (1928) and *Les notions de vitesse et de mouvement chez l'enfant (The Child's Conception of Movement and Speed)* (1946), and Heinz Werner's *Comparative Psychology of Mental Development* (1940).[106]

Two strands of thought were especially prominent in these notes. First, Kuhn pressed further his criticism of the idea that the relationship between theory and experiment, or between a concept and its application, could be accounted for in the terms of logical positivism or operationism. Assessing the "weaknesses of the positivist or operational position," Kuhn latched on to an observation that would animate other critics of these viewpoints: namely that, under the criterion of meaning offered in verification theory or operational analysis, most statements of the physical sciences would be meaningless, insofar as many currently accepted propositions in these fields did not lead directly to confirmation in immediate experience. For Kuhn, this was an issue about how language both conditioned thought and mapped onto experience. The second strand in Kuhn's notes followed out this perspective into the domain of the psychology of child learning and the methodology of the social sciences. At this stage in his education, Kuhn was taken with Piaget's account of the way in which children arrived at "conceptualizations" of their physical world that filtered its complexity into a framework that was "psychologically consistent or satisfying"—even if, especially at early developmental stages, it contained "logically contradictory" or "syncretic" elements. Kuhn came to think about this feature of developmental psychology as a model for scientific understanding. In particular, he used Piaget and Werner to think about the relations between a "physically visible world" and the "psychologically visible world" that mediated a human subject's appreciation of the former by means of words and the concepts they denoted. "Crudely and metaphorically," Kuhn wrote,

we may generalize as follows. There is the "physically visible world" consisting of what we can see with our available sensory and technical equipment. This phys[ically] v[isible] w[orld] is the pure raw flux, unorganized. We deal with this flux by forming Gestalts (conceptual-izations) (i.e. we isolate objects and situations) to which we gives words, phrases, etc. (i.e. symbols). This world of objects and situations is the "psychologically visible world". . . . To "see" a tree or a velocity dif-ference is to "see" something in the psych[ologically] vis[isible] world which is in turn a creation from the phys[ical] real world. . . . We are thus conscious of the phys[ical] r[eal] world, but can only "see" and can only *talk* about the psych[ological] r[eal] w[orld]. The psych[ological] real world contains only those elements expressible in the intentions [i.e., conceptual structure] of language. . . . In the broadest possible sense the scientific process consists of the attempt to minimize the verbal equip-ment implied by (or inherent in) the psych[ological] r[eal] w[orld]. . . . Thus science changes the psych[ological] r[eal] w[orld] Also, tho[ugh], science brings new features of the phys[ical] r[eal] w[orld] into [the] psych[ical] r[eal] w[orld] and ultimately expands . . . the phys[ical] r[eal] w[orld] itself. Thus the process is continuing.[107]

It was because of this view of science's reciprocal action upon the psy-chological and physical visible world through the medium of language and concepts that Kuhn reacted with excitement to Max Weber's "bril-liant" methodological essays. Weber's remarks on the importance of ideal types for understanding in the social sciences seemed to Kuhn an excellent approximation of how science provided economical conceptualizations of the physical world to human thinkers.[108] Around this time, Kuhn also seems to have read, and been impressed by, some of L. J. Henderson's work on the instrumental use of "conceptual schemes" across the sciences.[109]

From Pedagogy to History and Back Again: Kuhn's Road to *Structure*

By the time his tenure in the Society of Fellows drew to a close in 1951, Kuhn had travelled a considerable distance beyond the brief Conant had given him in the spring of 1947. We have seen that Conant's motives for engaging with the history of science were principally pedagogical. Kuhn, by contrast, very quickly seized upon historical investigations designed for pedagogical use and turned them into sources for a deeper understanding of the epistemological and metaphysical commitments of the natural sciences. In Kuhn's hands, the assessment of "the activity of the working scientist" as recorded in historical sources was transformed from a pedagogical tool into a vehicle for philosophical speculation.[110] Up to the end of the 1940s, Kuhn remained unclear on how to make sense of what kind of theory of knowledge the study of the history of science disclosed. To be sure, from the Aristotle moment onward, two elements of a theory of scientific discovery were fixed in Kuhn's mind. The first was a conviction in the reality of conceptual revolutions in science. The second element was a corollary of the first: for traditions of inquiry to shift a revolutionary manner, theories had to consist of more than the thin logical structures of the "positivists" or the static bodies of doctrine found in textbooks. Instead, such "conceptual structures" had to encompass human psychology, language, values, and social practices. But Kuhn possessed no settled account of the historical life of a scientific theory, as his eclectic appropriation of Piaget, Weber, and others made clear. What was it about science that made it prone to revolutions? And how could one make sense of the "nonrevolutionary" tradition or conceptual scheme being overturned in such episodes? These were the questions that Kuhn spent the next decade trying to answer.[111]

Some years after the publication of *Structure,* Kuhn himself presented an account of his intellectual development in this period. After writing up some of his ideas in a series of Lowell Lectures delivered in 1951, Kuhn recalled, he became convinced that he "did not yet know enough history or enough about my ideas to proceed toward publication."[112] Then, for "a period that I expected to be short but that lasted seven years, I set my more philosophical interests aside and worked straightforwardly at history."[113] As Kuhn told the story, he only returned to philosophy once he had received tenure (in the Department of History at Berkeley) in the late 1950s. A smattering of historical-cum-philosophical essays marked the final approach to the completion of the manuscript of *Structure* in 1962.[114] Kuhn insisted that the work he produced between 1951 and 1958 was "straightforwardly historical" insofar as the philosophical lessons he had learned about revolutions had to "vanish from historical writing." The reason for this was that, as Kuhn saw it, history was a narrative genre, and the finished narratives presented in historical essays tended to "disguise" the legwork of textual exegesis that produced them.[115] Throughout his post–*Structure* career, Kuhn refused to reprint the essays he wrote during these years precisely because, although they taught him "philosophy by example," they did not wear their epistemological significance on their sleeves. The same held for his first book, *The Copernican Revolution* (1957), which was for Kuhn a case history without a philosophical bent.[116] After a clarifying stint as a fellow of the Center for Advanced Study in the Behavioral Sciences in 1958–59, Kuhn was able to fill out the remaining details of his conception of the form that scientific revolutions had to take and moved toward publication, and into philosophy, at the beginning of the 1960s.[117]

Appropriately enough, however, the historical record discloses a more complex picture of the evolution of Kuhn's thinking than the one presented in his autobiographical writings. In the years following the Lowell Lectures, Kuhn did more than set himself a series of training exercises in historical scholarship. Rather, he continued to develop, albeit intermittently, his views about the nonlogical, nonformal elements that defined scientific theories and traditions of experimentation. These reflections are important because they show Kuhn trying to make sense of revolutions in terms of the social values and intellectual commitments shared by a community of scientific investigators. By the middle of the 1950s, Kuhn had come to understand the problem of revolutions and the ideological dimensions of theory in terms closest not to psychology or linguistics, as was the case in the late 1940s, but to intellectual history and sociology. A sociological notion of consensus informed his view of nonrevolutionary or "normal" science right up to the drafting of the penultimate version of *Structure* in 1960. Kuhn's own account of this crucial phase in the genesis of *Structure* highlights his growing recognition of the unworkability of consensus as *explanans* for normal science.[118] From our point of view, however, we can see Kuhn's late turn to paradigms as a return to his roots, both with regard to the linguistic-psychological approached tested during his junior fellowship, and to his earlier interest in science pedagogy.

As early as his Lowell Lectures, Kuhn had indicated that the gestalt switches he had found in the transition from Aristotelian to Newtonian mechanics had roots not just in psychology and linguistics but also in "social sources."[119] "I believe," he stated in the first lecture of the series, "that elements which, on [Karl] Pearson's description [of scientific methodology], can only be called prejudice and preconceptions are inextricably woven into the pattern of scientific research."[120] Hence psychological "predispositions" colored observation and experiment, and these predilections were, in some sense and to some degree, social. The following year, Kuhn responded to an invitation from Philipp Frank to join an Institute for the Unity of Science research project on the Sociology of Science with a yet more sociological take on the nonlogical aspects of scientific theory. In an unsent letter of reply, Kuhn took up a view of the social dimensions of the validation of theories that was yet more radical than Frank's, and in which not generic social values but professional mores were central. "I suspect," he noted,

> that in twentieth-century western science socially conditioned, implicit professional "faiths" have assumed many of the roles in the guidance of research and in the determination of the acceptability of scientific theories which religion and metaphysical systems played in the physical sciences of the seventeenth century. I believe this professional consensus has an important bearing upon the problems which a scientist considers worth attacking, the experiments which he employs to resolve his problems, the abstract aspects of his experiments which he considers relevant, and the logical and experimental criteria which he demands of "valid" argument.[121]

By the summer of 1953, perhaps as a result of this connection to Frank, Kuhn had been commissioned to write the entry on the history of science for the *International Encyclopedia of Unified Science*.[122] Asked by the editor of the *Encyclopedia* to provide an outline of his contribution, Kuhn expanded on his argument about the ways in which professional ideology shaped the conduct of scientific research. "For the professional group that employs it the content of a scientific theory is larger than the formal or formalizable content of the theory," Kuhn wrote. As a "profession ideology [*sic*] for the practicing scientist," theory served to "direct the scientist's attention to certain sorts of problems as 'useful' and to certain sorts of measurement as 'important'; it dictates preferred techniques of interpretation, and it sets standards of precision in experiment and of rigor in reasoning." Kuhn even moved toward a description of normal science in his observation that one of the "striking results of the 'ideological' portions of a professionally institutionalized theory is the relatively firm closure which it gives the field of scientific problems." Theories, in this expanded view, "preserve themselves by restricting the attention of the profession to problems which can in principal be solved within the theory and by inhibiting the recognition of important incongruities in the application of the theory to nature."[123] Two years later, Kuhn was still advocating in private correspondence the view that the "locus" of conceptual schemes or theories was not the domain of formal logic or objective doctrine but of "the professional scientific group (or some larger socio-cultural unit determined by common education, etc.)." For such groups, Kuhn claimed, "the research process can frequently be well described as a systematic attempt to make nature conform to the set of beliefs sanctioned by the group."[124]

This inclusive, sociological conception of scientific theories left its mark on Kuhn's first book, *The Copernican Revolution.* Although Kuhn would later characterize the monograph as "an extended case history" drawn from his teaching on Natural Sciences 4, the analysis it offered was much more wide-ranging and methodologically ambitious than anything Conant provided in the *Harvard Case Histories in Experimental Science.*[125] Throughout his study of the shift from Ptolemaic to Copernican astronomy, Kuhn insisted that a proper understanding of the embrace of the idea of a heliocentric universe required more than the consideration of internal changes in mathematical astronomy. The latter were in fact intimately connected to "conceptual changes in cosmology, physics, philosophy, and religion." As such, the episode known as "the Copernican Revolution" was not "singular" but "plural"; Copernicus's *De Revolutionibus,* technical though it was, "could only be assimilated by men able to create a new physics, a new conception of space, and a new idea of man's relation to God." Hence to understand scientific change one also had to grasp "intellectual history" and the interconnections between science, society, and culture. Consequently, in his treatment both of the Aristotelian tradition of cosmology and the Copernican framework that replaced it, Kuhn underscored the "extra-astronomical entanglements of astronomical theory," and especially the upheavals in theology and religious consciousness associated with the Reformation.[126]

What is most striking about all of this is that five years later, in *Structure,* Kuhn insisted that these wider sociological and cultural factors could *not* account for the kinds of revolutionary change in science he had been assessing since 1947. In *Structure,* Kuhn acknowledged that the description of some of a scientific community's "shared beliefs will present no problems," but that other candidates for consensually held beliefs that appeared plausible to the historian "would almost certainly have been rejected by some members of the group he studies." But if a consensus on certain beliefs or common rules for conducting inquiry were taken to be central to defining the coherence of a professional group, "some specification of common ground in the corresponding area is needed." And this challenge was just what Kuhn, in 1962, believed to be "a source of continual and deep frustration." What "the various research problems and techniques" of a scientific tradition had in common was "not that they satisfy some explicit or even some fully discoverable set of rules and assumptions that gives the tradition its character and its hold upon the scientific mind." All the community needed for coherence was a set of recognized "established achievements" that could be used to model further research, even if the members of the community would not have agreed about the defining features or proper application of the exemplars in question. Having a "paradigm" meant not needing to raise these issues; given the utility of paradigms for the actual practice of research, deliberating over their qualities was beside the point.[127]

What led Kuhn to abandon sociological talk of professional consensus in favor of model puzzle-solutions and paradigms? One explanation stands out. From 1947 onward, Kuhn wanted to use history and sociology for epistemological ends, but his thoughts about consensus and intellectual culture in the early to mid-1950s had led him into the history and sociology of knowledge and away from what he had long sought: a normative epistemology derived from science "conceived as an activity, as the thing which the scientist does."[128] In particular, Kuhn wanted to draw out the epistemology implicit in his Aristotle epiphany, in which revolutions involved "learning to live within a new viewpoint," and working within "a new world" of scientific exploration.[129] Kuhn's notorious remarks in *Structure* about revolutions as "changes of world view" were less the product of Kuhn's theory of science than its animating presupposition, for which he sought a conception of scientific development adequate to explain what had been intimated in his 1947 Aristotle experience.[130]

When we look at Kuhn's notes and drafts for *Structure,* most of which were composed between 1958 and 1960, we can see that he was stymied in his attempt to make "consensus" perform the philosophical work he desired.[131]

In an outline of the "Cyclic Stages in Scientific Development," Kuhn began with a "Period of consensus" in which scientists worked "within a single pattern of belief about valid theory, problems, cannons [*sic*] of explanation, instruments useful in science, etc." This was followed by the "Crisis stage" and then the "revolution" in which a new consensus would snap into place. Yet when Kuhn turned, in the same document, to the question of what was "meant by scientific revolution," he spoke not of agreement or consensual beliefs but of a change of "style or mental set" and of a new "way of seeing the world."[132] Although Kuhn still placed the notion of consensus at the center of his account of normal science until at least as late as the spring of 1959, his dissatisfaction with the concept was increasingly evident.[133] In several stabs at setting down the role of consensus in the course of scientific revolutions, Kuhn seemed to recognize that it could not serve as the criterion that could both mark science off from other social practices and account for the sharp, nonaccumulative shifts he saw in the Copernican revolution. In one note, Kuhn asked himself whether earlier stages in the development of a science would be characterized by "weaker consensus" and then crossed through most of his observations on the "Acquisition of Consensus."[134] Elsewhere, in a draft of Chapter 2 (then titled "The Normal Practice of Mature Science"), Kuhn defined revolutions as "periods in which a scientific community temporarily ceased to satisfy the criteria [for progress-producing consensus] and thus made the transition from consensus to dissensus to new consensus."[135] But he was beginning to see that the kind of basic common ground between practitioners in a profession was not enough to explain the mechanics of scientific discovery he wanted to extract from the historical record. Psychologists could be "professionalized" but psychology itself was not yet a "mature" sciences marked by consensus. "I must know more than I do now," he wrote to himself in regard to Chapter 2, "about the transition from non-consensus to consensus stage in the development of sciences": this could not be taken for granted, but rather had itself to be accounted for. While the "development from consensus to consensus" was "something sciences achieve with maturity," the transition to maturity "may vary so much from science to science that it defies treatment in this monograph."[136]

Kuhn's growing sense of the problems with consensus came to a head in the penultimate draft of *Structure,* which was completed sometime "before June 1960."[137] At this point, Kuhn had already hit upon the idea of the paradigm as the key to understanding normal science and the pre-cipitation of crises that led to revolutions.[138] But he had yet to abandon the view that normal science was characterized not only by common famil-iarity with a paradigm or established achievement—and thus by the pos-session of a model that obviated the need for explicit consensus—but also by shared rules and beliefs. Indeed, in the 1960 draft, the chapter dealing with these issues was titled "Normal Science as Rule-Determined"—not, as in *Structure,* "Normal Science as Puzzle-solving." Furthermore, the major architectonic difference between the 1960 draft and the published text was that the former did not contain what became in *Structure* Chapter 5: "The Priority of Paradigms." This passage of Kuhn's essay was the most exten-sively rewritten in the final manuscript, and it is not hard to see why. In "Normal Science as Rule-Determined," Kuhn struck a confessional note by admitting that his earlier view that "periods of normal science were periods of consensus, during which the entire scientific community agreed about the rules of the game" was "very probably wrong." The problem with investi-gating the shared beliefs and commitments of scientists was that "the num-ber of rules that can be educed by such study never seems sufficient to define the puzzles that scientists normally undertake, or to restrict scientific atten-tion to their pursuit." Having a shared puzzle allowed disagreements among the members of a community to be muted, at least until such time as the model puzzle-solution was itself thrown into question. Kuhn had mistaken the "absence of disagreement" for the presence of agreement. "Part of the consensus that characterizes the normal scientific community is only appar-ent," he wrote.[139]

It is clear from these remarks that Kuhn had all but forged his argument about the "priority of paradigms" in determining normal science and lay-ing the foundations for crisis. Some of Kuhn's famous claims about scientific development are contained in the penultimate draft, but, as we have already noted, there was no chapter on the primacy of paradigms. Although we cannot document in detail the reasons why Kuhn finally decided to push paradigms into the foreground in the published version of *Structure,* we can identify the conceptual breakthrough that helped Kuhn flesh out his understanding of how paradigms could allow for the coherence of a tradi-tion without consensus or rules. In a word, Kuhn returned to his pedagogical roots. As he put it in *Structure,* a "community's paradigms" are "revealed in its textbooks, lectures, and laboratory exercises." Education was the key to explaining how a paradigm could be provide an understanding of the salient problems and intelligible solutions in a given field, yet without relying on explicit methodological agreement on exactly what made the problems and solutions pertinent. On this point, Kuhn nodded toward Michael Polanyi's exploration of "tacit knowledge" in the conduct of science, but his argument was more concerned with the psychology of learning and the referential properties of words. According to Kuhn, what the focus on puzzle solving as the mark of normal science brought to light was that scientists "never learn concepts, laws, and theories by themselves. Instead, these intellectual tools are from the start encountered in a historically and pedagogically prior unit that displays them with and through their applications." In periods of revolution, a "new theory is announced together with applications to some concrete range of natural phenomena." If the theory was accepted, "those same applications or others accompany the theory into the textbooks from which the future practitioner will learn his trade." Kuhn was adamant that "the process of learning a theory depends upon the study of the applications, including practice problem solving both with a pencil and paper and with instruments in the laboratory." What the student learns, and what the successful scientist never forgets, is thus not simply a stock body of theory or rules, but rather how to model a problem and solve it using the theory they have, in a circular fashion, learned through model puzzle-solutions. Thus, "paradigms guide research by direct modeling as well as through abstract rules."[140]

The hero of Kuhn's account of the priority of paradigms over rules was Ludwig Wittgenstein, whose *Philosophical Investigations* was in the late 1950s beginning to attract significant attention within the intellectual circles to which Kuhn belonged at Harvard and Berkeley.[141] But his patron saints might as well have been L. J. Henderson and James Bryant Conant. For Kuhn's description of paradigm-based pedagogy was similar in many respects to the practitioner's logic of cases and instrumental, research-enabling conceptual schemes outlined by Henderson and Conant. Just how much Kuhn had relied on his own experiences in the Harvard complex was evident from the near total lack of historical or scholarly references to the actual pedagogical regimes in particular fields that Kuhn insisted were so vital in explaining the potency of paradigms.[142] In *Structure* and subsequent works, Kuhn made no systematic historical study of the training regimens—the textbooks, the laboratory exercises, the lectures and seminars—through which a Newton or a Boyle or an Einstein was initiated into the profession.[143] It would seem that Kuhn eschewed the history of science pedagogy and relied, in adding the novel fifth chapter of *Structure,* on his own familiarity with the pedagogy of the case history in the teaching of non-scientists.

By the early 1960s, this had become an intuitive familiarity. Upon completing his Junior Fellowship, Kuhn (along with Leonard Nash) took over Natural Sciences 4 from Conant.[144] His first full-time academic appointment was as assistant professor of general education in the Harvard College's General Education program. All of which is to say that, until he was denied tenure at Harvard in 1956—at which time he took a joint appointment in history and philosophy at Berkeley—Kuhn was a theorist of education as well a historian. As early as 1949, Kuhn participated in conferences on general education and the teaching of science to laymen.[145] These pedagogical involvements were not sidelines for Kuhn. His frustration with the textbook or positivist image of science had an educational as well as a philosophical aspect: in his teaching of non-scientists, his major concern was disabusing students of an unrealistic picture of science as an imposing body of knowledge or mechanical procedure and replacing it with historical case studies that would disclose science as a process.[146] Here his interests dovetailed with Conant's.

Ultimately, however, Kuhn went further than his mentor. To explain how paradigms rather than rules guided normal science, Kuhn reached for the case-based theory of learning he had both practiced as a teacher and applied to the knowledge-making activities of research scientists. Unlike Henderson or Conant, Kuhn used pedagogy to explain the historical cases of scientific advancement he employed in the education of non-scientists. This was not quite a closed circle or a feedback loop. What we see in *Structure* is yet another combination of pedagogy, research, and epistemology in the Harvard complex.

Epilogue

The Great Disembedding

The publication of *The Structure of Scientific Revolutions* in 1962 takes us back to where we began our investigation. In the Prologue, we surveyed the crisis of the Anglophone human sciences since World War II. The defining moment of this period of upheaval was the attack on the "behavioral science" program launched during the 1960s by a heterodox group of historicists, neopragmatists, Wittgensteinians, and critical theorists. *Structure* is often said to have led the charge. "The natural sciences, whose claims to objectivity had intimidated humanists and inspired philosophers and social scientists," writes one historian "fell before the historicist analysis of Thomas Kuhn." After *Structure*, the way was clear for "an influential band of scholars who challenged the ideal of objectivity" in the human sciences and expounded "variations on a revolutionary gospel of interpretation."[1] Leading philosophers and social scientists have claimed that *Structure* "cuts across [disciplinary] lines which many have thought fixed and sacred," and, in doing so, enabled the "demystification of scientific authority, its re-enclosure in time and society."[2] By helping to overturn a hitherto pervasive "positivism" and "objectivism," Kuhn is often said to have caused a revolution of his own in the philosophy of science and in the social sciences.[3]

An aim of this book has been to show just how much is missed when the battle lines are so sharply drawn. Before the image of a positivist/post-positivist divide gripped the historical imagination after the publication of *Structure,* key debates about epistemology in the human sciences were, as we have seen throughout this study, bound up with matters of pedagogy and research practice. Indeed, in the case of the Harvard complex, there were few stand-alone, purely "ideological" positions on the epistemic character of the human sciences. Instead, there were attempts, within specific constellations of disciplines and institutional networks, to identify epistemology with research practice, or to make pedagogy a reflection of the protocols of scientific inquiry, or to conflate epistemology with pedagogy. Regardless of whether we would now want to assign the theories of knowledge we find in Hendersonian social science, operationism, Quinean analytic philosophy, the general theory of action, or Kuhnian history of science to this or that ideological category, we must recognize that epistemological questions showed up for these practitioners of the human sciences as practical challenges that they had to face in order to conduct the kind of intellectual and professional lives they wished to live. That is why pedagogy and research practice seemed so closely aligned with the search for proxies for traditional epistemology in the Harvard complex. The "crisis" literature assessed in the Prologue, because it is neglectful of the history of the university or the tradition of scientific philosophy, has no means for gaining a purchase on epistemology as a practice, distinct from ideology or rhetoric. For this reason, the Harvard complex has been largely lost from view.

Since the early 1960s, in fact, the ideas, disciplinary projects, and institutional structures that composed the Harvard complex have been erased in both the historical memory and the academic practice of the human sciences. The epistemology that flourished in the interstitial academy has been gradually disembedded from the training regimes and research practices that defined it. *The Structure of Scientific Revolutions,* and its author, have been at the center of this process. We have seen how Kuhn drew a philosophy of science from the elements of what had been an eclectic education in the Harvard complex. Kuhn's solution to the problem of accounting for the absence of disagreement on fundamentals among scientific specialists—namely, a training regime in which a theory was given with a certain range of concrete applications—cannot be understood without knowing about the history of case-based reasoning and teaching at Harvard or about Kuhn's early career as an expert on general education. The marks of the Harvard complex, and especially the Harvard formula for balancing internal and external accounts of the history of science, are readily observable in *Structure,* not least in Kuhn's reliance on the *Harvard Case Histories in Experimental Science* and in his citations of the work of Bernard Barber, Henry Guerlac, Robert Merton, and others. Nevertheless, the final draft of *Structure* and the philosophical debates it sparked created a different perception of what kind of an intervention Kuhn's book was intended make.

The major revision following Kuhn's preparation of the penultimate draft of *Structure* in the first half of 1960 was the insertion of an extra chapter on "The Priority of Paradigms." In turn, the chief novelty of that chapter, when juxtaposed with the earlier draft, was the appeal to Wittgenstein's remarks in the *Philosophical Investigations* on the criteria according to which names or kind terms ("Moses" or "leaf") were applied to the world. Wittgenstein's point was that speakers used words in ways generally intelligible to their interlocutors without having exactly the same criterion in mind as all their fellow speakers and without being "equipped with rules for every possible application" of a given term.[4] Moreover, Wittgenstein's account of the loose weave of meaning, like Kuhn's claims about the "priority of paradigms," rested in the end on a particular conception of how words or concepts were learned through exposure to a discrete range of examples that would form the criteria for applying a word (or paradigm). As Wittgenstein put it in his exploration of the insufficiency of rules for determining meaning, one always had to ask "How did we *learn* the meaning of this word . . .? From what sort of examples?"[5] Wittgenstein was thus sensitive to the importance for guiding action of examples conceived as precedents or adjudged cases, much like the model of the English common law from which the Harvard case method had been forged.[6] Likewise, Kuhn tapped into this tradition of thinking about pedagogy and knowledge when he claimed in *Structure* that the paradigm, "like an accepted judicial decision in the common law . . . is an object for further articulation and specification under new or more stringent conditions."[7]

If Kuhn's embrace of Wittgenstein can be seen as a natural move for someone with Kuhn's education and interests, it rapidly came to seem a much more radical break with prevailing epistemological ideas than Kuhn had intended when he was drafting the book during the 1950s. Kuhn was at once a bystander and willing participant in this development. Somewhat unwittingly, Kuhn's invocation of the later Wittgenstein allowed *Structure* to catch a wave of new work in philosophy and the social sciences, work which placed in question the very idea of a "science" of behavior.[8] Toward the end of his long career, the anthropologist Clifford Geertz spoke of Wittgenstein's "enormous impact upon my sense of what I was about and what I hoped to accomplish." This involved for Geertz a generational shift in philosophy, "and that shift has been particularly congenial to those, like myself, who believe that the answers to our most general questions—why? how? what? whither?—to the degree they have answers, are to be found in the fine detail of lived life." Wittgenstein's relocation of philosophical puzzles in the realm of everyday practices of learning and agency seemed "almost custom designed to enable the sort of anthropological study I, and others of my ilk, do."[9] Geertz's rec- ollections neatly capture the intellectual atmosphere of the late 1950s and early 1960s. It was at this time that a group of Wittgenstein's disciples, gath- ered principally around Rush Rhees at the University of Swansea, began to publish the Studies in Philosophical Psychology, a series of attacks on the naturalistic approach to the study of human psychology and social action that offered distinctly Wittgensteinian takes on intentionality, motivation, the will, perception, and the role of rationality in explaining behavior.[10] Peter Winch's *The Idea of a Social Science* (1958), with its astringent rejec- tion of naturalistic accounts of human agency, is perhaps the most famous of the Swansea "red books" and remains a classic of the hermeneutic turn in the human sciences.[11] Meanwhile, across the Atlantic, philosophically inclined social scientists and historians like Geertz and Kuhn were becom- ing increasingly excited by the new vistas opened up by the *Investigations*. One of Kuhn's colleagues at Berkeley, the philosopher Stanley Cavell, would help to mold the reception of the later philosophy of Wittgenstein among American philosophers.[12] The nod to Wittgenstein in *Structure* put Kuhn at the forefront of this new anti-positivist, anti-behavioralist charge.

Riding the Wittgensteinian wave, *Structure* was rapidly pulled out of the Harvard context in which it had largely been forged. This distension of Kuhn's project in the history and philosophy of science was exacerbated by its reception. As John H. Zammito has perceptively observed, "the philo- sophical reception [of *Structure*] shifted everything Kuhn had been saying into the key of philosophy of language and focused all its energy on the dicey notion of meaning and its (in)variance."[13] Some of Kuhn's earliest critics, notably Dudley Shapere and Israel Scheffler, read his claim about paradigm change in revolutions as a claim about the way in which words changed their meaning.[14] Correspondingly, Kuhn's suggestion that terms preserved across a revolution would nonetheless apply to "different worlds" was interpreted as a radical thesis about the variability of meaning, such that the same word could, on either side of a revolutionary divide, have "incommensurable" meanings—a different sense and a different reference. This was the ground on which hostile readers like Shapere and Scheffler criticized Kuhn as a naïve philosopher of language. Scheffler's defense of the meaning invariance of words—and thus of the possibility of cumulative progress in science—rested on an account of the preservation of the denotation or reference of words even when very different senses or connotations were imposed upon them. Debates about Kuhn's alleged advocacy of meaning variance and incommensurability were soon filtered through the so-called causal theory of reference developed by the philosophers Saul Kripke, Hilary Putnam, and Keith Donnellan during the 1970s.[15]

What is most striking about the "semantic ascent" to which Kuhn's theory of science was subjected by the philosophers was that Kuhn himself, in subsequent essays, reframed his own arguments to fit these linguistic debates about reference, interpretation, and translation.[16] Kuhn even went so far as to try to meet the challenge posed by the causal theory of reference.[17] This apparent acquiescence makes sense when we recall that Kuhn had since the 1940s been desperate to make philosophy from his historical study of science; *Structure* was intended as a book for philosophers.[18] So it was that Kuhn spent much of the rest of his career attempting to meet the concerns of philosophers about linguistic incommensurability, rather than, for example, shedding more historical and sociological light on the claims about pedagogy that underpinned his insistence on the priority of paradigms in scientific research.

A final factor that helped to obscure the role of the Harvard complex in the formation of *Structure* belonged more explicitly to its author's intentions. Very early in his career, Kuhn felt dissatisfied with the "textbook" and "positivist" conceptions of science. What the young Kuhn objected to in these positions was their treatment of science as a finished body of knowledge produced by a settled formal method. In *Structure*, Kuhn pressed these objections yet further, and in doing so created the impression that his historical philosophy of science was a deathblow against logical positivism or operationism. What he called "early logical positivism" (no references were provided to characteristic works of this school) was responsible for obscuring the "necessity of revolutions" by reducing historically rich rival theories to ideologically neutral theoretical systems. Those operationists and positivists who insisted on laboratory manipulations or a purportedly "neutral" observation language as the basis of scientific knowledge mistook "the perceptual features that a paradigm so highlights that they surrender their regularities upon inspection," on the one hand, with "immediate experience" or "the given," on the other. At the root of these problems, Kuhn argued, was that, like the textbooks and "popularizations" of science on which they were "modeled," philosophical works in the positivist tradition "address themselves to an already articulated body of problems, data, and theory, most often to the particular set of paradigms to which the scientific community is committed at the time they are written."[19] Philosophers who today seek to rehabilitate logical empiricism are driven to distraction by Kuhn's vague understanding of the philosophy of science he sought to overturn.[20] But the power of his remarks, when coupled with the new wave of anti-behavioralist and anti-positivist studies in philosophy and the social sciences, is unmistakable.

Kuhn's iconoclastic intentions toward "logical positivism" allowed his book to be coupled with other purportedly postpositivist works in the philosophy of science from the same period. Retrospectively, *Structure* snapped into place alongside Norwood Russell Hanson's *Patterns of Discovery* (1958), Michael Polanyi's *Personal Knowledge* (1958), Stephen Toulmin's *Foresight and Understanding* (1961), and Paul Feyerabend's early essays as one component of a new postpositivist orientation in the philosophy of science.[21] This was, it is frequently said, a dispensation hostile to the formalism and verificationism of the Vienna Circle and its followers and committed to an understanding of science as an intellectual craft. It was, moreover, a short step to associate this "school" with insurgent movements against "positivism" elsewhere in the human sciences. After World War II, Carl Hempel's studies in the "logic" of scientific explanation made abstract what in logical empiricism had previously been engaged with the ongoing findings of logic, mathematics, and physics.[22] His extension of what became known as the covering law model of explanation into the philosophy of history helped to crystallize an emerging critique of the assimilation of history and the social sciences to the logic of the natural sciences.[23] In parallel to the postpositivist moment in the philosophy of science came an anti–Hempelian turn in the philosophy of history led by philosophers William Dray and Louis Mink and the structuralist literary critic Hayden White.[24] Around the same time, W. V. Quine's notorious dismantling of the "dogmas" of empiricism, together with the reception of the later Wittgenstein and of the ordinary language philosophy of J. L. Austin, was interpreted as a casus belli against the hegemony of "positivism" and especially against a science-centered philosophy. With the aid of the Studies in Philosophical Psychology series and other Wittgensteinian works, problems of intentionality, mental events, rationality, and social explanation came back onto the philosophical agenda, while at the same time opening up the social sciences in the ways that Geertz fondly recalled. A new "science of interpretation" was on the agenda.[25]

Or so it seemed in retrospect. Everything I have described in this book suggests that the idea of a revolution against positivism, or a tipping of the scales from positivism to historicism, misconstrues the epistemological stakes for those involved in the Harvard complex. Not only did Kuhn conjure up an image of positivism neither substantiated textually nor historically explicated; he also obscured the sources of his thinking in a milieu in which logical empiricism, operationism and the rhetoric of "conceptual schemes" mixed freely with discipline-building, institutional change, and pedagogical reform. Kuhn was not alone in this regard. The place of Quine in the scholarly imagination is equally striking: in the history of philosophy he is the figure who overturns Carnap's program of Logic of Science. But we have seen that he did not carry out this critique. Instead, he pursued his own "logical empiricism," which was shaped by the opportunities for engaging with scientific philosophy in Harvard's interstitial academy.[26] Even Clifford Geertz, for some the avatar of laissez-faire literary postmodernism in anthropology, was trained in the Department of Social Relations and was deeply marked by the eclectic approach of the Levellers, especially with respect to the independence of the cultural domain in human affairs.[27] The present way of conceiving of an anti-positivist break, in which one monolithic vision of knowledge was overthrown on epistemological grounds by various postpositivist doctrines, excludes the existence of the Harvard complex from the stories we tell about the contemporary human sciences.

There is one final element in this work of historical forgetting, which is principally institutional. The interstitial academy, as a fluid space within the research university that existed at the margins of more professionalized disciplines and schools, has itself become institutionalized. The forms of interdisciplinary discourse it fostered, in which pedagogy, practice, and epistemology were linked, have now been professionalized and incorporated into the modern academy. This has involved a two-track process. On the one hand, the interstitial academy itself melted away. Without exception, the Harvard enclaves either perished or were transformed. Several of these creations were, as we have seen, sustained by the will and authority of L. J. Henderson. None of his concerted attempts to reform the social sciences on the model of Pareto long survived his death in 1942. After the war, the Society of Fellows became part of a national system of postdoctoral fellowships, while the history of science provision at Harvard yielded, ultimately, to a Department of the History of Science.[28] Elsewhere, Philipp Frank's attempts to convert the Inter-Science Discussion Group into a permanent Institute for the Unity of Science, with ties to the general education programs of the postwar period, fell prey to the logic of professionalism. Unable to secure the cultural authority that Frank craved, the Institute gave way to the Boston Colloquium for Philosophy of Science, one among a host of new professional organizations catering for the academic field of "philosophy of science."[29] The Department of Social Relations endured a slow death after the failure of the Carnegie Project on Theory. Parsons became the subject of increasingly vituperative attacks from radical students and scholars, who blamed him for instilling a "consensus bias" in sociological theory. More damaging to his project was the growing dissatisfaction of his colleagues with the rationale for the Department. An asset in the heyday of Geertz, the Department's lack of discipline became a liability in the early 1960s. The notorious, LSD-fuelled "Prisons Project" of Timothy Leary and Richard Alpert gave Social Relations the unwanted mantle of a seedbed of the counterculture. Parsons's action-based view of the social realm became less attractive to a post-1960s generation of social scientists with new questions and critical frameworks. In 1970, the sociologists withdrew from the DSR, and by 1974 Social Relations had ceased to exist. Discrete departments of sociology, anthropology, and psychology took its place.[30] Finally, the General Education program in which Kuhn had taught before leaving for Berkeley in 1956 edged toward dissolution in the mid-1960s in the face of faculty criticism and undergraduate apathy.[31]

As the Harvard enclaves dropped away, they were replaced with more permanent structures in which "theory" could be pursued. The Center for Advanced Study in the Behavioral Sciences in Palo Alto, which opened its doors in 1954, was in the vanguard of the new institutions for extradepartmental inquiry in the human sciences. This was where Kuhn held a fellowship in 1958–59, during which time he hit upon the notion of the paradigm. The

Center was emphatically not the place where major experiments or empirical investigations were carried out. Rather, it was where existing knowledge was mulled and refigured. The scenes of scholarly labor were arranged accordingly: the major "physical resources" of the Center were "an individual study for each fellow, meeting rooms, a dining room, and a library." Technical "plant" was limited to "typewriters, calculating machines, the more common items of I.B.M. equipment, and a room with a one-way vision screen for the study of small group behavior."[32]

The Center for Advanced Study was part of a movement that created many new institutes for advanced study and humanities centers. These institutions proliferated across the United States over the next quarter of a century. Appointed to a professorship at the Institute for Advanced Study in Princeton in 1970, Clifford Geertz made its newly established School of Social Science a hothouse for theory, "a place for planning research, for reflecting upon it, or for writing it up, rather than a place for carrying it out."[33] Meanwhile, humanities centers, which replicated the Palo Alto model of offices, libraries, and seminar rooms, were opening their doors in universities across the nation. By the end of the century, the United States boasted at least three dedicated spaces for advanced study in the humanities and human sciences—the Center for Advanced Study in the Behavioral Sciences, the Institute for Advanced Study, and the National Humanities Center in North Carolina, with the Radcliffe Institute for Advanced Study waiting in the wings—and over one hundred campus-based advanced study or humanities centers.[34] Such centers belonged to a large cohort of what became known as "Organized Research Units" (ORUs), which after World War II were the preferred forum of organizing research outside of discrete departments or schools. An ORU might be an interdisciplinary graduate program or a center dedicated to a specific, cross-disciplinary area (international studies, robotics, etc.).[35] Or, like the Center for Advanced Study or the humanities centers, they would bring together scholars working on individual research projects, with the idea that serendipitous exchanges might result. Moreover, during the early Cold War, state-sponsored research produced a network of Federal Contract Research Centers that inhabited a "gray area" between the university and the world of private think tanks.[36] The interstitial spaces in which Harvard's human scientists had found themselves after World War I became less ad hoc and more professionalized from the 1950s onward.

All of this would be of merely historical interest were it not that the neglect of a phenomenon like the Harvard complex places unnecessary limitations on our understanding of what kind of tradition and field of possibilities the human sciences constitute. Recent studies have drawn "surprising" and apparently counterintuitive connections between Quine and Carnap, or Kuhn and the logical empiricists, or Geertz and the Parsonian behavioral scientists.[37] But these discoveries are only surprising because of the hold on our imaginations of the epic histories of epistemological conflict between "positivists" and "postpositivists."

One idea that the foregoing account of the development of the human sciences at Harvard helps us to recover is the notion that the history of epistemology need not be considered an all-or-nothing clash between distinct ideologies of knowledge, with "positivism," "scientism" or "objectivism" ranged against "postpositivism," "relativism," or "historicism." Not, of course, that one would want to downplay the very real conflict between rival traditions of inquiry during the twentieth century. Nor should it be thought that Harvard complex represents some lost golden age of epistemological comity in the human sciences. But it can teach us to be more realistic about the theory of knowledge and the modes it which it has historically been practiced. I use "realistic" here in a relatively undemanding sense. Questions of realism and anti-realism in philosophy are infamously complex, and I have no intention of stirring up that particular nest of hornets. Being "realistic" about the theory of knowledge and its recent history means recognizing that, in order to understand any particular ideology of knowledge, one must place overtly normative claims about the nature of human cognition and scientific inquiry within the social context of academic practices that at once produced and were shaped by those same normative, "philosophical" claims. Members of the Harvard complex, from Henderson to Kuhn, made these connections between epistemology and scientific practices intuitively. Within contemporary science studies, the turn to the study of scientific practices was designed precisely to provide an antidote to idealist and transcendental theories of scientific understanding.[38] Even the purest form of "objectivity," as one influential study has suggested, can itself be seen to rest on the everyday, nitty-gritty details of laboratory measurement, textbook illustration, and scientific calculation.[39] The point can be put in a slightly more philosophical register: what it is rational to believe and to do in the realm of scientific inquiry will depend both upon normative evaluation of what counts as a warranted knowledge claim and upon an assessment of how the process of inquiry actually goes in practice, in this or that place and time.[40] The example of the Harvard complex can remind us of these twin normative and empirical features of knowledge-making in the human sciences and thereby help us to put the various epistemological "isms" in their proper historical place.

In the Prologue, I described the balkanized intellectual landscape that has resulted from the stalled interpretive revolution against the behavioral sciences. What we see now is a scene in which diverse subcultures exist cheek by jowl—proponents of rational choice theories in political science here, social statisticians there, critical theorists in this institutional enclave, and so on. No single group holds the whip hand, even if, in particular cases, one group or another feels itself marginalized, besieged, or insurgent. Hence many normative or critical political theorists consider themselves a minority opposition to their quantitatively or formalistically inclined brethren in political science departments, or "continental" philosophers feel themselves outnumbered by their "analytic" colleagues in philosophy. The philosophical histories sketched by figures such as Richard Rorty and Charles Taylor have given these intramundane grievances an epochal character, setting one vision of the human against another. Equally, however, there are, as we have noted, some signs of at least a basis for discussion across the barricades. The discovery of affinities between those on either side of the positivist/postpositivist divide point the way to an appreciation of the sources of the contemporary human sciences that sees their major traditions not as implacably hostile but locked into a mutual process of innovation, struggle, and accommodation. Geertz may not be a Parsonian any more than Kuhn is a Carnapian, but by seeing them all as part of a middle ground of conflict, adjustment, and conceptual change, we can begin to find our way to a middle ground of our own, in which all exchanges between rival traditions need not be zero-sum games. Perhaps this kind of exercise will provide us with new narratives through which to define the salience of our research problems and justify our theoretical and methodological choices.

This is where the Harvard complex comes into the picture. At the historiographical level, it provides the necessary background against which the idea of the definitive split between Kuhn's generation and that of Parsons or Quine can be seen as a simplification of the historical record. It helps us to connect up the postwar crises of the human sciences with events of the interwar years and with the intellectual traditions forged at the turn of the twentieth century. Even more importantly, the Harvard complex itself furnishes one example of a practice-oriented theory of knowledge that contemporary practitioners of the human sciences may wish to ponder. Once again, this is not to romanticize Harvard's interstitial academy or the tradition of scientific philosophy. Reasonable people may disagree about the adequacy of the visions of the study of human thought and action offered by the primary movers in the Harvard complex. There are, for example, powerful arguments to be made about the conservative nature of the reception of logical empiricism and the history of science in the hands of Quine and Kuhn. My point, then, is not that the period stretching from Henderson and Parsons to Kuhn is a heyday of consensus or epistemological sophistication. The Harvard complex offers a paradigm for thinking about epistemology in the human sciences because it brought more than brute ideological conceptions of "knowledge" or "scientific explanation" into play. At stake in the debates held among Paretians, operationists, philosophical naturalists, and the others were matters of learning and research practice, as well as epistemology. There was much at issue in these exchanges, then, but also a wider, more varied terrain on which to debate. By tying concerns about knowledge-making to prosaic challenges of discipline formation, institutional recognition, research practice, and pedagogy, the Harvard complex provided those who moved within it a rich array of tools and concepts with which to develop disciplines dedicated to the scientific study of human affairs. For today's human scientists, this can serve as an instructive example of how sciences of human conduct can be made, and remade.

Notes

Abbreviations

CIBP   Chester I. Barnard Papers, Baker Library, Harvard Business School, Harvard University

HUSFP   Videotapes of interviews, Society of Fellows, HUC 1800.14, Harvard University Archives, Harvard University

LJHP   L. J. Henderson Papers, Baker Library, Harvard Business School, Harvard University

PDFASH   Papers of the Dean of the Faculty of Arts and Sciences, Harvard University Archives, Harvard University

SSSPH   Stanley Smith Stevens Papers, HUGFP 2.12, Harvard University Archives, Harvard University

TPPH   Talcott Parsons Papers, Harvard University Archives, Harvard University

TSKP   Thomas S. Kuhn Papers, MC240, Institute Archives and Special Collections, Massachusetts Institute of Technology

WVQPH   Willard Van Orman Quine Papers, Houghton Library, Harvard University.

Prologue

1. Thomas Kuhn, "The Structure of Scientific Revolutions," 24, 39, draft typescript, marked "penultimate draft of *Structure* before June 1960," box 4, TSKP.

2. Ibid., 41–43.

3. Thomas Kuhn, *The Structure of Scientific Revolutions*, vol. 2, no. 2, *International Encyclopedia of Unified Science* (Chicago: University of Chicago Press, 1962), 44–47.

4. Ibid., 9.

5. Clifford Geertz, "The Legacy of Thomas Kuhn: The Right Text at the Right Time," in Geertz, *Available Light: Anthropological Reflections on Philosophical Topics* (Princeton, NJ: Princeton University Press, 2000), 162.

6. Richard Rorty, *Philosophy and the Mirror of Nature* (1979; Oxford: Blackwell, 1980), 322–325.

7. Richard J. Bernstein, *The Restructuring of Social and Political Theory* (London: Metheun, 1979), 84–93.

8. There are historians and philosophers who offer more detailed models for combining commitments to naturalism and historicism than I can provide here, just as there are better examples than the Harvard complex of intellectual cultures intermediate between those two positions. For compelling accounts of how to acknowledge the social and historical nature of human knowledge—and especially knowledge of human affairs—while holding a place for a suitably redefined "objectivity" or "realism," see Thomas Haskell, "The Curious Persistence of Rights Talk in the Age of Interpretation," in Haskell, *Objectivity Is Not Neutrality: Explanatory Schemes in History* (Baltimore, MD: Johns Hopkins University Press, 1988), 115–144; John Dunn, "Practising History and Social Science on 'Realist' Assumptions," in Dunn, *Political Obligation in Its Historical Context: Essays in Political Theory* (Cambridge: Cambridge University Press, 1980), 81–111; John R. Searle, *Making the Social World: The Structure of Human Civilization* (Oxford: Oxford University Press, 2010); Pierre Bourdieu, *The Logic of Practice,* trans. Richard Nice (Cambridge: Polity, 1990), 30–65. For a salutary via media, see James T. Kloppenberg, *Uncertain Victory: Social Democracy and Progressivism in European and American Thought, 1870–1920* (New York: Oxford University Press, 1986).

9. For an overview of this Cold War "moment," see Ron Robin, *The Making of the Cold War Enemy: Culture and Politics in the Military-Industrial Complex* (Princeton, NJ: Princeton University Press, 2001); Steve J. Heims, *The Cybernetics Group* (Cambridge, MA: MIT Press, 1991); Dorothy Ross, "Changing Contours of the Social Science Disciplines," in *The Cambridge History of Science,* vol. 7: *The Modern Social Sciences,* ed. Theodore M. Porter and Dorothy Ross (Cambridge: Cambridge University Press, 2003), 229–234; Joel Isaac, "The Human Sciences in Cold War America," *Historical Journal* 50 (2007): 725–746.

10. Peter Mandler, "Deconstructing 'Cold War Anthropology,'" in *Uncertain Empire: American History and the Idea of the Cold War* ed. Joel Isaac and Duncan Bell (forthcoming).

11. Talcott Parsons, "The Prospects of Sociological Theory," in Parsons, *Essays in Sociological Theory,* rev. ed. (1950; New York: Free Press, 1954), 369.

12. James T. Patterson, *Grand Expectations: The United States, 1945–1974* (New York: Oxford University Press, 1996).

13. Nils Gilman, *Mandarins of the Future: Modernization Theory in Cold War America* (Baltimore, MD: Johns Hopkins University Press, 2003), 41–71; Howard Brick, *Transcending Capitalism: Visions of a New Society in Modern American Thought* (Ithaca, NY: Cornell University Press, 2006), 121–185; David Ciepley, *Liberalism in the Shadow of Totalitarianism* (Cambridge, MA: Harvard University Press, 2006).

14. Although the coinage is claimed by James Grier Miller, the adoption of the term in 1949 by the Ford Foundation for its program in the social and psychological sciences, along with its use for the newly created (and Ford Foundation–sponsored) Center for Advanced Study in the Behavioral Sciences (conceived in 1951), marks the entry of the term into the Anglophone human sciences. See H. Rowan Gaither et al., *Report of the Study for the Ford Foundation on Policy and Program* (Detroit, MI: Ford Foundation, 1949), 94; Ford Foundation, *Ford Foundation Annual Report 1953* (Detroit, MI: Ford Foundation, 1953), 65–67.

15. University of Minnesota, *The Social Sciences at Mid-Century: Papers Delivered at the Dedication of Ford Hall, April 19–21, 1951* (1951; Freeport, NY: Books for Libraries Press, 1968); Daniel Lerner and Harold D. Lasswell, *The Policy Sciences* (Stanford, CA: Stanford University Press, 1951); Leonard D. White, *The State of the Social Sciences* (Chicago: University of Chicago Press, 1956); Bernard Berelson, *The Behavioral Sciences Today* (New York: Basic Books, 1963). See also Hunter Crowther-Heyck, "Patrons of the Revolution: Ideals and Institutions in Postwar Behavioral Science," *Isis* 97 (2006): 420–446.

16. For a useful overview of the midcentury importance of culture-and-personality studies, see Peter Mandler, "Margaret Mead Amongst the Natives of Great Britain," *Past & Present* 204 (2009): 195–233; Joanne Meyerowitz, "'How Common Culture Shapes the Separate Lives': Sexuality, Race, and Mid-Twentieth-Century Social Constructionist Thought," *Journal of American History* 96 (2010): 1057–1084.

17. Ludwig Wittgenstein, *Philosophical Investigations,* trans. G. E. M. Anscombe, 3rd ed. (Oxford: Blackwell, 2001); Kuhn, *Structure.*

18. Isaiah Berlin, "Does Political Theory Still Exist?" in *Philosophy, Politics, and Society,* 2nd series, ed. Peter Laslett and W. G. Runciman (Oxford: Blackwell, 1962), 1–33; Alasdair MacIntryre, "The Idea of a Social Science," in MacIntyre, *Against the Self-Images of the Age* (1967; London: Duckworth, 1971), 211–229; C. Wright Mills, *The Sociological Imagination* (New York: Oxford, 1959).

19. Hubert L. Dreyfus, *What Computers Can't Do: A Critique of Artificial Intelligence* (New York: Harper & Row, 1972).

20. For critiques of the deductive-nomological model of scientific reasoning, see Norwood Russell Hanson, *Patterns of Discovery: An Inquiry into the Conceptual Foundations of Science* (Cambridge: Cambridge University Press, 1958); Stephen Toulmin, *Foresight and Understanding: An Enquiry into the Aims of Science* (London: Hutchinson, 1961); Kuhn, *Structure.* On the narrative turn in the philosophy of history, see Hayden White, *Metahistory: The Historical Imagination in Nineteenth-Century Europe* (Baltimore, MD: Johns Hopkins University Press, 1973); Louis O. Mink, *Historical Understanding,* ed. Brian Fay, Eugene O. Golob, and Richard T. Vann (Ithaca, NY: Cornell University Press, 1987).

21. Charles Taylor, "Interpretation and the Sciences of Man," *Review of Metaphysics* 25 (1971): 3–51.

22. Charles Taylor, "Neutrality in Political Science," in *Philosophy, Politics, and Society,* 3rd series, ed. Peter Laslett and W. G. Runciman (Oxford: Blackwell, 1967), 25–57; Taylor, "Cognitive Psychology," in Taylor, *Human Agency and Language: Philosophical Papers,* vol. 1 (Cambridge: Cambridge University Press, 1985), 187–212. For a historical account of Taylor's intervention in the human sciences, see Naomi Choi, "Defending Anti-Naturalism after the Interpretive Turn: Charles Taylor and the Human Sciences," *History of Political Thought* 30 (2009): 693–718.

23. See especially W. V. Quine, *From a Logical Point of View* (Cambridge, MA: Harvard University Press, 1953).

24. Carl G. Hempel and Paul Oppenheim, "Studies in the Logic of Explanation," *Philosophy of Science* 15 (1948): 135–175; Ernest Nagel, *The Structure of Science: Problems in the Logic of Scientific Explanation* (London: Routledge and Kegan Paul, 1961).

25. Geertz, "Preface," in *Available Light,* ix–xiv.

26. For an overview, see Paul Rabinow and William M. Sullivan, eds., *Interpretive Social Science: A Reader* (Berkeley: University of California Press, 1979).

27. Clifford Geertz, "Blurred Genres: The Refiguration of Social Thought," in Geertz, *Local Knowledge: Further Essays in Interpretive Anthropology* (1980; New York: Basic Books, 2000), 19–35; Bernstein, *Restructuring.*

28. Morris Dickstein, ed., *The Revival of Pragmatism: New Essays on Social Thought, Law, and Culture* (Durham, NC: Duke University Press, 1998).

29. Quentin Skinner, "Introduction: The Return of Grand Theory," in *The Return of Grand Theory in the Human Sciences,* ed. Skinner (Cambridge: Cambridge University Press, 1985), 6.

30. Thomas Kuhn, "The Natural and the Human Sciences," in Kuhn, *The Road Since Structure: Philosophical Essays, 1970–1993, with an Autobiographical Interview,* ed. James Conant and John Haugeland (Chicago: University of Chicago Press, 2000), 216–223; Clifford Geertz, "The Strange Estrangement: Charles Taylor and the Natural Sciences," in Geertz, *Available Light,* 143–159; Richard Rorty, "Science as Solidarity" and "Inquiry as Recontextualization: An Anti-Dualist Account of Interpretation," in Rorty, *Objectivity, Relativism, and Truth: Philosophical Papers,* vol. 1 (Cambridge: Cambridge University Press, 1991), 35–45, 93–110. Kuhn's essay was prepared in 1989 and published in 1991; Geertz's piece appeared in print in 1995; and Rorty's essays were written in the mid- to late 1980s.

31. Skinner's abandonment of a long-gestating book on hermeneutics may be seen as symptomatic of this impasse. See Kari Palonnen, *Quentin Skinner: History, Politics, Rhetoric* (Cambridge: Polity, 2003), 137. Portions of the abortive manuscript eventually appeared in Skinner, "A Reply to My Critics," in *Meaning and Context: Quentin Skinner and His Critics,* ed. James Tully (Cambridge: Polity, 1988), 231–288. See also Skinner, *Visions of Politics,* vol. 1, *Regarding Method* (Cambridge: Cambridge University Press, 2002).

32. S. M. Amadae, *Rationalizing Capitalist Democracy: The Cold War Origins of Rational Choice Liberalism* (Chicago: University of Chicago Press, 2003).

33. Gilman, *Mandarins of the Future*, 256–276.

34. For a forceful restatement of the second-order vision, see Scott Soames, *Philosophical Analysis in the Twentieth Century*, 2 vols. (Princeton: Princeton University Press, 2003).

35. Enthusiastic endorsements of this approach include Steven Pinker, *How the Mind Works* (New York: Norton, 1997) and Francis Fukuyama, *The Origins of Political Order* (London: Profile, 2011). A critical account of these developments is given generally in Raymond Tallis, *Aping Mankind: Neuromania, Darwinitis, and the Misrepresentation of Humanity* (Durham, UK: Acumen, 2011); and more specifically in the disciplines of international relations and political theory in Duncan Bell, "Beware False Prophets: Biology, Human Nature and the Future of International Relations Theory," *International Affairs* 82 (2006): 493–510.

36. Jerome Kagan, *The Three Cultures: Natural Sciences, Social Sciences, and the Humanities in the 21st Century* (Cambridge: Cambridge University Press, 2009), 130.

37. See, e.g., Donald P. Green and Ian Shapiro, *Pathologies of Rational Choice Theory: A Critique of Applications in Political Science* (New Haven, CT: Yale University Press, 1994).

38. Peter A. Hall, "The Dilemmas of Contemporary Social Science," *boundary 2* 34 (2007): 121–141.

39. Barbara Herrnstein Smith, "Cutting-Edge Equivocation: Conceptual Moves and Rhetorical Strategies in Contemporary Anti-Epistemology," *The South Atlantic Quarterly* 101 (2002): 187–212.

40. Kristin Renwick Monroe, ed., *Perestroika! The Raucous Rebellion in Political Science* (New Haven, CT: Yale University Press, 2005). For a skeptical take on the controversy from an avowed "quantified," see Stephen Earl Bennett, "'Perestroika' Lost: Why the Latest 'Reform' Movement in Political Science Should Fail," *PS: Political Science & Politics* 35 (2002): 177–179.

41. Much of the commentary on continuities in the thought of Carnap and Kuhn has turned on the affinities of Carnap's account of the construction of linguistic frameworks in the exact sciences (what Carnap called *Wissenschaftslogik*) with Kuhn's description of scientific paradigms. See George A. Reisch, "Did Kuhn Kill Logical Empiricism?" *Philosophy of Science* 58 (1991): 264–277; John Earman, "Carnap, Kuhn, and the Philosophy of Scientific Methodology," in *World Changes: Thomas Kuhn and the Nature of Science*, ed. Paul Horwich (Cambridge, MA: MIT Press, 1993), 9–36; Gürol Irzik and Teo Grünberg, "Carnap and Kuhn: Arch Enemies or Close Allies?" *British Journal for the Philosophy of Science* 46 (1995): 285–307; Peter Achinstein, "Subjective Views of Kuhn," *Perspectives on Science* 9 (Winter 2001): 423–432; Michael Friedman, "Kuhn and Logical Empiricism," in *Thomas Kuhn*, ed. Thomas Nickles (Cambridge: Cambridge University Press, 2003), 19–44. The philosopher of science Michael Friedman has thought the views of Carnap and Kuhn so complementary as to provide the platform for a renewed scientific philosophy, rooted in a formal yet historical account of relativized constitutive a priori principles in the formation of scientific knowledge. See

Michael Friedman, *Dynamics of Reason* (Stanford, CA: CSLI Publications, 2001); Alan W. Richardson, "Narrating the History of Reason Itself: Friedman, Kuhn, and a Constitutive A Priori for the Twenty-First Century," *Perspectives on Science* 10 (2002): 253–274. An alternative tactic for repudiating the notion of an absolute break between Carnap and Kuhn has been to demonstrate that *The Structure of Scientific Revolutions* cannot be read as a substantive critique of Carnapian logical empiricism. See Alan W. Richardson, "'That Sort of Everyday Image of Logical Positivism': Thomas Kuhn and the Decline of Logical Empiricist Philosophy of Science," in *The Cambridge Companion to Logical Empiricism,* ed. Alan Richardson and Thomas Uebel (Cambridge: Cambridge University Press, 2007), 346–369. Kuhn himself was party to some of these discussions before his death in 1996. Initially, he acknowledged similarities but emphasized deep methodological differences in Kuhn, "Afterwords," in Horwich, *World Changes,* 313–314. In an interview conducted shortly before he died, Kuhn went further by implying that he and Carnap had arrived at similar notions from different scholarly traditions: Aristedes Baltas, Kostas Gavroglu, and Vassiliki Kindi, "A Discussion with Thomas Kuhn," in Thomas Kuhn, *The Road Since Structure,* 305–306.

42. Geertz never made a secret of his debt to Parsons and the version of Weberian social theory Parsons promulgated. See, e.g., Richard Handler, "An Interview with Clifford Geertz," *Current Anthropology* 32 (1991): 604, 605; Clifford Geertz, "Commentary," in *Clifford Geertz by His Colleagues,* ed. Richard A. Schweder and Byron Good (Chicago: University of Chicago Press, 2005), 111, 114. This connection is given a fuller exposition in James L. Peacock, "Geertz's Concept of Culture in Historical Context: How He Saved the Day and Maybe the Century," in Schweder and Good, *Clifford Geertz by His Colleagues,* 52–62; Adam Kuper, "Culture, Identity and the Project of a Cosmopolitan Anthropology," *Man* 29 (1994): 540–541; Kuper, *Culture: The Anthropologists' Account* (Cambridge, MA: Harvard University Press, 1999), 70–72, 80.

43. See Charles Taylor, *Philosophy and the Human Sciences: Philosophical Papers,* vol. 2 (Cambridge: Cambridge University Press, 1985); Taylor, *Human Agency and Language;* Taylor, *Philosophical Arguments* (Cambridge, MA: Harvard University Press, 1995).

44. Taylor, "Preface," in *Philosophical Arguments,* vii.

45. Taylor, "Overcoming Epistemology," in *Philosophical Arguments,* 5–6.

46. Taylor, "Preface," in *Philosophical Arguments,* vii.

47. Taylor, "Overcoming Epistemology," in *Philosophical Arguments,* 3–4.

48. Ibid., 5.

49. Charles Taylor, *Sources of the Self* (Cambridge, MA: Harvard University Press, 1989); Jürgen Habermas, *Knowledge and Human Interests,* 2nd ed., trans. Jeremy J. Shapiro (Cambridge: Polity, 1987); Habermas, *The Philosophical Discourse of Modernity: Twelve Lectures,* trans. Frederick Lawrence (Cambridge: Polity, 1987); Michel Foucault, *The Order of Things: An Archaeology of the Human Sciences*

(London: Routledge, 2002); Alasdair MacIntyre, *After Virtue: A Study in Moral Theory,* 2nd ed. (London: Duckworth, 1985); Rorty, *Philosophy and the Mirror of Nature.* Dates given in the main text refer to the publication of first editions in the original language.

50. Richard Rorty, "The Historiography of Philosophy: Four Genres," in Rorty, *Truth and Progress: Philosophical Papers,* vol. 3 (Cambridge: Cambridge University Press, 1998), 256.

51. For a critique of "philosophical historiography," see Ian Hunter, "The History of Philosophy and the Persona of the Philosopher," *Modern Intellectual History* 4 (2007): 571–600.

52. Robert C. Bannister, *Sociology and Scientism: The American Quest for Objectivity, 1880–1940* (Chapel Hill: University of North Carolina Press, 1987); Dorothy Ross, *The Origins of American Social Science* (New York: Cambridge University Press, 1991); Thomas L. Haskell, *The Emergence of Professional Social Science: The American Social Science Association and the Nineteenth-Century Crisis of Authority* (Urbana: University of Illinois Press, 1977); Mary O. Furner, *Advocacy and Objectivity: A Crisis in the Professionalization of American Social Science, 1865–1905* (Lexington: University Press of Kentucky, 1975); Edward A. Purcell, Jr., *The Crisis of Democratic Theory: Scientific Naturalism and the Problem of Value* (Lexington: University Press of Kentucky, 1973); Mark C. Smith, *Social Science in the Crucible: The American Debate over Objectivity and Purpose, 1918–1941* (Durham, NC: Duke University Press, 1994); Bruce Kuklick, *The Rise of American Philosophy: Cambridge, Massachusetts, 1860–1930* (New Haven, CT: Yale University Press, 1977); David M. Ricci, *The Tragedy of Political Science: Politics, Scholarship, and Democracy* (New Haven, CT: Yale University Press, 1984); JoAnne Brown, *The Definition of a Profession: The Authority of Metaphor in the History of Intelligence Testing, 1890–1930* (Princeton, NJ: Princeton University Press, 1992).

53. See, e.g., George Steinmetz, "Introduction: Positivism and Its Others in the Social Sciences," in *The Politics of Method in the Human Sciences: Positivism and Its Epistemological Others,* ed. George Steinmetz (Durham, NC: Duke University Press, 2005), 31–33; Steinmetz, "Scientific Authority and the Transition to Post-Fordism: The Plausibility of Positivism in U.S. Sociology since 1945," in *The Politics of Method in the Human Sciences,* 275–323; Philip Mirowski, *Machine Dreams: How Economics Became a Cyborg Science* (Cambridge: Cambridge University Press, 2002); Mark Bevir, "Political Studies as Narrative and Science, 1880–2000," *Political Studies* 54 (2006): 583–606; John McCumber, *Time in the Ditch: American Philosophy and the Mccarthy Era* (Evanston, IL: Northwestern University Press, 2000).

54. Dorothy Ross, "A Historian's View of American Social Science," in *Scientific Authority and Twentieth-Century America,* ed. Ronald G. Walters (Baltimore, MD: Johns Hopkins University Press, 1997), 32–49; James T. Kloppenberg, "Institutionalism, Rational Choice, and Historical Analysis," *Polity* 28 (1995): 125–128.

55. Joseph Rouse, *Engaging Science: How to Understand Its Practices Philosophically* (Ithaca, NY: Cornell University Press, 1996), 65–67.56. See especially Alberto Coffa, *The Semantic Tradition from Kant to Carnap: To the Vienna Station* (Cambridge: Cambridge University Press, 1991); Friedman, *Dynamics of Reason*; Alan W. Richardson, *Carnap's Construction of the World: The* Aufbau *and the Emergence of Logical Empiricism* (Cambridge: Cambridge University Press, 1998).

57. On the Kantian sources of scientific philosophy, see Coffa, *The Semantic Tradition from Kant to Carnap;* Michael Friedman, *Reconsidering Logical Positivism* (Cambridge: Cambridge University Press, 1999); A. W. Carus, *Carnap and Twentieth-Century Thought: Explication as Enlightenment* (Cambridge: Cambridge University Press, 2007), 65–90.

58. Immanuel Kant, *Critique of Pure Reason,* ed. and trans. Paul Guyer and Allen W. Wood (Cambridge: Cambridge University Press, 1998), A717/B745. The letters "A" and "B" refer to the first (1781) and second (1787) editions of the *Critique.* The numbers that follow the edition marker refer to the section number.

59. Carl B. Boyer, *A History of Mathematics,* 2nd ed. (New York: Wiley, 1989), 519–522, 533–552.

60. Friedman, *Dynamics of Reason,* 27–28.

61. David Hilbert, *The Foundations of Geometry,* trans. E. J. Townsend (La Salle, IL: Open Court, 1950).

62. Michael Potter, *Reason's Nearest Kin: Philosophies of Arithmetic from Kant to Carnap* (Oxford: Oxford University Press, 2000), 56.

63. See Jean van Heijenoort, ed., *From Frege to Gödel: A Source Book in Mathematical Logic, 1879–1931* (Cambridge, MA: Harvard University Press, 1967); Peter Hylton, *Russell, Idealism and the Emergence of Analytic Philosophy* (Oxford: Oxford University Press, 1992), 167–236.

64. Quoted in Boyer, *History of Mathematics,* 562.

65. Gottlob Frege, "*Begriffsschrift,* A Formula Language, Modelled Upon That of Arithmetic, for Pure Thought," in van Heijenoort, *From Frege to Gödel,* 7, 12.

66. See Klaus Christian Köhnke, *The Rise of Neo-Kantianism: German Academic Philosophy between Idealism and Positivism,* trans. R. J. Hollingdale (Cambridge: Cambridge University Press, 1991), 252–256.

67. Carus, *Carnap and Twentieth-Century Thought,* 69–77.

68. On Mach's project in science and philosophy see John T. Blackmore, *Ernst Mach: His Work, Life, and Influence* (Berkeley: University of California Press, 1972).

69. Ernst Mach, *The Analysis of Sensations and the Relation of the Physical to the Psychical* (Chicago: Open Court, 1914), 22; Habermas, *Knowledge and Human Interests,* 81–90. On the Kantian dimensions of James's thought, see Murray G. Murphy, "Kant's Children: The Cambridge Pragmatists," *Transactions of the Charles S. Peirce Society* 4 (1968): 3–33.

70. On James's debt to Mach, see Gerald Holton, "Ernst Mach and the Fortunes of Positivism in America," *Isis* 83 (1992): 33–36.

71. Henri Poincaré, *Science and Hypothesis*, trans. W. J. Greenstreet (London: Scott, 1905). On the limits of Poincaré's conventionalism, in the fullest sense of the word, see Michael Friedman, "Poincaré's Conventionalism and the Logical Positivists," in Friedman, *Reconsidering Logical Positivism*, 71–86.

72. See Barbara S. Heyl, "The Harvard 'Pareto Circle,'" *Journal of the History of the Behavioral Sciences* 4 (1968): 316–334; Lawrence T. Nichols, "The Rise of Homans at Harvard: Pareto and the *English Villagers*," in *George C. Homans: History, Theory, and Method*, ed. A. Javier Treviño (Boulder, CO: Paradigm, 2006), 43–62; Joel Isaac, "W. V. Quine and the Origins of Analytic Philosophy in the United States," *Modern Intellectual History* 2 (2005): 205–234; Gary L. Hardcastle, "Debabelizing Science: The Harvard Science of Science Discussion Group, 1940–41," in *Logical Empiricism in North America*, ed. Gary L. Hardcastle and Alan W. Richardson (Minneapolis: University of Minnesota Press), 170–196; Gerald Holton, "On the Vienna Circle in Exile: An Eyewitness Report," in *The Foundational Debate: Complexity and Constructivity in Mathematics and Physics*, ed. W. DePauli-Schimanovich, E. Kohler, and F. Stadler (Dordrecht: Kluwer, 1995), 269–292; Peter Galison, "The Americanization of Unity," *Daedalus* 127 (1998): 45–71; Joy Harvey, "History of Science, History and Science, and Natural Science: Undergraduate Teaching of the History of Science at Harvard, 1938–1970," *Isis* 90 (Supplement, 1999): S281–282, S284; Jensine Andresen, "Crisis and Kuhn," *Isis* 90 (Supplement, 1999): S56–63; Jamie Cohen-Cole, "Instituting the Science of the Mind: Intellectual Economies and Disciplinary Exchanges at Harvard's Center for Cognitive Studies," *British Journal for the History of Science* 40 (2007): 567–597.

73. Kuhn, *Structure*, 1.

74. On the notion of subculture and its application to the history of science, see Peter Galison, *Image and Logic: A Material Culture of Microphysics* (Chicago: University of Chicago Press, 1997).

75. See, e.g., Ian Hacking, *Representing and Intervening: Introductory Topics in the Philosophy of Social Science* (Cambridge: Cambridge University Press, 1983); Steven Shapin and Simon Schaffer, *Leviathan and the Air-Pump: Hobbes, Boyle, and the Experimental Life* (Princeton, NJ: Princeton University Press, 1985); David Gooding, Trevor Pinch, and Simon Schaffer, eds., *The Uses of Experiment: Studies in the Natural Sciences* (Cambridge: Cambridge University Press, 1989).

76. David Kaiser, *Drawing Theories Apart: The Dispersion of Feynman Diagrams in Postwar Physics* (Chicago: University of Chicago Press, 2005); Andrew Warwick, *Masters of Theory: Cambridge and the Rise of Mathematical Physics* (Chicago: University of Chicago Press, 2003); Lorraine Daston and Peter Galison, *Objectivity* (New York: Zone Books, 2007).

77. I have explored these issues in greater depth in "Tangled Loops: Theory, History, and the Human Sciences in Modern America," *Modern Intellectual History* 6 (2009): 397–424; and "Tool Shock: Technique and Epistemology in the Postwar Social Sciences," *History of Political Economy* 42 (Supplement 1, 2010): 133–164.

78. For an introduction to this literature, see Charles Camic and Neil Gross, "The New Sociology of Ideas," in *The Blackwell Companion to Sociology*, ed. Judith R. Blau (Oxford: Blackwell, 2001), 236–249.

79. See especially Pierre Bourdieu, *Homo Academicus*, trans. Peter Collier (Cambridge: Polity Press, 1988).

80. For a helpful summary of Bourdieu's sociology of intellectuals, see Loïc Wacquant, "Sociology as Socioanalysis: Tales of *Homo Academicus*," *Sociological Forum* 5 (1990): 677–689.

81. Neil Gross, "Becoming a Pragmatist Philosopher: Status, Self-Concept, and Intellectual Choice," *American Sociological Review* 67 (2002): 52–76; Gross, *Richard Rorty: The Making of an American Philosopher* (Chicago: University of Chicago Press, 2008).

82. This is a point forcefully made in Hubert Dreyfus and Paul Rabinow, "Can There Be a Science of Existential Structure and Social Meaning?" in *Bourdieu: A Critical Reader*, ed. Richard Shusterman (Oxford: Blackwell, 1999), 84–93.

1. The Interstitial Academy

1. Technically, 1936 marked the three hundredth anniversary of the legislative act of the Great and General Court of Massachusetts that provided for the establishment of a "schoale or colledge" in the Bay Colony. Only in March 1639 was the newly established college in Cambridge renamed in honor of John Harvard, in recognition of the bequest of his library and half of his land to the college. Harvard was first officially referred to as "The University" in the 1780 Constitution of Massachusetts, at a time when the college was, in fact, the only department of the institution. (A school of medicine would be the first accretion to the University in 1782.) In any case, 1936 was treated as the de facto tercentenary of the University insofar as all departments were involved in the celebrations. For more information, see Samuel Eliot Morison, *Three Centuries of Harvard, 1636–1936* (Cambridge, MA: Harvard University Press, 1937), 5–9; Morison, "Introduction," in *The Development of Harvard University since the Inauguration of President Eliot, 1869–1929*, ed. Morison (Cambridge, MA: Harvard University Press, 1930), xxv–xxvii.

2. See David Hollinger, "The Two NYUs and 'The Obligation of the Universities to the Social Order' in the Great Depression," in Hollinger, *Science, Jews, and Secular Culture: Studies in Mid-Twentieth-Century American Intellectual History* (Princeton, NJ: Princeton University Press, 1996), 60–79.

3. Harvard Tercentenary Publications, *The Tercentenary of Harvard College: A Chronicle of the Tercentenary Year, 1935–1936* (Cambridge, MA: Harvard University Press, 1937), 169–186.

4. See, e.g., Laurence R. Veysey, *The Emergence of the American University* (Chicago: University of Chicago Press, 1965); Burton R. Clark, "The Organizational Conception," in *Perspectives on Higher Education: Eight Disciplinary and*

Comparative Views, ed. Clark (Berkeley: University of California Press, 1984), 106–131; Jonathan R. Cole, *The Great American University: Its Rise to Preeminence, Its Indispensable National Role, and Why It Must Be Protected* (New York: Public Affairs, 2009).

5. Ross B. Emmett, "Specializing in Interdisciplinarity: The Committee on Social Thought as the University of Chicago's Antidote to Compartmentalization in the Social Sciences," *History of Political Economy* 42 (Supplement 1, 2010): 267–287; Frank Tannenbaum, "The University Seminar Movement at Columbia University," in Tannenbaum, *The Balance of Power in Society and Other Essays* (London: Macmillan, 1969), 265–286; Ronald Gross, "Columbia's University Seminars— Creating a 'Community of Scholars,'" *Change* 14 (1982): 43–45.

6. Frederick Rudolph, *The American College and University: A History,* new ed. (Athens: University of Georgia Press, 1990), 462–464; Roger L. Geiger, *To Advance Knowledge: The Growth of American Research Universities, 1900–1940* (New York: Oxford University Press, 1986), 1–57; Veysey, *Emergence,* 268–341.

7. See Charles Franklin Thwing, *College Administration* (New York: The Century Co., 1900); Charles William Eliot, *University Administration* (Boston: Houghton Mifflin, 1909).

8. Robert H. Wiebe, *Businessmen and Reform: A Study of the Progressive Movement* (Cambridge, MA: Harvard University Press, 1962); Hugh Hawkins, *Between Harvard and America: The Educational Leadership of Charles W. Eliot* (New York: Oxford University Press, 1972), 215.

9. William K. Selden, "The Association of American Universities: An Enigma in Higher Education," *Graduate Journal* 8 (1968): 199–209.

10. David O. Levine, *The American College and the Culture of Aspiration, 1915– 1940* (Ithaca, NY: Cornell University Press, 1986); Rudolph, *American College,* 449–461.

11. Geiger, *To Advance Knowledge,* 140–245; Geiger, "Organized Research Units—Their Role in the Development of University Research," *Journal of Higher Education* 61 (1990): 1–19; Robert E. Kohler, *Partners in Science: Foundations and Natural Scientists, 1900–1945* (Chicago: University of Chicago Press, 1991).

12. A point emphasized in Veysey, *Emergence,* Part II. The eclectic institutional profile of the American university is also highlighted in Daniel Boorstin, "Universities in the Republic of Letters," *Perspectives in American History* 1 (1967): 369–379. For the argument that the discontinuity between the epoch of the college and the rise of the research university has been overstated, see Geiger, "Introduction to the Transaction Edition," in *To Advance Knowledge,* x–xi.

13. On the shifting ground of university organization and support for advanced research after World War II, see Rebecca S. Lowen, *Creating the Cold War University: The Transformation of Stanford* (Berkeley: University of California Press, 1997); Philip Mirowski, *Machine Dreams: Economics Becomes a Cyborg Science* (Cambridge: Cambridge University Press, 2002); Roger L. Geiger, *Research and*

Relevant Knowledge: American Research Universities since World War II (New York: Oxford University Press, 1993); Hunter Crowther-Heyck, "Patrons of the Revolution: Ideals and Institutions in Postwar Behavioral Science," *Isis* 97 (2006): 420–446.

14. For hagiographies of Eliot, see Eugen Kuehnemann, *Charles W. Eliot: President of Harvard University (May 19, 1869—May 19, 1909)* (Boston: Archibald Constable, 1909); Henry Hallam Saunderson, *Charles W. Eliot: Puritan Liberal* (New York: Harper & Brothers, 1928).

15. Morison, *Three Centuries,* 324.

16. Richard Norton Smith, *The Harvard Century: The Making of a University to a Nation* (Cambridge, MA: Harvard University Press, 1986), 27.

17. Henry Adams, *The Education of Henry Adams: An Autobiography* (1918; New York: Modern Library, 1996), 54–55.

18. Morison, *Three Centuries,* 324.

19. Charles William Eliot, *A Turning Point in Higher Education: The Inaugural Address of Charles William Eliot as President of Harvard College, October 19, 1869* (1869; Cambridge, MA: Harvard University Press, 1969).

20. Richard Hofstadter and Walter Metzger, *The Development of Academic Freedom in the United States* (New York: Columbia University Press, 1955); Hofstadter, *The Development and Scope of Higher Education in the United States* (New York: Columbia University Press, 1952); Hofstadter, "The Revolution in Higher Education," in *Paths of American Thought,* ed. Arthur M. Schlesinger and Morton White (London: Chatto & Windus, 1964), 269–290; Richard Storr, *The Beginning of Graduate Education in America* (Chicago: University of Chicago Press, 1953), 1–6; Hugh Hawkins, *Pioneer: A History of the Johns Hopkins University, 1874–1889* (Ithaca, NY: Cornell University Press, 1960); Rudolph, *American College.*

21. Indeed, Tewkesbury is cited as an authority in the recent Pulitzer Prize–winning book by Daniel Walker Howe, *What Hath God Wrought: The Transformation of America, 1815–1848* (New York: Oxford University Press, 2007), 459n40; 462n42; 870. For a criticism of Tewkesbury's analysis, see Natalie A. Naylor, "The Ante-Bellum College Movement: A Reappraisal of Tewkesbury's *Founding of American Colleges and Universities,*" *History of Education Quarterly* 13 (1973): 261–274.

22. Donald Tewksbury, *The Founding of American Colleges and Universities before the Civil War, with Particular Reference to the Religious Influences Bearing Upon the College Movement* (1932; Archon Books, 1965).

23. Hofstadter and Metzger, *Development of Academic Freedom,* 209–211.

24. George P. Schmidt, *The Liberal Arts College: A Chapter in American Cultural History* (1957; Westport, CT: Greenwood Press, 1975); Melvin I. Urofsky, "Reforms and Response: The Yale Report of 1828," *History of Education Quarterly* 5 (1965): 53–67.

25. Hofstadter and Metzger, *Development of Academic Freedom,* 209–274. See also Richard Storr, *Harper's University, the Beginnings: A History of the University of Chicago* (Chicago: University of Chicago Press, 1966); Hugh Hawkins, "Three

University Presidents Testify," *American Quarterly* 11 (1959): 99–119; Hawkins, "The University-Builders Observe the Colleges," *History of Education Quarterly* 11 (1971): 353–362.

26. George E. Peterson, *The New England College in the Age of the University* (Amherst, MA: Amherst College Press, 1964), 3.

27. Colin B. Burke, *American College Populations: A Test of the Traditional View* (New York: New York University Press, 1982); David F. Allmendinger, Jr., *Paupers and Scholars: The Transformation of Student Life in Nineteenth-Century New England* (New York: St. Martin's Press, 1975); W. Bruce Leslie, *Gentlemen and Scholars: College and Community in the "Age of the University," 1865–1917* (University Park: Pennsylvania State University Press, 1992); James Axtell, "The Death of the Liberal Arts College," *History of Education Quarterly* 11 (1971): 339–352; J. M. Opal, "The Making of the Victorian Campus: Teacher and Student at Amherst College, 1850–1880," *History of Education Quarterly* 42 (2002): 342–367.

28. James Turner and Paul Bernard, "The German Model and the Graduate School: The University of Michigan and the Origin Myth of the American University," in *The American College in the Nineteenth Century*, ed. Roger Geiger (Nashville, TN: Vanderbilt University Press, 2000), 221–241; Roger Geiger, "The Era of Multipurpose Colleges in American Higher Education, 1850–1890," in *American College in the Nineteenth Century*, 127–152.

29. For an early assessment of the local importance of colleges in nascent metropolitan cultures on the frontier, see Daniel J. Boorstin, *The Americans: The National Experience* (1965; London: Phoenix, 2000), 152–161.

30. David B. Potts, "American Colleges in the Nineteenth Century: From Localism to Denominationalism," *History of Education Quarterly* 11 (1971): 363–380.

31. Hofstadter and Metzger, *Development of Academic Freedom*, 212.

32. I use "metropolitan" here in its cultural and economic sense. That is to say, I do not claim that antebellum Boston was a metropolis defined by a large, high-density population, suburbs linked to the urban core by sophisticated transportation networks, and so on. Rather, pre–Civil War Boston was metropolitan in cultural development, economic power, and regional, national, and international links.

33. See E. Digby Baltzell, *Puritan Boston and Quaker Philadelphia: Two Protestant Ethics and the Spirit of Class Authority and Leadership* (New York: Free Press, 1979); Robert A. McCaughey, *Stand, Columbia: A History of Columbia University in the City of New York, 1754–2004* (New York: Columbia University Press, 2003), 96–97, 105.

34. Nonetheless, Dartmouth found itself at the center of a battle between political, religious, and quasi-metropolitan elites in the famous controversy surrounding James Marsh's bid to move the College away from Congregationalist orthodoxy. See Francis N. Stites, *Private Interest and Public Gain: The Dartmouth College Case, 1819* (Amherst: University of Massachusetts Press, 1972).

35. Seymour Martin Lipset and David Riesman, *Education and Politics at Harvard* (New York: McGraw-Hill, 1975), 51–52.

36. Hofstadter and Metzger, *Development of Academic Freedom,* 184–185; Lipset and Riesman, *Education and Politics,* 51; Morison, *Three Centuries,* 190–191.

37. See Daniel Walker Howe, *The Unitarian Conscience: Harvard Moral Philosophy, 1805–1861* (Middletown, CT: Wesleyan University Press, 1988), 14–15.

38. See Lipset and Riesman, *Education and Politics,* 51–53; Morison, *Three Centuries,* 187–191; Ronald Story, "Harvard Students, the Boston Elite, and the New England Preparatory System, 1800–1876," *History of Education Quarterly* 15 (Autumn 1975): 281–298.

39. G. A. Koch, *Republican Religion* (1933), quoted in Lipset and Riesman, *Education and Politics,* 53.

40. Sydney E. Ahlstrom, *A Religious History of the American People,* 2nd ed. (New Haven: Yale University Press, 2004), 389.

41. Howe, *Unitarian Conscience,* Part I; Bruce Kuklick, *The Rise of American Philosophy: Cambridge, Massachusetts, 1860–1930* (London: Yale University Press, 1977), 16–21.

42. Frederic Cople Jaher, "Nineteenth-Century Elites in Boston and New York," *Journal of Social History* 6 (Autumn 1972): 32–77; Cople Jaher, *The Urban Establishment: Upper Strata in Boston, New York, Charleston, Chicago, and Los Angeles* (Urbana: University of Illinois Press, 1982).

43. Ronald Story, *The Forging of an Aristocracy: Harvard and the Boston Upper Class, 1800-1870* (Middleton, CT: Wesleyan University Press, 1980), 32.

44. For the egalitarian view of the antebellum decades, see David M. Potter, *People of Plenty: Economic Abundance and the American Character* (Chicago: University of Chicago Press, 1954), esp. 142–165; Marvin Meyers, *The Jacksonian Persuasion: Politics and Belief* (Stanford, CA: Stanford University Press, 1957).

45. The flourishing democratic cultures of post-Revolutionary and antebellum America are surveyed in Sean Wilentz, *The Rise of American Democracy: Jefferson to Lincoln* (New York: Norton, 2005).

46. On the commercial classes in the revolutionary era, see T. H. Breen, *The Marketplace of Revolution: How Consumer Politics Shaped American Independence* (New York: Oxford University Press, 2004).

47. Jackson Turner Main, "Trends in Wealth Concentration before 1860," *Journal of Economic History* 31 (June 1971): 445–447. See also Edward Pessen, "The Egalitarian Myth and the American Social Reality: Wealth, Mobility, and Equality in the 'Era of the Common Man,'" *American Historical Review* 76 (October 1971): 989–1034. The importance of class formation in the making of nineteenth-century political culture in New York is examined in a pair of important books: Sean Wilentz, *Chants Democratic: New York City and the Rise of the American Working Class, 1788–1850,* new ed. (Oxford: Oxford University Press, 2004); Sven Beckert, *The Monied Metropolis: New York City and the Consolidation of the American Bourgeoisie, 1850–1896* (Cambridge: Cambridge University Press, 2001).

48. Story, *Forging of an Aristocracy,* 3–4.

49. See Cople Jaher, "Nineteenth-Century Elites," 33–34.

50. Shaw Livermore, *The Twilight of Federalism: The Disintegration of the Federalist Party, 1815–1830* (Princeton, NJ: Princeton University Press, 1962); Samuel Eliot Morison, *Harrison Gray Otis, 1765–1848: The Urbane Federalist* (Boston: Houghton Mifflin, 1969).

51. Morison, *Three Centuries.*

52. Story, *Forging of an Aristocracy.*

53. Howe, *Unitarian Conscience,* 301–305.

54. Yale College, *Reports on the Course of Instruction in Yale College by a Committee of the Corporation and the Academical Faculty* (New Haven, CT: Hezekiah Howe, 1828); Melvin I. Urofsky, "Reforms and Response: The Yale Report of 1828," *History of Education Quarterly* 5 (1965): 53–67; Jack C. Lane, "The Yale Report of 1828 and Liberal Education: A Neorepublican Manifesto," *History of Education Quarterly* 27 (1987): 325–338.

55. Philip Alexander Bruce, *History of the University of Virginia, 1819–1919: The Lengthened Shadow of One Man,* vol. 1 (New York: Macmillan, 1920), 324, 326–327. http://etext.virginia.edu/toc/modeng/public/BruHist.html, accessed September 6, 2010.

56. On Waylan's advocacy of practical instruction in the college, see Walter Cochrane Bronson, *The History of Brown University, 1764–1914* (Providence, RI: Brown University, 1914), 214–220, 259–279; "Francis Wayland's Report to the Brown Corporation, 1850," in *American Higher Education: A Documentary History,* ed. Richard Hofstadter and Wilson Smith, vol. 2 (Chicago: University of Chicago Press, 1961), 478–487.

57. David B. Tyack, *George Ticknor and the Boston Brahmins* (Cambridge, MA: Harvard University Press, 1967), 126–127; Paul Frothingham, *Edward Everett: Orator and Statesman* (Boston, 1925).

58. Tyack, *George Ticknor,* 107–23.

59. Charles William Eliot, *Harvard Memories* (Cambridge, MA: Harvard University Press, 1923), 7.

60. Robert A. McCaughey, *Josiah Quincy, 1772–1874: The Last Federalist* (Cambridge, MA: Harvard University Press, 1974), 142–162.

61. Josiah Quincy, *Report of the President of Harvard University, Submitting for Consideration a General Plan of Studies, Conformably to a Vote of the Board of Overseers of That Seminary, Passed February 4 1830* (Cambridge, MA: E. W. Metcalf, 1830), 3; McCaughey, *Josiah Quincy,* 167–169.

62. McCaughey, *Josiah Quincy,* 179–185.

63. On Peirce's career, see Sven R. Peterson, "Benjamin Peirce: Mathematician and Philosopher," *Journal of the History of Ideas* 16 (1955): 89–112.

64. Robert A. McCaughey, "The Transformation of American Academic Life: Harvard University, 1821–1892," *Perspectives in American History* 8 (1974): 260–262; McCaughey, *Josiah Quincy,* 170–176.

65. A. Hunter Dupree, *Asa Gray, 1810–1888* (Cambridge, MA: Harvard University Press, 1959), 118–131; Edward Lurie, *Louis Agassiz: A Life in Science*

(Chicago: University of Chicago Press, 1960); McCaughey, "Transformation of American Academic Life," 262.

66. See George H. Daniels, *American Science in the Age of Jackson* (Tuscaloosa: University of Alabama Press, 1994); Sally Gregory Kohlstedt, *The Formation of the American Scientific Community: The American Association for the Advancement of Science, 1848–1860* (Urbana: University of Illinois Press, 1976); Alexandra Oleson and Sanborn Brown, eds., *The Pursuit of Knowledge in the Early American Republic: American Scientific and Learned Societies* (Baltimore, MD: Johns Hopkins University Press, 1976); Ronald L. Numbers, "Together but Not Equal: Amateurs and Professionals in Early American Scientific Societies," *Reviews in American History* 4 (1976): 497–503.

67. Mary Ann James, "Engineering an Environment for Change: Bigelow, Peirce, and Early Nineteenth-Century Practical Education at Harvard," in *Science at Harvard University: Historical Perspectives,* ed. Clark A. Elliot and Margaret W. Rossiter (Bethlehem, GA: Lehigh University Press; London: Associated University Presses, 1992), 55–75.

68. See Thomas Bender, ed., *The University and the City: From Medieval Origins to the Present* (New York: Oxford University Press, 1988).

69. Frothingham, *Everett,* 266–301; Constance Blackwell, "Between Harvard and the Past," *History of Universities* 9 (1990): 240–241.

70. See also Eliot, *Harvard Memories,* 19.

71. Blackwell, "Between Harvard and the Past," 241.

72. Eliot, *Harvard Memories,* 20–24.

73. Henry James, *Charles W. Eliot: President of Harvard University, 1869–1909,* 2 vols. (London: Constable, 1930), 1:159–169.

74. See, e.g., Charles William Eliot, "The New Education; Its Organization, Part II" (1869), reprinted in *Charles William Eliot and Popular Education,* ed. Edward A. Krug (New York: Bureau of Publications, Teachers College, Columbia University, 1961), 29–46.

75. Hawkins, *Between Harvard and America,* 13–14; James, *Charles W. Eliot,* 1:55–66.

76. Blackwell, "Between Harvard and the Past," 242.

77. See McCaughey, "Transformation of American Academic Life," 275–280.

78. Arthur E. Sutherland, *The Law at Harvard: A History of Ideas and Men, 1817–1967* (Cambridge, MA: Belknap Press of Harvard University Press, 1967), 159, 166.

79. George Santayana, *The Middle Span* (London: Constable, 1947), 161–162.

80. Eliot, *Turning Point,* 11.

81. Ibid., 9.

82. Ibid., 11.

83. James, *Charles W. Eliot,* 1:259–261; Hawkins, *Between Harvard and America,* 80–119.

84. Morison, *Three Centuries,* 334–335.

85. For general accounts of the professionalization of American life in this period, see Burton J. Bledstein, *The Culture of Professionalism: The Middle Class and the Development of Higher Education in America* (New York: Norton, 1976); Thomas L. Haskell, *The Authority of Experts: Studies in History and Theory* (Bloomington: Indiana University Press, 1984).

86. McCaughey, "Transformation of American Academic Life": 282–294.

87. See Daniel T. Rodgers, *Atlantic Crossings: Social Politics in a Progressive Age* (Cambridge, MA: Harvard University Press, 1998), 84–89.

88. Geiger, *To Advance Knowledge,* 20–39; Thomas L. Haskell, *The Emergence of Professional Social Science: The American Social Science Association and the Nineteenth-Century Crisis of Authority* (Urbana: University of Illinois Press, 1977).

89. Geiger, "Organized Research Units."

90. James, *Charles W. Eliot,* 2:3–19; McCaughey, "Transformation of American Academic Life": 287–289.

91. Veysey, *Emergence,* 339–340; Geiger, *To Advance Knowledge,* 17–20.

92. For accounts of this dissipation at the end of Eliot's administration, see John Reed, "Almost Thirty," *The New Republic* 86 (April 29, 1936): 332; Virginia Spencer Carr, *Dos Passos: A Life* (Garden City, NY: Doubleday, 1984), 51–52.

93. Santayana, *Middle Span,* 168.

94. Ibid., 440–443; Smith, *Harvard Century,* ch. 2.

95. Marcia Graham Synnott, *The Half-Opened Door: Discrimination and Admissions at Harvard, Yale, and Princeton, 1900–1970* (Westport, CT: Greenwood Press, 1979); Nell Painter, "Jim Crow at Harvard: 1923," *New England Quarterly* 44 (1971): 627–634; Moshik Temkin, *The Sacco-Vanzetti Affair: America on Trial* (New Haven, CT: Yale University Press, 2009), ch. 4.

96. Harry Aaron Yeomans, *Abbott Lawrence Lowell, 1856–1943* (Cambridge, MA: Harvard University Press, 1948).

97. A point eloquently made in the case of philosophy by Kuklick, *Rise of American Philosophy.*

98. See *Development of Harvard University;* Kuklick, *Rise of American Philosophy.*

99. On James's resolutely pluralistic vision of knowledge making, see Francesca Bordogna, *William James at the Boundaries: Philosophy, Science, and the Geographies of Knowledge* (Chicago: University of Chicago Press, 2008).

100. Quoted in Hawkins, *Between Harvard and America,* 270.

101. Ibid., 263.

102. A. L. Lowell, *Colonial Civil Service: The Selection and Training of Colonial Officials in England, Holland, and France* (New York: Macmillan, 1900).

103. See A. L. Lowell, "The Art of Examination" (1920), reprinted in Lowell, *At War with Academic Traditions in America* (Cambridge, MA: Harvard University Press, 1934), 137–142; Lowell et al., "General Examination and Tutors in Harvard College" (1927), reprinted in Lowell, *At War with Academic Traditions,* 157–180. On the effects of Lowell's examination system on teaching in the College, see Arthur

Burkhard, "The Harvard Tutorial System in German," *Modern Language Journal* 14 (1930): 269–284.

104. On Wilson's attempted reforms at Princeton, see James Axtell, *The Making of Princeton University: From Woodrow Wilson to the Present* (Princeton, NJ: Princeton University Press, 2006), 14–18.

105. Lowell's programmatic statements on these topics can be found in Lowell, *At War with Academic Traditions.*

106. Synnott, *Half-Opened Door,* 106–110.

107. On Lowell's political thought, see Robert Adcock, "Liberalism, Progress, and Comparative Inquiry: Trans-Atlantic Exchanges and the Making of the American Science of Politics" (PhD Dissertation, University of California at Berkeley, 2007), 148–164.

108. Lawrence T. Nichols, "The Establishment of Sociology at Harvard: A Case of Organizational Ambivalence and Scientific Vulnerability," in Elliot and Rossiter, *Science at Harvard University,* 203–204, 208–209.

109. Rodney G. Triplet, "Harvard Psychology, the Psychological Clinic, and Henry Murray: A Case in the Establishment of Disciplinary Boundaries," in Elliot and Rossiter, *Science at Harvard University,* 223–250; Sheldon M. Stern, "William James and the New Psychology," in *Social Sciences at Harvard, 1860–1920: From Inclucation to the Open Mind,* ed. Robert L. Church (Cambridge, MA: Harvard University Press, 1965), 175–222; Curtis M. Hinsley, "The Museum Origins of Harvard Anthropology, 1866–1915," in Elliot and Rossiter, *Science at Harvard University,* 121–145.

110. Morton Keller and Phyllis Keller, *Making Harvard Modern: The Rise of America's University* (New York: Oxford, 2001), 87.

111. See Clark A. Elliot, "The Tercentenary of Harvard University in 1936: The Scientific Dimension," *Osiris,* 2nd series, 14 (1999): 153–175.

112. Off to one side slightly was the Engineering School, which accepted, as it still does, undergraduates as well as graduates.

113. Morison, "Introduction," *Development of Harvard University,* xxv–xxxviii.

114. James Bryant Conant, "Harvard Present and Future" (1935), in Tercentenary Publications, *The Tercentenary of Harvard College,* 67–68.

115. Lipset and Riesman, *Education and Politics at Harvard,* 154–155.

116. James Bryant Conant, *My Several Lives: Memoirs of a Social Inventor* (New York: Harper & Row, 1970), 83. On Conant's election to the presidency of the University, see James G. Hershberg, *James B. Conant: From Harvard to Hiroshima and the Making of the Nuclear Age* (Stanford, CA: Stanford, 1993), 65–75.

117. Keller and Keller, *Making Harvard Modern,* 64–71.

118. On the formation of the American social sciences, see Dorothy Ross, *The Origins of American Social Science* (Cambridge: Cambridge University Press, 1991).

119. See Thomas Bender, "E. R. A. Seligman and the Vocation of Social Science," in Bender, *Intellect and Public Life: Essays on the Social History of Academic*

Intellectuals in the United States (Baltimore, MD: Johns Hopkins University Press, 1993), 50–52.

120. Edward S. Mason, "The Harvard Department of Economics from the Beginning to World War II," *Quarterly Journal of Economics* 97 (1982): 419–430.

121. See Harvard University, *Scientific Papers Contributed to the Harvard Tercentenary Conference of Arts and Sciences, August 31–September 12, 1936, Other than Those Related to the Three Collaborative Symposia Published by the University* (Cambridge, MA: Harvard University Press, 1938).

122. See Harvard Tercentenary Publications, *Tercentenary of Harvard College*, 87–95; Keller and Keller, *Making Harvard Modern*, 3–9.

123. The proceedings of the three symposia of the Tercentenary Conference of Arts and Sciences were duly published under their original titles: Harvard Tercentenary Publications, *Factors Determining Human Behavior* (Cambridge, MA: Harvard University Press, 1937); Harvard Tercentenary Publications, *Authority and the Individual* (Cambridge, MA: Harvard University Press, 1937); Harvard Tercentenary Publications, *Independence, Convergence, and Borrowing in Institutions, Thought, and Art* (Cambridge, MA: Harvard University Press, 1937).

124. E. B. Wilson to W. G. Land, November 30, 1935, carton 3, LJHP.

125. E. B. Wilson to L. J. Henderson, January 27, 1936, carton 3, LJHP.

126. L. J. Henderson to E. D. Adrian, January 3, 1935, carton 3, LJHP.

127. Conant quoted in Hershberg, *James B. Conant*, 94.

128. James B. Conant to L. J. Henderson, October 24, 1938, LJHP.

129. L. J. Henderson, "Prospects of Liberty in the Modern World," unpublished ms., March 1, 1930, box 5, LJHP.

130. L. J. Henderson, "Sociology 23 Lectures (1941–42)," in *On the Social System*, ed. Bernard Barber (Chicago: University of Chicago Press, 1970), 69–70.

131. E. B. Wilson to L. J. Henderson, March 9, 1935, carton 3, LJHP.

132. Faculty Committee, *The Behavioral Sciences at Harvard* (Cambridge, MA: Harvard University Press, 1954), 4.

2. Making a Case

1. Charles P. Curtis, Jr., to Chester I. Barnard, December 1, 1954, carton 2, CIBP; Harvard University Society of Fellows, "Conversations: 6. George Casper Homans," 1998, HUSFP; *Harvard University Catalogue, November 1932* (Cambridge, MA: Harvard University Press, 1932), 309. The full entry in the course catalogue reads: "Seminary in Sociology. Thursday, 4.15–6.15. Subject: Pareto and Methods of Scientific Investigation. Professor L. J. Henderson and Mr. Charles P. Curtis, Jr. This course requires a reading knowledge of French or of Italian."

2. See L. J. Henderson to Julian A. Ripley, Jr., October 11, 1932, carton 4, LJHP. The French edition of the *Trattato* was published as Vilfredo Pareto, *Traité de sociologie générale*, ed. Pièrre Boven, 2 vols. (Lausanne: Payot et Cie, 1917–1919).

3. To my knowledge, no comprehensive list of the seminar's membership exists. I have drawn on George C. Homans, *Coming to My Senses: The Autobiography of a Sociologist* (New Brunswick, NJ: Transaction Books, 1984), 105; L. J. Henderson, untitled memorandum, September 1932, carton 4, LJHP; Alvin W. Gouldner, *The Coming Crisis of Western Sociology* (London: Heinemann, 1971), 149; Richard Gillespie, *Manufacturing Knowledge: A History of the Hawthorne Experiments* (Cambridge: Cambridge University Press, 1991), 195–196; Society of Fellows, "Conversation 6," HUSF. I have excluded the eminent sociologist David Riesman from this list even though he took a physiology course from Henderson and was introduced by Henderson to another major figure in his circle, Elton Mayo. See Seymour Martin Lipset and David Riesman, *Education and Politics at Harvard* (New York: McGraw-Hill, 1975), 296.

4. Parsons (1949), Merton (1957), Davis (1959), Homans (1964) and Sorokin (1965) served as presidents of the ASA. Brinton (1963) was president of the AHA, Kluckhohn (1947) of the AAA, and Schumpeter (1948) of the AEA.

5. Malcolm Cowley, "A Handbook for Demagogues," *The New Republic* 80 (September 12, 1934): 134. The reference to DeVoto would seem to invoke his vigorous response to a hostile piece on Pareto published in the *New Republic* in 1933. See Bernard DeVoto, "Pareto and Fascism," *The New Republic* 76 (October 11, 1933): 244–245.

6. Vilfredo Pareto, *The Mind and Society*, vol. 1, *Non-Logical Conduct*, trans. Andrew Bongiorno and Arthur Livingston (London: Jonathan Cape, 1935), v. For a further reference to Henderson's Pareto seminar, see Harold A. Larrabee, "Pareto and the Philosophers," *Journal of Philosophy* 32 (September 1935): 507.

7. Gouldner, *Coming Crisis,* 149–151, 173.

8. See, e.g., Peter Hamilton, *Talcott Parsons* (London: Tavistock, 1983), 59–61; Geoffrey Hawthorn, *Enlightenment and Despair: A History of Social Theory,* 2nd ed. (Cambridge: Cambridge University Press, 1987), 213–216.

9. Milorad M. Novicevic, Thomas J. Hench, and Daniel A. Wren, "'Playing By Ear' . . . 'in an Incessant Din of Reasons': Chester Barnard and the History of Intuition in Management Thought," *Management Decision* 40 (2002): 992–1002, esp. 993–996; Hunter Crowther-Heyck, *Herbert A. Simon: The Bounds of Reason in Modern America* (Baltimore, MD: Johns Hopkins University Press, 2005), 68–69; Steve Fuller, *Thomas Kuhn: A Philosophical History for Our Times* (Chicago: University of Chicago Press, 2000), 163–169; Stephen Turner, "Merton's 'Norms' in Political and Intellectual Context," *Journal of Classical Sociology* 7 (2007): 161–178.

10. See Barbara S. Heyl, "The Harvard 'Pareto Circle,'" *Journal of the History of the Behavioral Sciences* 4 (1968): 316–334; Cynthia Eagle Russett, *The Concept of Equilibrium in American Social Thought* (New Haven, CT: Yale University Press, 1966); Lawrence T. Nichols, "The Rise of Homans at Harvard: Pareto and the *English Villagers,*" in *George C. Homans: History, Theory, and Method,* ed. A. Javier

Treviño (Boulder, CO: Paradigm, 2006), 43–62; Gouldner, *Coming Crisis,* 148; Fuller, *Thomas Kuhn,* 163–169.

11. Fritz Roethlisberger, *The Elusive Phenomena: An Autobiographical Account of My Work in the Field of Organizational Behavior at the Harvard Business School,* ed. George F. F. Lombard (Boston: Graduate School of Business Administration, Harvard University, 1977), 61.

12. Jürg Niehans, *A History of Economic Theory: Classic Contributions, 1720–1980* (Baltimore, MD: Johns Hopkins University Press, 1990), 259–267, 329, 360.

13. See Stuart L. Campbell, "The Four Paretos of Raymond Aron," *Journal of the History of Ideas* 47 (1986): 287–298.

14. William David Jones, "Toward a Theory of Totalitarianism: Franz Borkenau's Pareto," *Journal of the History of Ideas* 53 (1992): 455–466.

15. Max Lerner, "Pareto's Republic," *The New Republic* 83 (June 12, 1935): 137.

16. Schumpeter unconvincingly defended Pareto from the charge of fascism by insisting that there was "no point in judging his action [in accepting Mussolini's blandishments] . . . from the standpoint of Anglo-American tradition." See Schumpeter, "Vilfredo Pareto," 153. More sophisticated exculpations are offered in Raymond Aron, *Main Currents in Sociological Thought,* vol. 2, trans. Richard Howard and Helen Weaver (1967; London: Penguin, 1990), 171–175; Renato Cirillo, "Was Vilfredo Pareto Really a Precursor of Fascism?" *American Journal of Economics and Sociology* 42 (1983): 235–245. But others have convincingly shown that Pareto was at best ambiguous about, if not outright sympathetic to, the fascist usurpation and that his social theory adumbrated and endorsed many motifs of fascist thought. See H. Stuart Hughes, *Consciousness and Society: The Reorientation of European Social Thought, 1890–1930* (Brighton, UK: Harvester Press, 1979), 270–274; James W. Vander Zanden, "Pareto and Fascism Reconsidered," *American Journal of Economics and Sociology* 19 (1960): 399–411; Lerner, "Pareto's Republic."

17. Early responses in the academic journals included Andrew Bongiorno, "A Study of Pareto's Treatise on General Sociology," *American Journal of Sociology* 3 (1930): 349–370; Max Milikan, "Pareto's Sociology," *Econometrica* 4 (1936): 324–337; Harold A. Larrabee, "Pareto and the Philosophers," *Journal of Philosophy* 32 (1935): 505–515; Henry J. Bitterman, "Pareto's Sociology," *Philosophical Review* 45 (1936): 303–313; Symposium on Pareto's Significance for Social Theory, *Journal of Social Philosophy* 1 (1935): 36–89. Due largely to the efforts of DeVoto, Pareto was briefly an object of interest in literary magazines. See Bernard DeVoto, "Sentiment and Social Order: Introduction to the Teachings of Pareto," *Harpers* 167 (October 1933): 569–581; and a special issue on Pareto in *The Saturday Review of Literature* 12 (May 23, 1935).

18. The one exception to the undertheorized response to Pareto in the United States was Joseph Schumpeter, whose "elite conception of democracy" was indebted to elite theorists such as Pareto and Mosca. But *Capitalism, Socialism, and Democracy* would not appear until 1942, and his profile at Harvard was that

of a rogue economist. See John Medearis, *Joseph Schumpeter's Two Theories of Democracy* (Cambridge, MA: Harvard University Press, 2001), 107–113.

19. See John Parascandola, "Organismic and Holistic Concepts in the Thought of L. J. Henderson," *Journal of the History of Biology* 4 (1971): 63–113; Parascandola, "L. J. Henderson and the Mutual Dependence of Variables: From Physical Chemistry to Pareto," in *Science at Harvard University: Historical Perspectives*, ed. Clark A. Elliot and Margaret W. Rossiter (London: Associated University Presses, 1992), 167–190.

20. See L. J. Henderson, *The Fitness of the Environment* (New York: Macmillan, 1913); Henderson, *The Order of Nature: An Essay* (Cambridge, MA: Harvard University Press, 1917).

21. L. J. Henderson, "Memories," unpublished manuscript, September 21, 1936, box 7, LJHP; Henderson, "Memories," unpublished manuscript, 1936, box 7, LJHP.

22. See, e.g., Lawrence J. Henderson, "Review: A Philosophical Interpretation of Nature," *Quarterly Review of Biology* 1 (1926): 289–294.

23. Russett, *Concept of Equilibrium*, 118.

24. Henderson, "Memories," 1936.

25. Henderson, *Pareto's General Sociology: A Physiologist's Interpretation* (Cambridge, MA: Harvard University Press, 1935), 7.

26. Ibid., 20

27. See esp. Pareto, *Mind and Society*, 2:516–519.

28. To be sure, Henderson reproduced Pareto's classification of the residues, but he treated it not as a canon but as a "tolerably serviceable" "first attempt." See Lawrence J. Henderson, "Sociology 23 Lectures," *L. J. Henderson on the Social System*, ed. Bernard Barber (Chicago: University of Chicago Press, 1970), 120–125.

29. Ibid., 88.

30. Henderson, *Pareto's General Sociology*, 17–18.

31. Ibid., 112–113.

32. See Lawrence J. Henderson, "Henderson Paper, February 5, 1934," box 5, LJHP.

33. The works most often cited are: George C. Homans and Charles C. Curtis, *An Introduction to Pareto: His Sociology* (New York: Knopf, 1934); Crane Brinton, *French Revolutionary Legislation on Illegitimacy, 1789–1804* (Cambridge, MA: Harvard University Press, 1936); Brinton, *The Anatomy of Revolution* (London: G. Allen & Unwin, 1939); Talcott Parsons, *The Structure of Social Action: A Study in Social Theory With Special Reference to a Group of Recent European Writers (Alfred Marshall, Vilfredo Pareto, Émile Durkheim, Max Weber)* (New York: McGraw-Hill, 1937); George C. Homans, *English Villagers of the Thirteenth Century* (Cambridge, MA: Harvard University Press, 1942). For historical studies that focus on these works, see Heyl, "Harvard 'Pareto Circle'"; Fuller, *Thomas Kuhn*.

34. *Official Register of Harvard University* 32 (September 18, 1935), 145–146; *Official Register of Harvard University* 34 (October 1, 1937), 157.

35. "Sociology 23: Schedule of Lectures, 1939," carton 4, LJHP; Bernard Barber, "L. J. Henderson: An Introduction," in *L. J. Henderson on the Social System,* 40–41; Nichols, "The Rise of Homans," 53–54.

36. Roethlisberger, *Elusive Phenomena,* 66.

37. See "Sociology 23: Schedule of Lectures."

38. Bernard Barber, "Preface," in *L. J. Henderson on the Social System,* vii.

39. Roethlisberger, *Elusive Phenomena,* 66.

40. William Foote Whyte, *Participant Observer: An Autobiography* (Ithaca, NY: ILR Press, 1994), 55. See also George C. Homans and Orville T. Bailey, "The Society of Fellows, Harvard University, 1933–1947," in *The Society of Fellows,* ed. Crane Brinton (Cambridge, MA: Society of Fellows, 1959), 2–7.

41. See, e.g., Harvard University Society of Fellows, "Conversations: 3. Harry Tuchman Levin," HUSFP; Harvard University Society of Fellows, "Conversations: 2. Willard Van Oram [*sic*] Quine," HUSFP; Whyte, *Participant Observer,* 56. Noted Levin of the Society under Henderson's leadership: "For self-protection if for no other reason we all had to read Pareto, and argue about it with [Henderson]."

42. See Homans, *Coming to My Senses,* 121.

43. Whyte, *Participant Observer,* 65.

44. See Conrad M. Arensberg and Solon T. Kimble, *Family and Community in Ireland* (Cambridge, MA: Harvard University Press, 1940), 309–316; William Foote Whyte, *Street Corner Society,* 2nd ed. (Chicago: University of Chicago Press, 1954), 284–288.

45. James Grier Miller, *Living Systems* (New York: McGraw Hill, 1978).

46. Whyte, *Participant Observer,* 56.

47. See, e.g., Quine's admission that his use of the phrase came from Henderson: W. V. Quine, "On the Very Idea of a Third Dogma," in Quine, *Theories and Things* (Cambridge, MA: Harvard University Press, 1981), 41. On the pervasiveness of the idea of a "conceptual scheme" in history and philosophy of science, see Peter Galison, *Image and Logic: A Material Culture of Microphysics* (Chicago: University of Chicago Press, 1997), 787–790.

48. For brief summaries of the early works of these Junior Fellows, see their entries in Brinton, *Society of Fellows,* 143–146, 164–165, 259–260.

49. See Whyte, *Participant Observer,* 63.

50. Steven M. Horvath and Elizabeth C. Horvath, *The Harvard Fatigue Laboratory: Its History and Contributions* (Englewood Cliffs, NJ: Prentice-Hall, 1973).

51. Jeffrey L. Cruikshank, *A Delicate Experiment: The Harvard Business School, 1908–1945* (Boston, MA: Harvard Business School Press, 1987), 165.

52. William C. Buxton, "Snakes and Ladders: Parsons and Sorokin at Harvard," in *Sorokin & Civilization: A Centennial Assessment,* ed. Joseph B. Ford, Michael P. Richard, and Palmer C. Talbutt (New Brunswick, NJ: Transaction, 1996), 31–43.

53. See Wallace B. Donham, "Essential Groundwork for a Broad Executive Theory," *Harvard Business Review* 1 (1922): 1–10; Rakesh Khurana, *From Higher*

Aims to Hired Hands: The Social Transformation of American Business Schools and the Unfulfilled Promise of Management as a Profession (Princeton, NJ: Princeton University Press, 2007), 137–192; Cruikshank, *Delicate Experiment,* 154–155.

54. Parascandola, "L. J. Henderson," 184; Horvath and Horvath, *Fatigue Laboratory,* 30–31.

55. This orientation is clear in Lawrence J. Henderson, "Science, Logic, and Human Intercourse," *Harvard Business Review* 12 (1934): 317–327.

56. Roethlisberger, *Elusive Phenomena,* 60–71.

57. See, e.g., T. North Whitehead, *Leadership in a Free Society,* (London: Oxford University Press, 1936).

58. On the "Paretian turn" in the interpretation of the Hawthorne experiments, see Gillespie, *Manufacturing Knowledge,* 194–196.

59. Conant's father-in-law, the dean of Harvard physical chemistry Theodore Richards, was Henderson's brother-in-law.

60. Joy Harvey, "History of Science, History and Science, and Natural Sciences: Undergraduate Teaching of the History of Science at Harvard, 1938–1970," *Isis* 90 (Supplement 1990): S271–S272.

61. See, e.g., L. J. Henderson to E. D. Adrian, January 3, 1935, carton 3, LJHP; E. B. Wilson to L. J. Henderson, January 27, 1936, carton 3, LJHP; Harvard Tercentenary Publications, *Factors Determining Human Behavior* (Cambridge, MA: Harvard University Press, 1937); Clark A. Elliot, "The Tercentenary of Harvard in 1936: The Scientific Dimension," *Osiris* 14, 2nd series (1999): 153–175.

62. Henderson "A Relation of Physiology to the Social Sciences," *Erkenntnis: The Journal of Unified Science* 8 (1939/1940): 370.

63. See Gary L. Hardcastle, "Debabelizing Science: The Harvard Science of Science Discussion Group, 1940–41," in *Logical Empiricism in North America,* ed. Gary L. Hardcastle and Alan Richardson (Minneapolis: University of Minnesota Press, 2003), 170–196.

64. On Langdell's reforms, see Arthur E. Sutherland, *The Law at Harvard: A History of Ideas and Men, 1817–1967* (Cambridge, MA: Harvard University Press, 1967), 166–174; Bruce A. Kimball, "The Proliferation of Case Method Teaching in American Law Schools: Mr. Langdell's Emblematic 'Abominaton', 1890–1915," *History of Education Quarterly* 46 (2006): 193–195.

65. See Bruce A. Kimball, "'Warn Students That I Entertain Heretical Opinions, Which They Are Not to Take as Law': The Inception of Case Method Teaching in the Classrooms of the Early C. C. Langdell, 1870–1883," *Law and History Review* 17 (1999): 57–140.

66. John Forrester, "If P, Then What? Thinking in Cases," *History of the Human Sciences* 9 (1996): 14–15.

67. A point underscored with respect to case pedagogy at Harvard in Wallace B. Donham, "Business Teaching by the Case System," *American Economic Review* 12 (1922): 54–55. See also Galison, *Image and Logic,* 57, n. 66.

68. On Langdell's search for an adequate basis for legal knowledge and pedagogy, see Bruce A. Kimball, *The Inception of Modern Professional Education: C. C. Langdell, 1826–1906* (Chapel Hill: University of North Carolina Press, 2009).

69. This reading of Langdell is rebutted along with other shibboleths in Bruce A. Kimball, "The Langdell Problem: Historicizing the Century of Historiography, 1906–2000s," *Law and History Review* 22 (2004): 277–337.

70. C. C. Langdell, *A Selection of Cases on the Law of Contracts* (Boston: Little, Brown, 1871), vi–vii.

71. C. C. Langdell, "Professor Langdell"[Speech at the Quarter-Millennial Celebration of Harvard University, November 5, 1886], *Law Quarterly Review* 3 (1887): 124.

72. W. B. Cannon, *The Way of an Investigator: A Scientist's Experiences in Medical Research* (New York: W. W. Norton, 1945), 85; Cannon, "The Case Method of Teaching Systematic Medicine," *Boston Medical and Surgical Journal* 142 (1900): 31–36.

73. Cannon, *Way of an Investigator*, 85.

74. Benjamin Baker, "Teaching the Profession of Business at Harvard" (1915), quoted in Khurana, *From Higher Aims*, 114.

75. On Donham, see Cruikshank, *Delicate Experiment*, 92.

76. Khurana, *From Higher Aims*, 173.

77. Harvard University Graduate School of Business Administration, *Harvard Business Reports*, vol. 1 (New York: McGraw-Hill, 1925).

78. See Kenneth R. Andrews, ed., *The Case Method of Teaching Human Relations and Administration: An Interim Statement* (Cambridge, MA: Harvard University Press, 1953); Roethlisberger, *Elusive Phenomena*, ch. 9.

79. See James Bryant Conant, *On Understanding Science: An Historical Approach* (London: Oxford University Press, 1947); Conant, ed., *Harvard Case Histories in Experimental Science* (Cambridge, MA: Harvard University Press, 1957).

80. Henderson, "Memories," LJHP, 28–29, 34.

81. Ibid., 118–119.

82. Ibid., 4.

83. Ibid., 152.

84. Ibid., 152, 5.

85. Ibid., 154.

86. Ibid., 164–170; 172–173.

87. Ernst Mach, *The Science of Mechanics: A Critical and Historical Account of Its Development*, 6th ed., trans. Thomas J. McCormack (La Salle, IL: Open Court, 1960), 1.

88. On Mach's "historicism," see John T. Blackmore, *Ernst Mach: His Life, Work, and Influence* (Berkeley: University of California Press, 1972), 33, 133–134; Laurence D. Smith, *Behaviorism and Logical Positivism: A Reassessment of the Alliance* (Stanford, CA: Stanford University Press, 1986), 264–275.

89. Pierre Duhem, *The Aim and Structure of Physical Theory,* trans. Philip P. Wiener (1914; Princeton, NJ: Princeton University Press, 1954).

90. On Richards's course, see James Bryant Conant, *My Several Lives: Memoirs of a Social Inventor* (New York: Harper and Row, 1970), 30.

91. Arnold Thackray, "The Pre-History of an Academic Discipline: The Study of the History of Science in the United States, 1891–1941," in *Transformation and Tradition in the Sciences: Essays in Honor of I. Bernard Cohen,* ed. Everett Mendelsohn (Cambridge: Cambridge University Press, 1984), 400.

92. Ibid., 399–401.

93. I. Bernard Cohen, "A Harvard Education," *Isis* 75 (1984): 13.

94. John T. Edsall, "Lawrence J. Henderson and George Sarton," *Isis* 75 (1984): 11.

95. L. J. Henderson, "Memories," box 7, LJHP.

96. Cohen, "Harvard Education," 14.

97. Harvey, "History of Science, History and Science, and Natural Sciences," S271–S272.

98. Robert K. Merton, "George Sarton: Episodic Recollections by an Unruly Apprentice," *Isis* 76 (1985): 470–471.

99. Henderson, "Memories," LJHP, 210–211.

100. Ibid.

101. See I. Bernard Cohen, "A Harvard Education," *Isis* 75 (1984): 13; John T. Edsall, "Lawrence J. Henderson and George Sarton," *Isis* 75 (1984): 11.

102. Henderson, "Sociology 23," 71.

103. Ibid., 72.

104. Ibid., 72.

105. Ibid., 68.

106. Ibid., 73.

107. Ibid., 73–74.

108. Ibid., 74.

109. See, e.g., Brinton, *Anatomy of Revolution;* Parsons, Homans, *English Villagers;* Clyde Kluckhohn, "A Navaho Personal Document with a Brief Paretian Analysis," *Southwestern Journal of Anthropology* 1 (1945): 260–283; Abbott Lawrence Lowell, "Sociology 23: Summary of Lecture by A. L. Lowell," Carton 4, LJHP.

110. Letter dated October 6, 1937, Carton 4, LJHP.

111. Homans, *English Villagers,* 3–4.

112. Homans, *Coming to My Senses,* 168.

113. Homans, *English Villagers,* 4.

114. Sarah Igo, *The Averaged American: Surveys, Citizens, and the Making of a Mass Public* (Cambridge, MA: Harvard University Press, 2007).

115. Although not, of course, later, when he became the scourge of Parsonian visions of theory. See George C. Homans, "Fifty Years of Sociology," *Annual Review of Sociology* 12 (1986): xiii–xxx.

116. Homans, *English Villagers,* ch. 25. It is worth noting that Homans's remarks on these topics did not refer to Henderson. Homans instead cited the *Trattato* and Chester I. Barnard's *The Functions of the Executive* (Cambridge, MA: Harvard University Press, 1938) as his principal inspirations. See Homans, *English Villagers,* 459, n. 1. But the lineages were clear; indeed, Barnard's book was written under the influence of Henderson and his Paretian orientations.

117. Homans's most influential book was *Social Behavior: Its Elementary Forms* (New York: Harcourt, Brace, and World, 1961). The impact of this work is most evident in Robert Hamblin and John H. Kunkel, eds., *Behavioral Theory in Sociology: Essays in Honor of George C. Homans* (New Brunswick, NJ: Transaction, 1977).

118. George C. Homans, "A Conceptual Scheme for the Study of Social Organization," *American Sociological Review* 12 (1947): 13–26; Homans, *The Human Group* (New York: Harcourt, Brace, 1950).

119. For a more detailed discussion of these themes, see Joel Isaac, "Tool Shock: Technique and Epistemology in the Postwar Social Sciences," *History of Political Economy* 42 (Supplement 1, 2010): 151–154.

120. Whyte, *Street Corner Society.*

121. On Kluckhohn's view of the significance of the Navajo reservation at Rimrock, see Willow Roberts Powers, "The Harvard Study of Values: Mirror for Postwar Anthropology," *Journal of the History of the Behavioral Sciences* 36 (2000): 15–29.

122. Dickson and White, *Management and the Worker,* 551.

123. Ibid., 564.

3. What Do the Science-Makers Do?

1. Morton Keller and Phyllis Keller, *Making Harvard Modern: The Rise of America's University* (Oxford: Oxford University Press, 2001), 89.

2. See, e.g., John J. Cerullo, "Skinner at Harvard: Intellectual or Mandarin?" in *B. F. Skinner and Behaviorism in American Culture,* ed. Laurence D. Smith and William R. Woodward (Bethlehem, PA: Lehigh University Press, 1996), 226.

3. Forrest G. Robinson, *Love's Story Told: A Life of Henry A. Murray* (Cambridge, MA: Harvard University Press, 1992).

4. Richard Gillespie, "The Hawthorne Experiments and the Politics of Experimentation," in *The Rise of Experimentation in American Psychology,* ed. Jill G. Morawski (New Haven, CT: Yale University Press, 1988), 115. Emphasis added.

5. Lawrence J. Henderson, *Pareto's General Sociology: A Physiologist's Assumption* (Cambridge, MA: Harvard University Press, 1937), 63–64.

6. On "indigenous epistemologies" in interwar experimental psychology, see Laurence D. Smith, *Behaviorism and Logical Positivism: A Reassessment of the Alliance* (Stanford, CA: Stanford University Press, 1986).

7. I follow here Deborah J. Coon's important study "Standardizing the Subject: Experimental Psychologists, Introspection, and the Quest for a Technoscientific Ideal," *Technology and Culture* 34 (1993): 757–783.

8. Bruno Latour, *Science in Action: How to Follow Scientists and Engineers Through Society* (Cambridge, MA: Harvard University Press, 1987), 174–175.

9. John M. O'Donnell, *The Origins of Behaviorism: American Psychology, 1870–1920* (New York: New York University Press, 1985), 7.

10. Ibid., 2–4. See also Dorothy Ross, *G. Stanley Hall: The Psychologist as Prophet* (Chicago: University of Chicago Press, 1972); Matthew Hale, *Human Science and the Social Order: Hugo Münsterberg and the Origins of Applied Psychology* (Philadelphia, PA: Temple University Press, 1980); Thomas M. Camfield, "The Professionalization of American Psychology, 1870–1917," *Journal of the History of the Behavioral Sciences* 9 (1973): 66–75.

11. O'Donnell, *Origins of Behaviorism,* 9–14; E. G. Boring, *A History of Experimental Psychology,* 2nd ed. (New York: Appleton-Century-Crofts, 1950), 505–543; Mitchell G. Ash, "Psychology," in *The Cambridge History of Science,* vol. 7, *The Modern Social Sciences,* ed. Theodore M. Porter and Dorothy Ross (Cambridge: Cambridge University Press, 2003), 255–259; Franz Samelson, "Struggle for Scientific Authority: The Reception of Watson's Behaviorism, 1913–1920," *Journal of the History of the Behavioral Sciences* 17 (1981): 399–425.

12. See Rebecca Lemov, *World as Laboratory: Experiments with Mice, Mazes, and Men* (New York: Hill and Wang, 2005), ch. 2; Coon, "Standardizing," 758; Hale, *Human Science,* 60–69 and *passim.*

13. Michael M. Sokal, "The Origins of the Psychological Corporation," *Journal of the History of the Behavioral Sciences* 17 (1981): 54–67.

14. JoAnne Brown, "Mental Measurements and the Rhetorical Force of Numbers," in *The Estate of Social Knowledge,* ed. JoAnne Brown and David K. van Keuren (Baltimore, MD: Johns Hopkins University Press, 1991), 134–152; Leila Zenderland, *Henry Herbert Goddard and the Origins of American Intelligence Testing* (Cambridge: Cambridge University Press, 1998); Henry L. Minton, *Lewis M. Terman: Pioneer in Psychological Testing* (New York: New York University Press, 1988); John Carson, *The Measure of Merit: Talents, Intelligence, and Inequality in the French and American Republics* (Princeton, NJ: Princeton University Press, 2007).

15. Brown, "Mental Measurements," 138–144.

16. Roosevelt quoted in ibid., 139; Charles McCarthy, *The Wisconsin Idea* (New York: Macmillan, 1912).

17. See John Carson, "Army Alpha, Army Brass, and the Search for Army Intelligence," *Isis* 84 (1993): 278–309; Richard T. von Mayrhauser, "Making Intelligence Functional: Walter Dill Scott and Applied Psychological Testing in World War I," *Journal of the History of Ideas* 25 (1989): 60–72.

18. Loren Baritz, *The Servants of Power: A History of the Use of Social Science in American Industry* (Westport, CT: Greenwood Press, 1974); Richard Gillespie,

Manufacturing Knowledge: A History of the Hawthorne Experiments (Cambridge: Cambridge University Press, 1991); Anson Rabinbach, *The Human Motor: Energy, Fatigue, and the Origins of Modernity* (New York: Basic Books, 1990).

19. Jill G. Morawski and Gail A. Hornstein, "Quandary of the Quacks: The Struggle for Expert Knowledge in American Psychology, 1890–1940," in *Estate of Social Knowledge,* 106–133.

20. For a general account, see Nathan G. Hale, *The Rise and Crisis of Psychoanalysis in the United States: Freud and the Americans, 1917–1985* (New York: Oxford University Press, 1995).

21. Morawski and Hornstein, "Quandary of the Quacks," 110–116.

22. James H. Capshew, *Psychologists on the March: Science, Practice, and Professional Identity in America, 1929–1969* (Cambridge: Cambridge University Press, 1999), 22, 28–29.

23. Mary S. Morgan and Malcolm Rutherford, eds., *From Interwar Pluralism to Postwar Neoclassicism: Annual Supplement to Volume 30, History of Political Economy* (Durham, NC: Duke University Press, 1998); Stephen P. Turner and Jonathan H. Turner, *The Impossible Science: An Institutional Analysis of American Sociology* (Newbury Park, CA: Sage, 1990), 65–75.

24. Capshew, *Psychologists on the March,* 32–33.

25. Robert Woodworth, *Contemporary Schools of Psychology* (New York: Ronald, 1931); Capshew, *Psychologists on the March,* 33.

26. Michael M. Sokal, "The Gestalt Psychologists in Behaviorist America," *American Historical Review* 89 (1984): 1240–1263; Jean Matter Mandler and George Mandler, "The Diaspora of Experimental Psychology: The Gestaltists and Others," in *The Intellectual Migration: Europe and America, 1930–1960,* ed. Donald Fleming and Bernard Bailyn (Cambridge, MA: Belknap Press of Harvard University Press, 1969), 371–419.

27. Edna Heidbredder, *Seven Psychologies* (New York: Century, 1933); Capshew, *Psychologists on the March,* 34–35.

28. J. G. Morawski, "Organizing Knowledge and Behavior at Yale's Institute of Human Relations," *Isis* 77 (1986): 219–242; Lemov, *World as Laboratory,* ch. 4.

29. Heidbredder, *Seven Psychologies,* 200–233, 287–327.

30. G. Stanley Hall, *Life and Confessions of a Psychologist* (1923), quoted in Ralph Barton Perry, "Psychology, 1876–1929," in *The Development of Harvard University since the Inauguration of President Eliot, 1869–1929,* ed. Samuel Eliot Morison (Cambridge, MA: Harvard University Press, 1930), 218.

31. William James, *Principles of Psychology,* 2 vols. (London: Macmillan, 1890).

32. Bruce Kuklick, *The Rise of American Philosophy: Cambridge, Massachusetts, 1860–1930* (New Haven, CT: Yale University Press, 1977), 186–189; Perry, "Psychology," 218–219.

33. Bruce Kuklick, *Josiah Royce: An Intellectual Biography* (Indianapolis, IN: Hackett, 1985), 83–98.

34. Perry, "Psychology," 219; Kuklick, *Rise of American Philosophy,* ch. 22.

35. Rodney G. Triplet, "Harvard Psychology, the Psychological Clinic, and Henry A. Murray: A Case Study in the Establishment of Disciplinary Boundaries," in *Science at Harvard University,* ed. Clark A. Elliot and Margaret W. Rossiter (Bethlehem, PA: Lehigh University Press/Associated University Presses, 1992), 224; Perry, "Psychology," 219.

36. Hale, *Human Science and Social Order.*

37. Kuklick, *Rise of American Philosophy,* 432–434.

38. Edwin G. Boring, *Psychologist at Large: An Autobiography and Selected Essays* (New York: Basic Books, 1961), 55.

39. Triplet, "Harvard Psychology," 224–225.

40. On McDougall, see Egil Asprem, "A Nice Arrangement of Heterodoxies: William McDougall and the Professionalization of Psychical Research," *Journal of the History of the Behavioral Sciences* 46 (2010): 129–140; Allport, "Gordon Allport," 12.

41. Perry, "Psychology," 219; Triplet, "Harvard Psychology," 225.

42. Triplet, "Harvard Psychology," 225.

43. Gordon Allport, "Gordon Allport," in *A History of Psychology in Autobiography,* vol. 5, ed. Edwin G. Boring and Gardner Lindzey (New York: Appleton-Century-Crofts, 1967), 8–15.

44. Triplet, "Harvard Psychology," 226–228; Boring, *Psychologist at Large,* 43–45.

45. Boring, *Psychologist at Large,* 55–56.

46. Keller and Keller, *Making Harvard Modern,* 90.

47. Boring, *Psychologist at Large,* 56.

48. Triplet, "Harvard Psychology," 238–241.

49. P. W. Bridgman, *The Logic of Modern Physics* (New York: Macmillan, 1927), 5.

50. Ibid., 7–9. My exposition of Einstein's redefinition of simultaneity relies both on Bridgman's account and on the helpful explication in Peter Galison, *Einstein's Clocks, Poincaré's Maps: Empires of Time* (New York: Norton, 2003), 14–24.

51. Quoted in Galison, *Einstein's Clocks,* 18–19.

52. Bridgman, *Logic,* 8.

53. Quoted in Maila L. Walter, *Science and Cultural Crisis: An Intellectual Biography of Percy Williams Bridgman (1882–1961)* (Stanford, CA: Stanford University Press, 1990), 14.

54. Gerald Holton, "B. F. Skinner, P. W. Bridgman, and the Lost Years," in Holton, *Victory and Vexation in Science: Einstein, Bohr, Heisenberg, and Others* (Cambridge, MA: Harvard University Press, 2005), 73.

55. Bridgman *Logic,* 1.

56. Ibid., 2.

57. Ibid., v.

58. P. W. Bridgman, "Operational Analysis," *Philosophy of Science* 5 (1938), reprinted in Bridgman, *Reflections of a Physicist* (New York: Philosophical Library, 1955), 11.

59. Bridgman, *Logic,* 2.

60. Bridgman, "Operational Analysis," 2.

61. Ibid.

62. See the next section for references to the critical reaction to the *Logic.*

63. Bridgman, *Logic,* 9–10.

64. Quoted in Walter, *Science and Cultural Crisis,* 34.

65. Bridgman, "Operational Analysis," 26.

66. D. M. Newitt, "Percy Williams Bridgman. 1882–1961," *Biographical Memoirs of Fellows of the Royal Society* 8 (1962): 27–29.

67. Walter, *Science and Cultural Crisis,* 34.

68. Bridgman, "Operational Analysis," 25.

69. Henderson, "An Approximate Definition of Fact," *University of California Publications in Philosophy* 14 (1932), reprinted in *L. J. Henderson on the Social System,* ed. Bernard Barber (Chicago: University of Chicago Press, 1970), 167, 174.

70. Parsons, *The Structure of Social Action,* 2nd ed. (Glencoe, IL: Free Press, 1949), 37.

71. Paul Samuelson, *Foundations of Economic Analysis* (Cambridge, MA: Harvard University Press, 1947), 4–5.

72. Paul Samuelson, "How *Foundations* Came to Be," *Journal of Economic Literature* 36 (1998): 1375–1386; Harvard University Society of Fellows, "Conversations: 8. Paul A. Samuelson," 1998, HUSFP.

73. For claims about the anticipation of operationism, see Tim B. Rogers, "Operationism in Psychology: A Discussion of Contextual Antecedents and an Historical Interpretation of Its Longevity," *Journal of the History of the Behavioral Sciences* 25 (1989): 146; Smith, *Behaviorism,* 87; Boring, *History,* 656; S. S. Stevens, "Psychology and the Science of Science," *Psychological Bulletin* 36 (1939): 226.

74. Smith, *Behaviorism,* 86–87; E. G.Boring, "Intelligence as the Tests Test It," *New Republic* 36 (1923): 35–37; Edward Tolman, "A Behavioristic Theory of Ideas," *Psychological Review* 33 (1926): 369.

75. See Clark Hull, "Knowledge and Purpose as Habit Mechanisms," *Psychological Review* 37 (1930): 511–525; E. G.Boring, *The Physical Dimensions of Consciousness* (New York: Century, 1933); Edward Tolman, *Purposive Behavior in Animals and Men* (New York: Appleton-Century, 1932).

76. See, e.g., Christopher D. Green, "Of Immortal Mythological Beasts: Operationism in Psychology," *Theory and Psychology* 2 (1992): 297; Kuklick, *Rise of American Philosophy,* 461.

77. Boring, *History,* 656.

78. Herbert Feigl, "The *Wiener Kreis* in America," in Fleming and Bailyn, *Intellectual Migration,* 645.

79. B. F. Skinner, "B. F. Skinner . . . An Autobiography," in *Festschrift for B. F. Skinner,* ed. P. B. Dews (New York: Appleton-Century-Crofts, 1970), 10; Smith, *Behaviorism,* 376, n. 100.

80. Smith, *Behaviorism,* 284; B. F. Skinner, Review of Laurence D. Smith, *Behaviorism and Logical Positivism: A Reassessment of the Alliance, Journal of the History of the Behavioral Sciences* 23 (1987): 206–209.

81. B. F. Skinner, "The Concept of the Reflex in the Description of Behavior," (PhD Dissertation, Harvard University, 1930), 2, 55.

82. Skinner himself gave precedence to Harry M. Johnson. See Smith, *Behaviorism,* 343, fn. 48.

83. Boring, *History,* 656.

84. Sigmund Koch, "Psychology's Bridgman vs. Bridgman's Bridgman," *Theory and Psychology* 2 (1992): 269; Green, "Of Immortal Mythological Beasts," 297; Smith, *Behaviorism,* 343, n. 48.

85. S. S. Stevens, "The Operational Basis of Psychology," *American Journal of Psychology* 47 (1935): 323–330; Stevens, "The Operational Definition of Psychological Concepts," *Psychological Review* 42 (1935): 517–527.

86. On the provenance of the essays, see Gary L. Hardcastle, "S. S. Stevens and the Origins of Operationism," *Philosophy of Science* 62 (1995): 413.

87. S. S. Stevens, "Psychology: The Propaedeutic Science," *Philosophy of Science* 3 (1936): 90–103; Stevens, "Psychology and the Science of Science."

88. John A. McGeoch, "Learning as an Operationally Defined Concept," *Psychological Bulletin* 32 (1935): 688.

89. John A. McGeoch, "A Critique of Operational Definition," *Psychological Bulletin* 34 (1937): 703–704.

90. Two years earlier—that is, before Stevens published his seminal articles—Boring was invoking Bridgman's operational viewpoint, and explaining the parallels between his enterprise in *The Physical Dimensions of Consciousness* and operationism. See Boring to Yerkes, 1934, published in Boring, *Psychologist at Large,* 108–110.

91. E. G. Boring, "Temporal Perception and Operationism," *American Journal of Psychology* 48 (1936): 519–522.

92. E. G. Boring, "An Operational Restatement of G. E. Müller's Psychophysical Axioms," *Psychological Review* 48 (1941): 457–464.

93. E. G. Boring, "Human Nature vs. Sensation: William James and the Psychology of the Present" (1942), reprinted in *Psychologist at Large,* 194–209.

94. Edward Tolman, "An Operational Analysis of 'Demands,'" *Erkenntnis* 6 (1936): 383–390; Tolman, "The Intervening Variable," in *Psychological Theory,* ed. M. Marx (1936; New York: Macmillan, 1951), 87–102.

95. Percy Bridgman, *The Nature of Physical Theory* (Princeton, NJ: Princeton University Press, 1936); Bridgman, "Operational Analysis," *Philosophy of Science* 5 (1938), reprinted in *Reflections of a Physicist,* 1–26.

96. Stevens, "Psychology and the Science of Science," 256.

97. R H. Seashore and B. Katz, "An Operational Definition and Classification of Mental Mechanisms," *Psychological Record* 1 (1937): 3–24; A. C. Benjamin, "The Operational Theory of Meaning," *Philosophical Review* 46 (1937): 644–649;

Carroll Pratt, *The Logic of Modern Psychology* (New York: Macmillan, 1939); *Lundberg, "Quantitative Methods in Social Psychology," *American Sociological Review* 1 (1936): 38–54; J R. Kantor, "The Operational Principle in the Physical and Psychological Sciences," *Psychological Record* 2 (1938): 3–32; Sigmund Koch, "The Logical Character of the Motivation Concept: I," *Psychological Review* 48 (1941): 15–38; Koch, "The Logical Character of the Motivation Concept: II," *Psychological Review* 48 (1941): 127–154.

98. Clark Hull, *Principles of Behavior: An Introduction to Behavior Theory* (New York: Appleton-Century-Crofts, 1943). For a discussion of Hull's theory of science, and his engagement (or lack thereof) with operationism, see Green, "Of Immortal Mythological Beasts": 302–304; Smith, *Behaviorism*, ch. 6–8.

99. William Malisoff, "The Universe of Operations," *Philosophy of Science* 3 (1936): 360–304.

100. R. B. Lindsay, "A Critique of Operationalism in Physics," *Philosophy of Science* 4 (1937): 456–470.

101. R. H. Waters and L. A. Pennington, "Operationism in Psychology," *Psychological Review* 45 (1938): 414–423; A. G. Bills, "Changing Views on Psychology as a Science," *Psychological Review* 45 (1938): 377–394.

102. H. Israel and B. Goldstein, "Operationism in Psychology," *Psychological Review* 51 (1944): 177–188.

103. E. G. Boring et al., Symposium on Operationism, *Psychological Review* 52 (1945): 241–294.

104. Percy Bridgman, "The Nature of Some of Our Physical Concepts. I," *British Journal for the Philosophy of Science* 1 (1951): 257–272; Bridgman, "The Nature of Some of Our Physical Concepts. II," *British Journal for the Philosophy of Science* 2 (1951): 25–44; Bridgman, "The Nature of Some of Our Physical Concepts. III," *British Journal for the Philosophy of Science* 2 (1951): 142–160.

105. Green, "Of Immortal Mythological Beasts": 311–315.

106. Philip G. Frank, ed., *The Validation of Scientific Theories* (Boston: Beacon Press, 1956), ch. 2.

107. See esp. Rogers, "Operationism in Psychology."

108. Skinner, "Concept of the Reflex," 1–2; Gerald Holton, "Ernst Mach and the Fortunes of Positivism in America," *Isis* 83 (1992): 40.

109. Boring, "Temporal Perception," 521.

110. Boring, "Operational Restatement."

111. Stevens, "Operational Definition," 517.

112. Stevens, "Operational Basis," 323, 330.

113. For a general account of this process, see Ash, "Psychology," 251–267. On the American context, see Morawski and Hornstein, "Quandary of the Quacks."

114. Rogers, "Operationism in Psychology."

115. On Bridgman and Skinner's shared debt to Mach, see Holton, "Ernst Mach": 39–41; Holton, "Skinner," 74.

116. E. G. Boring et al., "Rejoinders and Second Thoughts," *Psychological Review* 52 (1945): 282.

117. Skinner, "The Operational Analysis of Psychological Terms," *Psychological Review* 52 (1945): 271. In one of his contributions to the 1945 symposium, Carroll Pratt expressed "surprise" at Skinner's apparent concession that "traditional or classical psychology," with its mentalistic bent, "was sound enough in its choice of subject-matter and experimental procedures, but went astray only in the realm of logic and definition." A New Light behaviorist of zealous convictions, Skinner meant nothing of the sort. See Boring et al., "Rejoinders," 288–289.

118. Skinner, "Operational Analysis": 272–274.

119. Boring et al., "Rejoinders," 283.

120. Holton, "Skinner."

121. Walter, *Science and Cultural Crisis,* 192.

122. Hardcastle, "S. S. Stevens,: 417–418.

123. Skinner to Stevens, n.d., quoted in ibid., 418.

124. Ibid.

125. Skinner, "Operational Analysis," 271.

126. Boring et al., "Rejoinders," 291–293.

127. See, e.g., E. G. Boring, "The Use of Operational Definitions in Science," *Psychological Review* 52 (1945): 243.

128. Boring, *History,* 657.

129. On the distinctiveness of Stevens' operationism, see Hardcastle, "S. S. Stevens": 423–422.

130. Stevens, "Operational Basis," 327. See also Stevens, "Operational Definition," 517–518; Stevens, "Psychology," 97; Stevens, "Psychology and the Science of Science," 227–228.

131. Stevens, "Operational Definition," 518.

132. Stevens, "Psychology and the Science of Science," 224. On the autonomy of Stevens's views, see Hardcastle, "S. S. Stevens," 415–417.

133. Quoted in Holton, "Skinner," 78.

134. Stevens, "Psychology and the Science of Science," 227.

135. On Skinner's "lost year" between graduating college and beginning graduate school, during which time he tried, and failed, to make a living as a writer in New York, see B. F. Skinner, *Particulars of My Life* (London: Jonathan Cape, 1976), 254–277.

136. Quoted in Daniel N. Wiener, *B. F. Skinner: Benign Anarchist* (Boston: Allyn and Bacon, 1996), 42.

137. Cerullo, "Skinner at Harvard, 219; Wiener, *B. F. Skinner,* 42.

138. Skinner, "B. F. Skinner," 8–9.

139. Bertrand Russell, *Philosophy* (New York: Norton, 1927). Published in the UK as Bertrand Russell, *An Outline of Philosophy* (London: George Allen & Unwin, 1927).

140. John Broadus Watson, *Behaviorism* (London: Kegan Paul, Trench, Trubner, 1925).

141. Wiener, *B. F. Skinner,* 29.

142. Quoted in Smith, *Behaviorism and Logical Positivism,* 263.

143. Skinner, "B. F. Skinner," 10.

144. Smith, *Behaviorism and Logical Positivism,* 264–275; Holton, "Ernst Mach," 40.

145. Skinner, Review of Laurence D. Smith, *Behaviorism and Logical Positivism: A Reassessment of the Alliance, Journal of the History of the Behavioral Sciences* 23 (1987): 209.

146. Skinner, "B. F. Skinner," 10.

147. See Alexandra Rutherford, "B. F. Skinner and the Auditory Inkblot," *History of Psychology* 6 (2003): 362–378.

148. Skinner, "B. F. Skinner," 10–11.

149. See Hardcastle, "S. S. Stevens," 416.

150. On the Psychological Roundtable, see Cohen-Cole, "Thinking about Thinking," 45–46; Gary L. Hardcastle, "The Cult of Experiment: The Psychological Roundtable, 1936–1941," *History of Psychology* 3 (2000): 334–370.

151. Gordon Allport, "The Psychologist's Frame of Reference," *Psychological Bulletin* 37 (1940): 1–28.

152. Cohen-Cole, "Thinking about Thinking," ch. 1.

153. Ibid., 64–72.

154. The report of the visiting committee was published as Alan Gregg et al., *The Place of Psychology in an Ideal University: The Report of the University Commission to Advise on the Future of Psychology at Harvard* (Cambridge, MA: Harvard University Press, 1947). On the division of the Department, see Gordon Allport and E. G. Boring, "Psychology and Social Relations at Harvard University," *American Psychologist* 1 (1946): 119–122.

155. Skinner, Review of Smith, 208.

4. Radical Translation

1. Bruce Kuklick, *The Rise of American Philosophy: Cambridge, Massachusetts, 1860–1930* (New Haven, CT: Yale University Press, 1977), 139.

2. George Herbert Palmer and Ralph Barton Perry, "Philosophy, 1870–1929," in *The Development of Harvard University since the Inauguration of President Eliot, 1869–1929,* ed. Samuel Eliot Morison (Cambridge, MA: Harvard University Press, 1930), 3–4. Palmer's authorship of this section of the essay is confirmed on page 3, n. 1.

3. Daniel Walker Howe, *The Unitarian Conscience: Harvard Moral Philosophy, 1805–1861* (Cambridge, MA: Harvard University Press, 1970).

4. Daniel J. Wilson, *Science, Community, and the Transformation of American Philosophy, 1860–1930* (Chicago: University of Chicago Press, 1990); Louis Menand, *The Metaphysical Club: A Story of Ideas in America* (London: Flamingo, 2001); Kuklick, *Rise of American Philosophy.*

5. George P. Adams and William Pepperell Montague, eds., *Contemporary American Philosophy: Personal Statements,* 2 vols. (London: George Allen & Unwin, 1930). The Harvard contingent was composed of G. H. Palmer—to whom the volumes were dedicated— William Ernest Hocking, Lewis, Montague, Perry, and Santayana. Some of the other contributors had also trained at Harvard.

6. Hans Hahn, Otto Neurath, and Rudolf Carnap, *Wissenschaftliche Weltauffassung: Der Wiener Kreis* ("The scientific conception of the world: the Vienna Circle") in Otto Neurath, *Empiricism and Sociology,* trans. Paul Foulkes and Marie Neurath, ed. Marie Neurath and Robert S. Cohen (Dordrecht: Reidel, 1973), 303–304.

7. Ibid., 309.

8. For systematic assessments of the reception of logical empiricism in America, see George A. Reisch, *How the Cold War Transformed Philosophy of Science: To the Icy Slopes of Logic* (Cambridge: Cambridge University Press, 2005); Gary L. Hardcastle and Alan W. Richardson, eds., *Logical Empiricism in North America* (Minneapolis: University of Minnesota Press, 2003); Peter Galison, "Constructing Modernism: The Cultural Location of *Aufbau,*" in *Origins of Logical Empiricism,* ed. Ronald N. Giere and Alan W. Richardson (Minneapolis: University of Minnesota Press, 1996), 35–38; George Reisch, "On the *International Encyclopedia,* the Neurath-Carnap Disputes, and the Second World War," in *Logical Empiricism: Historical and Contemporary Perspectives,* ed. Paolo Parrini, Wesley C. Salmon, and Merrilee H. Salmon (Pittsburgh, PA: University of Pittsburgh Press, 2003), 94–108.

9. On philosophical naturalism, see Mario de Caro and David Macarthur, eds., *Naturalism in Question* (Cambridge, MA: Harvard University Press, 2004); Dagfinn Føllesdal, ed., *Philosophy of Quine,* vol. 2, *Naturalism and Ethics* (New York: Garland, 2000). The literature on the history of analytic philosophy and logical empiricism is already vast, ranging from creative redescriptions of the tradition by philosophers like Richard Rorty and Hilary Putnam to popular biographies of seminal figures such as Russell, Wittgenstein, and Ayer. For recent expository introductions to the analytic tradition as a whole, see Avrum Stroll, *Twentieth-Century Analytic Philosophy* (New York: Columbia University Press, 2000); Scott Soames, *Philosophical Analysis in the Twentieth Century,* 2 vols. (Princeton, NJ: Princeton University Press, 2003). The study of the history of analytic philosophy has been opened up most of all by the late Burton Dreben. A close colleague of Quine's at Harvard, Dreben served for many years as the éminence grise of Emerson Hall, publishing little but exercising an enormous influence on analytic philosophers like Quine, Putnam, John Rawls, Morton White, Stanley Cavell, T. M. Scanlon, Warren Goldfarb, and Charles Parsons. In his teaching, Dreben emphasized the importance of examining the historical development of the analytic tradition, thus inspiring a new generation of philosophers to follow in his footsteps. See Burton Dreben, "Quine," in *Perspectives on Quine,* ed. Robert B. Barrett and Roger F. Gibson (Oxford: Basil Blackwell, 1990), 81–95; Dreben, "Putnam, Quine—and the Facts" (1992), reprinted in Føllesdal, *Philosophy of Quine,* vol. 4, *Ontology,* 305–327;

Dreben, "In *Mediis Rebus*" (1994), reprinted in Føllesdal, *Philosophy of Quine,* vol. 1, 39–45; Dreben, "Quine and Wittgenstein: The Odd Couple" (1996), reprinted in Føllesdal, *Philosophy of Quine,* vol. 1, 79–101. The work of his students is showcased in Juliet Floyd and Sanford Shieh, eds., *Future Pasts: The Analytic Tradition in Twentieth-Century Philosophy* (Oxford: Oxford University Press, 2001).

10. W. V. Quine, *From Stimulus to Science* (Cambridge, MA: Harvard University Press, 1995), 16.

11. The canonical statement of Quine's conception of naturalized epistemology is "Epistemology Naturalized," in W. V. Quine, *Ontological Relativity and Other Essays* (New York: Columbia University Press, 1969), 69–90. For a survey of Quine's position, see Robert J. Fogelin, "Aspects of Quine's Naturalized Epistemology," in *The Cambridge Companion to Quine,* ed. Roger F. Gibson (Cambridge: Cambridge University Press, 2004), 19–46; Barry Stroud, "The Significance of Naturalized Epistemology" (1981), reprinted in Føllesdal, *Philosophy of Quine,* vol. 2, 165–181.

12. Rudolf Carnap, *The Logical Structure of the World and Pseudoproblems in Philosophy,* trans. Rolf A. George (1967; Chicago: Open Court, 2003), 7.

13. Ibid., 107–109.

14. Ibid., 7, 5, 106–107. See Gary Hardcastle, "Logical Empiricism and the Philosophy of Psychology," in *The Cambridge Companion to Logical Empiricism,* ed. Alan Richardson and Thomas Uebel (Cambridge: Cambridge University Press, 2007), 235.

15. Quine has been called "the most distinguished American recruit to logical empiricism," "the living philosopher who has commanded the greatest influence among his colleagues," and the most eminent "living systematic philosopher." See, respectively, Anthony Quinton, "The Importance of Quine" (1967), reprinted in Føllesdal, *Philosophy of Quine,* vol. 1, *General Reviews and Analytic/Synthetic,* 157; A. J. Ayer, *Philosophy in the Twentieth Century,* (London: Weidenfeld and Nicolson, 1982), 242; Stuart Hampshire quoted in Roger F. Gibson, Jr., *The Philosophy of W. V. Quine: An Expository Essay* (Tampa: University of South Florida Press, 1982), xvii.

16. W. V. Quine, "Truth By Convention," (1935), reprinted in Quine, *The Ways of Paradox and Other Essays,* rev. ed. (Cambridge, MA: Harvard University Press, 1976), 77–106; Quine, "Two Dogmas of Empiricism," *Philosophical Review* 60 (1951): 20–43; Quine, "Carnap and Logical Truth," in *The Philosophy of Rudolf Carnap,* ed. Paul Arthur Schilpp (1954; La Salle, IL: Open Court, 1963), 385–406; Quine, "Epistemology Naturalized"; Quine, *From Stimulus to Science.*

17. In "Carnap and Logical Truth," 385, Quine acknowledged that his "dissent from Carnap's philosophy of logical truth is hard to state and argue in Carnap's terms."

18. My treatment of these issues is highly condensed. For helpful general accounts of the evolution of Carnap's thought in this period, see Michael Friedman, "Introduction: Carnap's Revolution in Philosophy," in *The Cambridge Companion to Carnap,* ed. Michael Friedman and Richard Creath (Cambridge: Cambridge

University Press, 2007), 1–18; Thomas Ricketts, "Carnap: From Logical Syntax to Semantics," in Giere and Richardson, *Origins of Logical Empiricism,* 231–250.

19. A point emphasized in Alan W. Richardson, "From Epistemology to Logic of Science: Carnap's Philosophy of Empirical Knowledge in the 1930s," in Giere and Richardson, *Origins of Logical Empiricism,* 309–332.

20. The classic examples of this part of Carnap's project are Rudolf Carnap, *The Logical Syntax of Language,* trans. Amethe Smeaton (London: Routledge and Kegan Paul, 1937); Carnap, "Testability and Meaning," *Philosophy of Science* 3 (1936): 419–471.

21. I have taken this term from Friedman, "Introduction," 9.

22. Carnap, "Task of the Logic of Science," 56, 60. Emphasis in original.

23. Alan W. Richardson, "Logical Empiricism, American Pragmatism, and the Fate of Scientific Philosophy in North America," in Hardcastle and Richardson, *Logical Empiricism in North America,* 9.

24. Quine, "Epistemology Naturalized," 74.

25. Richard Creath, "The Linguistic Doctrine and Conventionality: The Main Argument in 'Carnap and Logical Truth,'" in Hardcastle and Richardson, *Logical Empiricism in North America,* 255. On this criticism of "Two Dogmas" see also Gary Gutting, *What Philosophers Know: Case Studies in Recent Analytic Philosophy* (Cambridge: Cambridge University Press, 2009), 11–30.

26. Thomas Ricketts, "Languages and Calculi," in Hardcastle and Richardson, *Logical Empiricism in North America,* 258.

27. Michael Friedman, "Coordination, Constitution, and Convention: The Evolution of the A Priori in Logical Empiricism," in Richardson and Uebel, *The Cambridge Companion to Logical Empiricism,* 111–116; Richard Creath, "Quine's Challenge to Carnap," in Friedman and Creath, *Cambridge Companion to Carnap,* 316–335; Richardson, "Logical Empiricism," 8–10; Creath, "Linguistic Doctrine," 251.

28. Friedman, "Coordination," 113. Quine's enduring epistemological debts to logic and the foundations of mathematics are on vivid display even in "Epistemology Naturalized," 69–70.

29. John Dewey, "From Absolutism to Experimentalism," in Adams and Montague, *Contemporary American Philosophy,* vol. 2, 13–19; C. I. Lewis, "Logic and Pragmatism," in Adams and Montague, *Contemporary American Philosophy,* vol. 2, 31; Morris Cohen, "The Faith of a Logician," in Adams and Montague, *Contemporary American Philosophy,* vol. 1, 222–223.

30. Kuklick, *Rise of American Philosophy,* 451–480.

31. "Amateur" would be a more accurate, but also misleadingly pejorative, term.

32. W. V. Quine, *The Time of My Life: An Autobiography* (Cambridge, MA: MIT Press, 1985), 18–19, 30, 36–37.

33. W. V. Quine, "On Knowledge as a Unit," [unpublished essay] February 23, 1927, *2001M-7(b), box 1, WVQPH. The Quine papers have recently been recatalogued. Here I use the original series codes as these were used to organize the papers before they were processed.

34. John Barnard, *From Evangelicalism to Progressivism at Oberlin College, 1866–1917* (Columbus: Ohio State University Press, 1969), 127.

35. W. V. Quine quoted in Daniel Isaacson, "Quine and Logical Positivism," in Gibson, *The Cambridge Companion to Quine,* 249. See also Quine, *The Time of My Life,* 58.

36. On the centrality of behaviorism in Quine's later philosophy, see Gibson, *Philosophy of W. V. Quine.*

37. Quine, *Time of My Life,* 51–52.

38. Gottlob Frege, *The Foundations of Arithmetic: A Logico-Mathematical Enquiry into the Concept of Number,* trans. J. L. Austin, 2nd ed. (Oxford: Basil Blackwell, 1980), 99.

39. Bertrand Russell, "On Denoting," (1905), reprinted in Russell, *Logic and Knowledge: Essays 1901–1950,* ed. Robert Charles Marsh (1956; Nottingham: Spokesman, 2007), 41–56

40. Bertrand Russell, *Our Knowledge of the External World, As a Field for Scientific Method in Philosophy* (Chicago: Open Court, 1914), 33.

41. Ibid., 42–43.

42. Ibid., 58–59.

43. See the notes collected in the folder marked "Logic and Psychology Notes: Lectures and Reading," *2001M-0007(b), box 1, WVQPH. These reading notes of Quine's very likely do not cover all of his reading material, but they show that, in addition to the texts already mentioned, Quine consulted the following works (authors and titles as transcribed by Quine): John Venn, *Symbolic Logic;* Johann [sic] Reichenbach, *Relitivitätstheorie und Erkenntnis apriori;* C. K. Irvine, *Elements of Logic;* Georg Cantor, *Contributions to the Founding of the Theory of Transfinite Numbers;* E. V. Huntingdon, "The Fundamental Propositions of Algebra"; Stanlislaw Jaskowski, *Concerning the Rules of Suppositions in Formal Logic.*

44. Quine glosses Wittgenstein's Tractarian position on logical truth in W. V. Quine, "The Validity of Deduction," unpublished note, April 11, 1930, *2001M-0007(b), box 1, WVQPH. It may well be, however, that Quine's knowledge of the *Tractatus* came solely through his reading of Russell.

45. Quine, *Time of the My Life,* 72–73.

46. See Peter Hylton, *Russell, Idealism, and the Emergence of Analytic Philosophy* (Oxford: Clarendon Press, 1990), 72–101.

47. W. V. Quine, "Mathematics as a Mode of Thought," [unpublished manuscript] January 10, 1930, *1991M-0068, box 11, WVQPH.

48. C. I. Lewis, *A Survey of Symbolic Logic* (Berkeley: University of California Press, 1918). On Sheffer, see Michael Scanlan, "The Known and Unknown H. M. Sheffer," *Transactions of the Charles S. Peirce Society* 36 (2000): 193–224.

49. *Official Register of Harvard University* 27 (September 17, 1930), 130–135.

50. Quine, *Time of My Life,* 82.

51. Quine details the courses he took in graduate school in W. V. Quine, "General Lecture Notes in Philosophy and Philosophy Reading" *2001M-0007(b), box 1,

WVQPH; and Quine, *Time of My Life,* 82. His essays from graduate school document his participation in the Lewis and Whitehead courses. It is important to note that Quine's recorded itinerary does not completely match the course catalogues. No Prall course on Leibniz is listed in the *Official Register of Harvard University* for either 1930–31 or 1931–32. (No professor is listed for the Leibniz course in 1931–32, so it is possible that Prall taught this course and that Quine took it that year; in 1930–31, Donald C. Williams taught a course on Descartes and Leibniz.)

52. Quine, *Time of My Life,* 84–86.

53. See W. V. Quine, "General Lecture Notes in Philosophy and Philosophy Reading" and "Kant," [unpublished notes] *2001M-0007(b), box 1, WVQPH.

54. Quine, *The Time of My Life,* 82–83; Dreben, "Quine," 82–84.

55. Warren D. Goldfarb, "Logic in the Twenties: The Nature of the Quantifier," *Journal of Symbolic Logic* 44 (1979): 351–368.

56. W. V. Quine, "Notes on an Interview with A. N. Whitehead in his study, March 16, 1931," [transcript], *1991M-0068, Box 9, WVQPH.

57. Dreben, "Quine," 83–84.

58. Quine, *Time of My Life,* 78–79.

59. Lucien Price, *Dialogues of Alfred North Whitehead, As Recorded by Lucien Price* (New York: Mentor, 1954).

60. Alfred North Whitehead, *Science and the Modern World* (1925; New York: Free Press, 1967).

61. C. I. Lewis, "Autobiography," in *The Philosophy of C. I. Lewis,* ed. Paul Arthur Schilpp (La Salle, IL: Open Court, 1968), 5–6, 9–13.

62. C. I. Lewis, *Mind and the World-Order: Outline of a Theory of Knowledge* (London: Scribner's, 1929), vii.

63. Ibid., ix.

64. W. V. Quine, "Concepts and Working Hypotheses," [unpublished essay] March 10, 1931, *2001M-0007(b), box 1, WVQPH.

65. Ibid.

66. Ibid. Italics added.

67. Quine, "Two Dogmas," 42.

68. For Quine's orthodox reading of Lewis, see W. V. Quine, "On the Validity of Singular Empirical Judgements," [unpublished essay] March 17, 1931, *2001M-0007(b), box 1, WVQPH.

69. W. V. Quine, "Futurism and the Conceptual Pragmatist," [unpublished essay] May 6, 1931, *2001M-0007(b), box 1, WVQPH.

70. Quine, "Concepts and Working Hypotheses."

71. Ibid.

72. W. V. Quine, "Autobiography of W. V. Quine," in *The Philosophy of W. V. Quine,* ed. Lewis Edwin Hahn and Paul Arthur Schilpp (La Salle, IL: Open Court, 1986), 10–11; Quine, *The Time of My Life,* 84–86. A helpful summary of Quine's work on this topic can be found in W. V. Quine, "On the Quantitative Aspect of

Propositions," [unpublished manuscript] April 15, 1932, *1991M-0068, box 11, WVQPH.

73. Quotations are taken from *Official Register of Harvard University* 33 (September 23, 1936), 158–165.

74. *Official Register of Harvard University* 34 (October 1, 1937), 170. The course description runs as follows: "A critical survey of the views of the Vienna Circle and related authors."

75. Malachi Haim Hacohen, *Karl Popper—The Formative Years, 1902–1945: Politics and Philosophy in Interwar Vienna* (Cambridge: Cambridge University Press, 2000), 189 191.

76. Quine, *Time of My Life*, 94–95.

77. Ibid., 98.

78. Ibid.

79. For a historical reconstruction of these debates in connection with Carnap's philosophy, see Steve Awodey and A. W. Carus, "The Turning Point and the Revolution: Philosophy of Mathematics in Logical Empiricism from *Tractatus* to *Logical Syntax*," in Richardson and Uebel, *The Cambridge Companion to Logical Empiricism*, 165–192; Friedman, "Introduction," 5–9.

80. Carnap, *Logical Syntax*, 7.

81. Quine, "Autobiography" 11–13; Quine, *The Time of My Life*, 86–108; Donald Davidson, "On Quine's Philosophy," and Donald Davidson and W. V. Quine, "Exchange Between Donald Davidson and W. V. Quine," in Føllesdal, *Philosophy of Quine*, vol. 1, 48–50 and 58–60.

82. Quine, "Autobiography" 12.

83. W. V. Quine "Talk at the Faculty Reception of the Department of Philosophy" [unpublished manuscript], October 20, 1933. *1991M-0068, box 11, WVQPH.

84. W. V. Quine, "Towards a Calculus of Concepts," *The Journal of Symbolic Logic* 1 (March 1936): 2. See also Quine, "A Theory of Classes Presupposing No Canons of Type," *Proceedings of the National Academy of Sciences of the United States of America* 22 (May 1936): 320–326; Quine, "On the Axiom of Reducibility," *Mind* 45 (October 1936): 498–500.

85. W. V. Quine, "Ontological Remarks on the Propositional Calculus" (1934), reprinted in *Ways of Paradox*, 265–271.

86. See W. V. Quine "A System of Logistic Based Upon Classes," October 22, 1933; "Conference on 'The Paradoxes of Set Theory and Semantics,'" November 10, 1936; "The Theory of Types and Its Alternatives," April 1, 1937; "Remarks on the Nature & Purposes of Math'l Logic," May 4, 1937; "On the Theory of Logical Types," February 24, 1938; "Math. Colloquium Apr. 14, 1938," April 14, 1938. All in *1991M-0068, box 11, WVQPH.

87. W. V. Quine, "Reply to Hao Wang," *The Philosophy of W. V. Quine*, 644–645.

88. See W. V. Quine, *Mathematical Logic* (New York: Norton, 1940); Quine, *Elementary Logic* (Boston: Ginn, 1941); Quine, *Methods of Logic* (New York: Holt, 1950).

89. Davidson and Quine, "Exchange between Donald Davidson and W. V. Quine," 59.

90. Isaacson, "Quine and Logical Positivism," 239.

91. Neil Tennant, "Carnap and Quine," in *Logic, Language, and the Structure of Scientific Theories,* ed. Wesley Salmon and Gereon Wolters (Pittsburgh: University of Pittsburgh Press, 1994), 314.

92. Richard Creath, "The Initial Reception of Carnap's Doctrine of Analyticity" (1987), reprinted in Føllesdal, *Philosophy of Quine,* vol. 1, 321–343; Peter Hylton, "'The Defensible Province of Philosophy': Quine's 1934 Lectures on Carnap," in Floyd and Shieh, *Future Pasts,* 257–275.

93. W. V. Quine, "Lectures on Carnap. Lecture III: *Philosophy as Syntax,*" [manuscript] November 22, 1934, *1991M-0068, box 11, WVQPH. Quine's 1934 lectures on Carnap have been published in W. V. Quine and Rudolf Carnap, *Dear Carnap, Dear Van: The Quine-Carnap Correspondence,* ed. Richard Creath (Berkeley: University of California Press, 1990), 47–103.

94. W. V. Quine, "Introducing Carnap," [unpublished manuscript] April 8, 1936, *1991M-0068, box 11, WVQPH.

95. See, for example, W. V. Quine, "Nominalism," [unpublished manuscript] October 25, 1937, *1991M-0068, box 11, WVQPH; "Dr. Van O. Quine, Instructor in Logic, Guest of Math Club," *Northeastern News,* April 21, 1937, *1991M-0068, Box 11, WVQPH.

96. Quine, *Time of My Life,* 140.

97. A helpful overview of the mechanisms of cultural transfer in logical empiricism can be found in Freidrich Stadler, "Transfer and Transformation of Logical Empiricism: Quantitative and Qualitative Aspects," in Hardcastle and Richardson, *Logical Empiricism in North America,* 216–233.

98. Gerald Holton, "On the Vienna Circle in Exile: An Eyewitness Report," in *The Foundational Debate: Complexity and Constructivity in Mathematics and Physics,* ed. Werner Depauli-Schimanovich et al. (Dordrecht: Kluwer, 1995), 269–292; Holton, "From the Vienna Circle to Harvard Square: The Americanization of a European World Conception," in *Scientific Philosophy: Origins and Development,* ed. Freidrich Stadler (Dordrecht: Kluwer, 1993). An extended version of the latter is Holton, "Ernst Mach and the Fortunes of Positivism in America," *Isis* 83 (1992): 27–60.

99. Peter Galison, "The Americanization of Unity," *Daedalus* 127 (1998): 45–71. See also Galison, "The Ontology of the Enemy: Norbert Weiner and the Cybernetic Vision," *Critical Inquiry* 21 (1994): 228–266.

100. Reisch,* *How the Cold War Transformed Philosophy of Science;* Don Howard, "Two Left Turns Make a Right: On the Curious Political Career of North American Philosophy of Science at Midcentury," in Hardcastle and Richardson, *Logical Empiricism in North America,* 25–93. The purported depoliticization of American philosophy during the Cold War is examined, with mixed success, in John McCumber, *Time in the Ditch: American Philosophy and the McCarthy Era* (Evanston, IL: Northwestern University Press, 2001).

101. On the connections between pragmatism, secular intellectual culture, and the rise of Jewish intellectuals in American public and academic life, see David Hollinger, "Jewish Intellectuals and the De-Christianization of American Public Culture in the Twentieth Century" and "The 'Tough-Minded' Justice Holmes, Jewish Intellectuals, and the Making of An American Icon" in Hollinger, *Science, Jews, and Secular Culture* (Princeton, NJ: Princeton University Press, 1996), 17–59.

102. George Reisch, "From 'the Life of the Present' to the 'Icy Slopes of Logic': Logical Empiricism, the Unity of Science Movement, and the Cold War," in Richardson and Uebel, *The Cambridge Companion to Logical Empiricism,* 60–65; Reisch, *How the Cold War Transformed Philosophy of Science,* 57–72.

103. Ernest Nagel, "Impressions and Appraisals of Analytic Philosophy in Europe I," *Journal of Philosophy* 33 (January 1936): 5–24; John Dewey, "Unity of Science as a Social Problem," in *International Encyclopedia of Unified Science,* vol. 1, ed. Otto Neurath, Rudolf Carnap, and Charles Morris (Chicago: University of Chicago Press, 1955); William Gruen, "What is Logical Empiricism?" *Partisan Review* 6 (1939): 64–77; Albert Wohlstetter and Morton White, "Who Are the Friends of Semantics?" *Partisan Review* 6 (1939): 50–57.

104. On Morris, see Reisch, *How the Cold War Transformed Philosophy of Science,* 38–47; Galison, "Cultural Location," 36–40.

105. See Otto Neurath, Rudolf Carnap, and Charles Morris, eds., *International Encyclopedia for Unified Science,* 2 vols. (Chicago: University of Chicago Press, 1955).

106. Reisch, *How the Cold War Transformed Philosophy of Science,* 41–42.

107. On Deweyan reservations about Neurath's Unity of Science program, see David Hollinger, "The Unity of Knowledge and the Diversity of Knowers: Science as an Agent of Cultural Integration in the United States between the Two World Wars," *Pacific Historical Review* 80 (2011): 211–230; Reisch, *How the Cold War Transformed Philosophy of Science,* 167–190.

108. Galison, "Cultural Location," 37–39; George Reisch, "Disunity in the *International Encyclopedia of Unified Science,*" in Hardcastle and Richardson, *Logical Empiricism in North America,* 197–215; Reisch, "On the *International Encyclopedia,* the Neurath-Carnap Disputes, and the Second World War," in Parrini, Salmon, and Salmon, *Logical Empiricism,* 94–108.

109. See Herbert Feigl, "The *Wiener Kreis* in America" (1969), in Feigl, *Inquiries and Provocations: Selected Writings, 1929–1974,* ed. Robert S. Cohen (Dordrecht: D. Reidel, 1981), 60–67.

110. Paul K. Feyerabend, "Herbert Feigl: A Biographical Sketch," in *Mind, Matter, and Method: Essays in Philosophy and Science in Honor of Herbert Feigl,* ed. Paul K. Feyerabend and Grover Maxwell (Minneapolis: University of Minnesota Press, 1966), 8.

111. Feigl, "*Wiener Kreis,*" 69–70.

112. Albert E. Blumberg and Herbert Feigl, "Logical Positivism: A New Movement in European Philosophy," *Journal of Philosophy* 28 (May 1931): 281–296.

113. Of particular importance were Feigl's essays on the mind-body problem and his role in the establishment of the Minnesota Center for the Philosophy of Science in the 1950s. For a general overview, see Rudolf Haller, "On Herbert Feigl," in Hardcastle and Richardson, *Logical Empiricism in North America*, 115–128.

114. Quine, *Time of My Life*, 116.

115. L. J. Henderson, *Pareto's General Sociology: A Physiologist's Interpretation* (Cambridge, MA: Harvard University Press, 1937), 23, n. 1.

116. See Rudolf Carnap, "Logic," in Harvard Tercentenary Publications, *Factors Determining Human Behavior* (Cambridge, MA: Harvard University Press, 1937), 106–118.

117. On the reaction of the philosophers to Quine's promotion of Carnap, see Quine, *Time of My Life*, 117. Lewis's early, critical responses to "logical positivism" include "Experience and Meaning" (1934), "Verification and Types of Truth" (1936–37), and "Logical Positivism and Pragmatism" (1941), in *Collected Papers of Clarence Irving Lewis*, ed. John D. Goheen and John L. Motherhead, Jr. (Stanford, CA: Stanford University Press, 1970). Prall was keen to translate the *Logische Syntax*, but was beaten to the punch: see Quine and Carnap, *Dear Carnap, Dear Van*, 149–155.

118. On Henderson's harangues, see Harvard University Society of Fellows, "Conversations: 2. Willard Van Oram [*sic*] Quine," 1992, HUSFP. Quine's debt to Henderson on the rhetoric of conceptual schemes is recorded in W. V. Quine, "On the Very Idea of a Third Dogma" in Quine, *Theories and Things* (Cambridge, MA: Belknap Press of Harvard University Press, 1981), 41.

119. Harvard Society of Fellows, "Conversation 2"; Quine, *Time of My Life*, 110.

120. Harvard Society of Fellows, "Conversation 2."

121. Harvard University Society of Fellows, "Conversations: 5. Burrhus Fredric Skinner," 1998, HUSFP.

122. For the introduction of the course on Logical Positivism, see *Official Register of Harvard University* 34 (October 1, 1937), 170.

123. Lewis's considered but generally negative response to logical empiricism is discussed in Murray G. Murphey, *C. I. Lewis: The Last Great Pragmatist* (Albany: State University of New York Press, 2005), 214–244. On Quine in these years, see Joel Isaac, "Missing Links: W. V. Quine, the Making of 'Two Dogmas,' and the Analytic Roots of Postanalytic Philosophy," *History of European Ideas* 37 (2011): 267–279.

124. See Bessie Zaban Jones, "To the Rescue of the Learned: The Asylum Fellowship Plan at Harvard, 1938–1940," *Harvard Library Bulletin* 32 (1984): 205–238.

125. Feigl, *"Weiner Kreis,"* 73; Philipp Frank, *Modern Science and Its Philosophy* (Cambridge, MA: Harvard University Press, 1949), 50.

126. Charles Morris, "The Unity of Science Movement and the United States," *Synthese* (1938): 25–29.

127. "Fifth International Congress for the Unity of Science," *Journal of Unified Science/Erkenntnis* 8 (1939–1940): 191; Morris, "Unity of Science": 25–26.

128. "Fifth International Congress for the Unity of Science," 369–371.

129. For a perceptive study of this important moment in the history of scientific philosophy, see Gary L. Hardcastle, "Debabelizing Science: The Harvard Science of Science Discussion Group, 1940–41," in Hardcastle and Richardson, *Logical Empiricism in North America*, 170–196.

130. Bertrand Russell, *An Inquiry into Meaning and Truth* (London: Allen and Unwin, 1940).

131. Quine, *Time of My Life*, 149–150.

132. S. S. Stevens, "Quantifying the Sensory Experience," in Feyerabend and Maxwell, *Mind, Matter, and Method*, 215–216.

133. Rudolf Carnap, "Intellectual Autobiography of Rudolf Carnap," in Schilpp, *Philosophy of Rudolf Carnap*, 34–35.

134. Hardcastle, "Debabelizing Science," 174.

135. Quoted in ibid., 172.

136. Names of those involved drawn from a folder marked "Sci of Sci—Discussion Group," box 4, SSSPH. See also Hardcastle, "Debabelizing Science."

137. Hardcastle, "Debabelizing Science," 175.

5. The Levellers

1. On World War II as a scientific revolution, see M. Fortun and S. S. Schweber, "Scientists and the Legacy of World War II: The Case of Operations Research," *Social Studies of Science* 23 (1993): 595–642.

2. James Hershberg, *James B. Conant: Harvard to Hiroshima and the Making of the Nuclear Age* (New York: Knopf, 1993), ch. 8–13.

3. Roger L. Geiger, *Research and Relevant Knowledge: American Research Universities since World War II* (New York: Oxford University Press, 1993), 6–13; Rebecca Lowen, *Creating the Cold War University: The Transformation of Stanford* (Berkeley: University of California Press, 1997), 43–66.

4. As the authors of *The Behavioral Sciences at Harvard* (Report by a Faculty Committee, June 1954) admitted, among lone researchers, there was "a belief that the large research project, especially if dressed up as 'interdisciplinary', gets first call on foundation funds."

5. On Ruml, see Martin Bulmer and Joan Bulmer, "Beardsley Ruml and the Laura Spelman Rockefeller Memorial," *Minerva* 19 (1981): 347–407.

6. A point eloquently made in Rebecca Lemov, *World as Laboratory: Experiments with Mice, Mazes, and Men* (New York: Hill and Wang, 2005), 46–67.

7. See J. G. Morawski, "Organizing Knowledge and Behavior at Yale's Institute of Human Relations," *Isis* 77 (1986): 219–242.

8. Lemov, *World as Laboratory*, 125–144.

9. Howard Brick, "The Reformist Dimension of Talcott Parsons's Early Social Theory," in *The Culture of the Market: Historical Essays,* ed. Thomas L. Haskell and Richard F. Teichgraeber III (Cambridge: Cambridge University Press, 1993), 357–396; Joel Isaac, "Theories of Knowledge and the American Human Sciences, 1920–1960" (PhD Thesis, University of Cambridge, 2005), ch. 2.

10. Talcott Parsons, "'Capitalism' in Recent German Literature: Sombart and Weber" (1928–29), in Parsons, *Talcott Parsons: The Early Essays,* ed. Charles Camic (Chicago: University of Chicago Press, 1991), 3–37.

11. A point made forcefully, if controversially, in Charles Camic, "Reputation and Predecessor Selection: Parsons and the Institutionalists," *American Sociological Review* 57 (1992): 421–445. For a critical response, see Jeffrey C. Alexander and Giuseppe Sciortino, "On Choosing One's Intellectual Predecessors: The Reductionism of Camic's Treatment of Parsons and the Institutionalists," *Sociological Theory* 14 (1996): 154–171.

12. Charles Camic, "Introduction: Talcott Parsons before *The Structure of Social Action,*" in Parsons, *The Early Essays,* lxvii. See also Camic, "The Making of a Method: A Historical Reinterpretation of the Early Parsons," *American Sociological Review* 52 (1987): 421–439.

13. Talcott Parsons, "Pareto's Central Analytical Scheme" (1936), in Parsons, *The Early Essays,* 135.

14. Ibid., 141–142.

15. Talcott Parsons to Paul H. Buck, April 3, 1944, section HUGFP 15.2, box 6, TPPH.

16. See Lawrence T. Nichols, "The Establishment of Sociology at Harvard: A Case of Organizational Ambivalence and Scientific Vulnerability," in *Science at Harvard University: Historical Perspectives,* ed. Clark A. Elliot and Margaret W. Rossiter (Bethlehem, GA: Lehigh University Press/London: Associated University Presses, 1992), 191–222.

17. Talcott Parsons, "Clyde Kluckhohn and the Integration of Social Science," in *Culture and Life: Essays in Memory of Clyde Kluckhohn,* ed. Walter W. Taylor, John L. Fisher, and Evon Z. Vogt (Carbondale: Southern Illinois University Press, 1973) 32.

18. Ibid., 32.

19. J. T. Dunlop et al., "Toward a Common Language for the Area of Social Science" (Unpublished Memorandum, Harvard University, 1941), 1–5.

20. Ibid., 3.

21. On abstractions as tools, see ibid., 4, 5.

22. Ibid., 5–10.

23. Ibid., 11–16.

24. Ibid., 4.

25. James Bryant Conant, "President's Report," *Official Register of Harvard University* 41 (September 28, 1944), 8–9.

26. Paul H. Buck, "Faculty of Arts and Sciences," *Official Register* 41 (September 28, 1944), 30.

27. James Bryant Conant, "President's Report," *Official Register of Harvard University* 41 (September 26, 1944), 15; Peter Galison, "The Americanization of Unity," *Daedalus* 127 (1998): 59.

28. James Bryant Conant, "President's Report," *Official Register* 41 (September 28, 1944), 6–8.

29. Ibid., 6. On the ASTP, see Louis E. Keefer, *Scholars in Foxholes: The Story of the Army Specialized Training Program in World War II* (Jefferson, NC: McFarland, 1988); David C. Engerman, *Know Your Enemy: The Rise and Fall of America's Soviet Experts* (New York: Oxford University Press, 2009), 17–21.

30. James Bryant Conant, "President's Report," *Official Register* 41 (September 28, 1944), 6–8; Conant, "President's Report," *Official Register* 41 (September 26, 1944), 6, 17.

31. James Bryant Conant, "President's Report," *Official Register of Harvard University* 44 (July 7, 1947), 15–16.

32. Paul H. Buck, "Faculty of Arts and Science," *Official Register of Harvard University* 45 (December 1, 1948), 32.

33. Ibid., 32; Galison, "Americanization of Unity," 59.

34. Galison, "Americanization of Unity," 62–63.

35. Paul H. Buck, "Faculty of Arts and Science," *Official Register* 45 (December 1, 1948), 32; "Appendix 2," in Elliot and Rossiter, *Science at Harvard University,* 357.

36. On MIT's "Rad Lab," see Pater Galison, *Image and Logic: A Material Culture of Microphysics* (Chicago: Chicago University Press, 1997), 288–291.

37. On Aiken and the Mark I, see I. Bernard Cohen, "Howard H. Aiken, Harvard University, and IBM," in Elliot and Rossiter, *Science at Harvard University,* 251–284.

38. Morton Keller and Phyllis Keller, *Making Harvard Modern: The Rise of America's University* (New York: Oxford University Press, 2001), 163.

39. Galison, *Image and Logic,* ch. 4; Galison, "Americanization of Unity"; Fortun and Schweber, "Scientists and the Legacy of World War II"; Philip Mirowski, *Machine Dreams: Economics Becomes a Cyborg Science* (Cambridge: Cambridge University Press, 2002), 153–190; Jamie Cohen-Cole, "Instituting the Science of Mind: Intellectual Economies and Disciplinary Exchange at Harvard's Center for Cognitive Studies," *British Journal for the History of Science* 40 (2007): 567–597; Hunter Crowther-Heyck, "George A. Miller, Language, and the Computer Metaphor of Mind," *History of Psychology* 2 (1999): 37–64.

40. Galison, "Americanization of Unity."

41. Paul H. Buck, "Faculty of Arts and Sciences," *Official Register* 44 (July 7, 1947), 27–29.

42. Galison, "Americanization of Unity," 57; Warren McCulloch and Walter Pitts, "A Logical Calculus of the Ideas Immanent in Nervous Activity," *Bulletin of Mathematical Biophysics* 5 (1943): 115–133; Jean-Pierre Dupuy, *On the Origins of*

Cognitive Science: The Mechanization of Mind, trans. M. B. deBevoise (Cambridge, MA: MIT Press, 2009), 49–63; Howard Gardner, *The Mind's New Science: A History of the Cognitive Revolution,* new ed. (New York: Basic Books, 1984), 10–16.

43. Michael A. Bernstein, *A Perilous Progress: Economists and Public Purpose in Twentieth-Century America* (Princeton, NJ: Princeton University Press, 2001), 73–114; Mirowski, *Machine Dreams.*

44. Steve J. Heims, *The Cybernetics Group* (Cambridge, MA: MIT Press, 1991); Dupuy, *On the Origins of Cognitive Science,* ch. 3.

45. Two useful surveys are Ellen Herman, *The Romance of American Psychology: Political Culture in the Age of Experts* (Berkeley: University of California Press, 1995), ch. 1–4; Peter Buck, "Adjusting to Military Life: The Social Sciences Go to War, 1941–1950," in *Military Enterprise and Technological Change: Perspectives on the American Experience,* ed. Merritt Roe Smith (Cambridge, MA: MIT Press, 1985), 203–252.

46. Buck, "Adjusting to Military Life," 212–214.

47. On the sources of behavioral science, see Ron Robin, *The Making of the Cold War Enemy: Culture and Politics in the Military-Intellectual Complex* (Princeton, NJ: Princeton University Press, 2001), 24–33; Bernard Berelson, "Introduction to the Behavioral Sciences," in *The Behavioral Sciences Today,* ed. Bernard Berelson (New York: Basic Books, 1963), 1–11.

48. Stouffer cited in Buck, "Adjusting to Military Life," 222.

49. On the *TAS* studies, see Robin, *Making of the Cold War Enemy,* 19–23.

50. Forrest G. Robinson, *Love's Story Told: A Life of Henry A. Murray* (Cambridge, MA: Harvard University Press, 1992), 281–285.

51. Barry M. Katz, *Foreign Intelligence: Research and Analysis in the Office of Strategic Services, 1942–1945* (Cambridge, MA: Harvard University Press, 1989).

52. William J. Buxton and Lawrence T. Nichols, "Talcott Parsons and the 'Far East' at Harvard, 1941–48: Comparative Institutions and National Policy," *American Sociologist* 31 (2000): 5–17.

53. Gordon Allport, Clyde Kluckhohn, O. H. Mowrer, Henry Murray, and Talcott Parsons to Dean Paul H. Buck, June 10, 1943, section UAIII 5.55.26, box Soc Sci-Z, PDFASH.

54. Gordon Allport, Clyde Kluckhohn, O. H. Mowrer, Henry Murray, and Talcott Parsons to Paul H. Buck, August 31, 1943, section HUGFP 15.2, box 1, TPPH.

55. "Basic Social Science" [among the notes of the Allport Committee on the Social Sciences at Harvard]. n.d. (c. 1943), section HUGFP 15.2, box 1, TPPH.

56. Talcott Parsons to Paul H. Buck, April 3, 1944, TPPH.

57. Lawrence T. Nichols, "Social Relations Undone: Disciplinary Divergence and Departmental Politics at Harvard, 1946–1970," *American Sociologist* 29 (1998): 87.

58. Paul H. Buck, "Faculty of Arts and Sciences," *Official Register* 45 (May 20, 1948), 36–37.

59. Talcott Parsons, "Report: Organized Social Research in the Urban Community," n.d., section 5.55.26, box "Soc Sci-Z," PDFASH.

60. Ellen Condliffe Lagemann, *The Politics of Knowledge: The Carnegie Corporation, Philanthropy, and Public Policy* (Chicago: University of Chicago Press, 1992), 168–170.

61. George Homans, *Coming to My Senses: The Autobiography of a Sociologist* (New Brunswick, NJ: Transaction Books, 1984), 301.

62. On the origins and early organization of the Department of Social Relations, see Parsons, "On Building Social System Theory," 841; Benton Johnson and Miriam M. Johnson, "The Integrating of the Social Sciences: Theoretical and Empirical Research and Training in the Department of Social Relations at Harvard," in *The Nationalization of the Social Sciences,* ed. Samuel Z. Klausner and Victor M. Lidz (Philadelphia: University of Pennsylvania Press, 1986), 131–139; Rebecca M. Lemov, "The Laboratory Imagination: Experiments in Human and Social Engineering," PhD dissertation (University of California at Berkeley, 2000), 297–311.

63. Johnson and Johnson, "Integrating of the Social Sciences," 132.

64. On the BASR's role as a training ground for sociology graduate students, see Craig Calhoun and Jonathan VanAntwerpen, "Orthodoxy, Heterodoxy, and Hierarchy: 'Mainstream' Sociology and Its Challengers" in *Sociology in America: A History,* ed. Craig Calhoun (Chicago: University of Chicago Press, 2007), 390–396.

65. Lagemann, *Politics of Knowledge,* 171.

66. Arthur J. Vidich, "The Department of Social Relations and 'Systems Theory' at Harvard: 1948–50," *International Journal of Politics, Culture, and Society* 13 (June 2000): 616.

67. George Mandler, *Interesting Times: An Encounter with the Twentieth Century 1924–* (Mahwah, NJ: Erlbaum, 2002), 159.

68. David M. Schneider, *Schneider on Schneider: The Conversion of the Jews and Other Anthropological Stories,* ed. Richard Handler (Durham, NC: Duke University Press, 1995), 79, 81, 76.

69. Clifford Geertz, *After the Fact: Two Countries, Four Decades, One Anthropologist* (Cambridge, MA: Harvard University Press, 1995), 101.

70. Nichols, "Social Relations Undone," 94.

71. Parsons, "Preface," v.

72. David L. Szanton, "The Origin, Nature, and Challenge of Area Studies in the United States," in *The Politics of Knowledge: Area Studies and the Disciplines,* ed. David L. Szanton (Berkeley, CA: GAIA Books, Global, Area, and International Archive, UC Berkeley, 2002), available at <http://escholarship.org/uc/item/59n2d2n1> (accessed December 19, 2008); Cohen-Cole, "Instituting the Science of Mind," 575–576; Ralph Tyler, "Study Center for Behavioral Scientists," *Science* 123 (1956): 406–407.

73. Talcott Parsons, "Carnegie Project on Theory," n.d., box 6, section 42.8.4, TPPH.

74. Talcott Parsons to John W. Gardner, January 19, 1949, section 42.8.4, box 6, TPPH.

75. Ibid.

76. Parsons had broached the possibility of collaboration with Shils in early 1949. See Talcott Parsons to Edward Shils, April 6, 1949, section 42.8.4, box 6, TPPH.

77. The differences between Homans and Parsons are part of the professional lore of American sociology, but they have not been examined in great detail. Aspects of their early relationship are touched on in Lawrence T. Nichols, "The Rise of Homans at Harvard: Pareto and the *English Villagers,*" in *George C. Homans: History, Theory, and Method,* ed. A. Javier Treviño (Boulder, CO: Paradigm, 2006), 43–62.

78. See Jamie Cohen-Cole, "Thinking About Thinking in Cold War America," (PhD Dissertation, Princeton University, 2003), 24–73.

79. Parsons, "Preface," v.

80. With the exception of Sears, who joined the Department in 1949.

81. Homans reserved some especially acid remarks for these seminars in his memoirs. See Homans, *Coming to My Senses,* 302.

82. On the significance of this interpersonal dimension of postwar scientific research, see Steven Shapin, *The Scientific Life: A Moral History of a Late Modern Vocation* (Chicago: University of Chicago Press, 2008).

83. On this reflexive aspect of Cold War intellectual culture, see Jamie Cohen-Cole, "The Creative American: Cold War Salons, Social Science, and the Cure for Modern Society," *Isis* 100 (2009): 219–262.

84. Carnegie Project Minutes, "Meeting of Group 2—Oct. 3, 1949" box 6, section 42.8.4, TPPH.

85. Ibid.

86. Ibid.

87. Ibid.

88. Carnegie Project Minutes, "Meeting of Group 2. October 10, 1949," box 6, section 42.8.4, TPPH.

89. Carnegie Project Minutes, "Meeting of Monday Group—October 24, 1949," box 6, section 42.8.4, TPPH.

90. Ibid.

91. Homans, *Coming to My Senses,* 302. Although we cannot date precisely when Group 2 ceased to meet, the absence of transcripts from the seminar after October 24 in the Carnegie Project files in the Parsons Papers suggests that it was discontinued at that time.

92. Talcott Parsons, "The Carnegie Project on Theory," n.d., box 6, section 42.8.4, TPPH.

93. Ibid.

94. Talcott Parsons et al., "Some Fundamental Categories of the Theory of Action," in *Toward a General Theory of Action: Theoretical Foundations for the Social Sciences,* ed. Talcott Parsons and Edward A. Shils (1951; New York: Harper & Row, 1962), 3.

95. See Talcott Parsons and Edward A. Shils, "Values, Motives, and Systems of Action," in Parsons and Shils, *Toward a General Theory of Action,* 247, "Figure 1: The Action Frame of Reference."

96. Ibid., 48–49.

97. For an early trial of the pattern variables, see Talcott Parsons, "The Professions and Social Structure," (1939), in Parsons, *Essays in Sociological Theory,* rev. ed. (Glencoe, IL: Free Press, 1954), 34–49.

98. Parsons and Shils, "Values, Motives, and Systems of Action," 48.

99. Ibid., 49.

100. Ibid., 77.

101. Talcott Parsons to John W. Gardner, December 13, 1949, box 6, section 42.8.4, TPPH.

102. Talcott Parsons, "A Narrative Account of Theoretical Developments: Dec. 3–18, 1949," n.d., box 6, section 42.8.4, TPPH.

103. Carnegie Project Minutes, "December 10, 1949," box 6, section 42.8.4, Parsons Papers; Carnegie Project Minutes, "December 16, 1949," box 6, section 42.8.4, TPPH.

104. Parsons, *The Social System,* x.

105. It appears that Robert F. Bales (an assistant professor) and M. Brewster Smith were approached carry out the small group studies. For Smith's refusal to undertake such a study, see Smith to Talcott Parsons, January 3, 1950, box 6, section 42.8.4, TPPH.

106. My account of this meeting is drawn from Homans, *Coming to My Senses,* 303.

107. Parsons and Shils, "Values, Motives, and Systems of Action," in *Toward a General Theory of Action,* 50–51.

108. See George C. Homans, *The Human Group* (London: Routledge & Kegan Paul, 1951).

109. Homans, *Coming to My Senses,* 303. Parsons himself continued to believe in the imminent dawn of a general theory. In 1954 he told a committee appointed to examine the behavioral sciences at Harvard that his research program "aims at no less than a unification of theory in all the fields of the behavioral sciences." See Faculty Committee, *The Behavioral Sciences at Harvard,* 114.

110. See Nichols, "Social Relations Undone," 90–91; Bernard Barber, "Parsons' Second Project: The Social System. Sources, Development, and Limitations," *The American Sociologist* 29 (1998): 81; Howard Brick, "Talcott Parsons' 'Shift Away from Economics', 1937–1946," *Journal of American History* 87 (2000): 498. Representative works from this productive but eclectic period in the history of DSR include Talcott Parsons, Robert F. Bales, and Edward A. Shils, *Working Papers in the Theory of Action* (New York: Free Press, 1953) and Clyde M. Kluckhohn and Henry A. Murray, *Personality in Nature, Society, and Culture,* 2nd ed. (New York: Knopf, 1953).

111. Department and Laboratory of Social Relations, Harvard University, *Department and Laboratory of Social Relations, Harvard University: The First Decade, 1946–1956* (Official Report, Harvard University, 1956), 34.

112. Parsons, "Clyde Kluckhohn," 32; Parsons, "On Building Social System Theory: A Personal History," *Daedalus* 99 (1970): 843.

6. Lessons of the Revolution

1. See Robert K. Merton, "The Sociology of Science: An Episodic Memoir," in *The Sociology of Science in Europe,* ed. Robert K. Merton and Jerry Gaston (Carbondale: Southern Illinois University Press, 1977), 3–141.

2. See, e.g., Thomas Kuhn, *The Structure of Scientific Revolutions,* vol. 2, no. 2, *International Encyclopedia of Unified Science* (Chicago: University of Chicago Press, 1962), vii–xiv.

3. Steve Fuller, *Thomas Kuhn: A Philosophical History for Our Times* (Chicago: University of Chicago Press, 2000).

4. For a survey of Sarton's academic career, see Arnold Thackray and Robert K. Merton, "On Discipline Building: The Paradoxes of George Sarton," *Isis* 63 (1972): 473–495; Lewis Pyenson, *The Passion of George Sarton: A Modern Marriage and Its Discipline* (Philadelphia, PA: American Philosophical Society, 2007).

5. Joy Harvey, "History of Science, History and Science, and Natural Sciences: Undergraduate Teaching of the History of Science at Harvard, 1938–1970," *Isis* 90, Supplement (1999): S271–S279.

6. Thackray and Merton, "On Discipline Building," 476; Gerald Holton, "George Sarton, His *Isis,* and the Aftermath," *Isis* 100 (2009): 79–88.

7. Lewis Pyenson and Christophe Verbruggen, "Ego and the International: The Modernist Circle of George Sarton," *Isis* 100 (2009): 60–78.

8. James Bryant Conant, "George Sarton and Harvard University," *Isis* 48 (1957): 303.

9. Harvey, "History of Science," S273–S274.

10. George Sarton, "An Institute for the History of Science and Civilization (Third Article)," *Isis* 28 (1938): 13.

11. *Official Register of Harvard University* 31 (September 20, 1934), 106.

12. *Official Register of Harvard University* 38 (September 18, 1941), 136–137.

13. Harvey, "History of Science," S275–S277.

14. James Hershberg, *James B. Conant: Harvard to Hiroshima and the Making of the Nuclear Age* (New York: Knopf, 1993), 258–278; Paul S. Boyer, *By the Bomb's Early Light: American Thought and Culture and the Dawn of the Atomic Age* (New York: Pantheon, 1985), 33–45.

15. On this novel form of scaled-up science, see Peter Galison and Bruce Hevly, eds., *Big Science: The Growth of Large-Scale Research* (Stanford, CA: Stanford University Press, 1992).

16. See Roger L. Geiger, *Research and Relevant Knowledge: American Research Universities since World War II* (New York: Oxford University Press, 1992), 6–19; Stuart W. Leslie, *Cold War and American Science: The Military-Industrial-Academic*

Complex at MIT and Stanford (New York: Columbia University Press, 1993), 6–7, 20–25.

17. For a survey of these laboratory cultures, see Peter Galison, *Image and Logic: A Material Culture of Microphysics* (Chicago: University of Chicago Press, 1997), 239–293.

18. M. Fortun and S. S. Schweber, "Scientists and the Legacy of World War II: The Case of Operations (OR)," *Social Studies of Science* 23 (1993): 598.

19. See Kai Bird and Martin Sherwin, *American Prometheus: The Triumph and Tragedy of J. Robert Oppenheimer* (London: Atlantic, 2005); Jessica Wang, *American Science in an Age of Anxiety: Scientists, Anticommunism, and the Cold War* (Chapel Hill: University of North Carolina Press, 1999).

20. Thomas F. Gieryn, *Cultural Boundaries of Science: Credibility on the Line* (Chicago: University of Chicago Press, 1999), 65–84; Mark Solovey, "Riding Natural Scientists' Coattails onto the Endless Frontier: The SSRC and the Quest for Scientific Legitimacy," *Journal of the History of the Behavioral Sciences* 40 (2004): 393–422.

21. Fuller, *Thomas Kuhn*, 153–157, 161–162.

22. David Kaiser, "The Postwar Suburbanization of American Physics," *American Quarterly* 56 (2004): 851–888.

23. David Kaiser, "Scientific Manpower, Cold War Requisitions, and the Production of American Physicists after World War II," *Historical Studies in the Physical and Biological Sciences* 33 (2002): 131–159.

24. On the 1931 International Congress, see Andrew Brown, *J. D. Bernal: The Sage of Science* (Oxford: Oxford University Press, 2005), 105–107.

25. J. D. Bernal, *The Social Function of Science* (London: Routledge, 1939).

26. Peter J. Kuznick, *Beyond the Laboratory: Scientists as Political Activists in 1930s America* (Chicago: University of Chicago Press, 1987).

27. Philip Mirowski, *Machine Dreams: Economics Becomes a Cyborg Science* (Cambridge: Cambridge University Press, 2002), 182–184.

28. Michael Polanyi, *The Contempt of Freedom: The Russian Experiment and After* (London: Watts, 1940); Polanyi, *Personal Knowledge: Towards a Post-Critical Philosophy* (Chicago: University of Chicago Press, 1958). See also Brown, *J. D. Bernal*, 162–163; Charles Thorpe, "Community and Market in Michael Polanyi's Philosophy of Science," *Modern Intellectual History* 6 (2009): 59–89.

29. Peter Novick, *That Noble Dream: The "Objectivity Question" and the American Historical Profession* (Cambridge: Cambridge University Press, 1988), 293–294.

30. I. Bernard Cohen, *Science, Servant of Man* (1948; London: Sigma, 1949).

31. Robert K. Merton, "George Sarton: Episodic Reflections by an Unruly Apprentice," *Isis* 76 (1985): 471. On the roots of Merton's sociology of science in the Harvard complex, see Jonathan Turner, "Merton's 'Norms' in Political and Intellectual Context," *Journal of Classical Sociology* 7 (2002): 161–178; Lawrence T. Nichols, "Merton as Harvard Sociologist: Engagement, Thematic Continuities, and

Institutional Linkages," *Journal of the History of the Behavioral Sciences* 46 (2010): 72–95.

32. Robert K. Merton, "A Life of Learning," American Council of Learned Societies Occasional Paper no. 25 (1994).

33. Robert K. Merton, "Science, Technology, and Society in Seventeenth Century England," *Osiris* 4 (1938): 360–632.

34. Ibid., 501–502, n. 24; 531, n. 38; 544, n. 10.

35. See David Hollinger, "The Defense of Democracy and Robert K. Merton's Formulation of the Scientific Ethos," in Hollinger, *Science, Jews, and Secular Culture* (Princeton, NJ: Princeton University Press, 1996), 80–96; Novick, *That Noble Dream*, 296–297.

36. Robert K. Merton, "Science and the Social Order," in Merton, *Social Theory and Social Structure*, enlarged ed. (1938; New York: Free Press, 1968), 591–603.

37. Robert K. Merton, "A Note on Science and Democracy" (1942), reprinted in revised form as "Science and Democratic Social Structure," in Merton, *Social Theory and Social Structure*, 604–615.

38. See esp. Talcott Parsons, "Social Science: A Basic National Resource" (1948), reprinted in *The Nationalization of the Social Sciences*, ed. Samuel Z. Klausner and Victor M. Lidz (Philadelphia: University of Pennsylvania Press, 1986), 43–48.

39. Bernard Barber, *Science and the Social Order* (1952; New York: Collier, 1962), 93–121.

40. Peter Galison, "The Americanization of Unity," *Daedalus* 127 (1998): 45–71.

41. Philipp Frank, "The Institute for the Unity of Science: Its Background and Purpose," *Synthese* 6 (1947): 161.

42. See George Reisch, *How the Cold War Transformed Philosophy of Science: To the Icy Slopes of Logic* (Cambridge: Cambridge University Press, 2005), 294–306.

43. Quoted in Gerald Holton, "On the Vienna Circle in Exile: An Eyewitness Report," in *The Foundational Debate: Complexity and Constructivity in Mathematics and Physics*, ed. Werner Depauli-Schimanovich et al. (Dordrecht: Kluwer, 1995), 272–273.

44. See esp. Philipp Frank, "Institute for the Unity of Science," 160–167; Frank, "The Logical and Sociological Aspects of Science," Special Issue on *Contributions to the Analysis and Synthesis of Knowledge*, *Proceedings of the American Academy of Arts and Sciences* 80 (1951): 16–30; Frank, "The Variety of Reasons for the Acceptance of Scientific Theories," in *The Validation of Scientific Theories*, ed. Frank (Boston: Beacon Press, 1956), 3–18.

45. Frank, "Institute for the Unity of Science," 160.

46. Ibid., 162.

47. Ibid., 163–164.

48. Ibid., 165–166.

49. Ibid., 167.

50. Philipp Frank, "Science Teaching and the Humanities," *Synthese* 6 (1947–1948): 383.

51. Philipp Frank, "Introduction," in Frank, *The Validation of Scientific Theories*, vii.

52. Frank, "Science Teaching," 404.

53. This was the title of a collection of Frank's essay published soon after his arrival in the United States. See Philipp Frank, *Between Physics and Philosophy* (Cambridge, MA: Harvard University Press, 1941).

54. Frank, "Science Teaching," 406.

55. *Official Register of Harvard University* 45 (September 15, 1948), 25.

56. *Official Register of Harvard University* 46, No. 49 (September 1949), 27.

57. See Thomas E. Uebel, "Philipp Frank's History of the Vienna Circle: A Programmatic Retrospective" and Gary L. Hardcastle, "Debabelizing Science: The Harvard Science of Science Discussion Group, 1940–41," in *Logical Empiricism in North America,* ed. Gary L. Hardcastle and Alan W. Richardson (Minneapolis: University of Minnesota Press, 2003), 156–164, 188–189.

58. Hans Hahn, Otto Neurath, and Rudolf Carnap, *Wissenschaftliche Weltauffassung: Der Wiener Kreis* (The scientific conception of the world: the Vienna Circle) in in Otto Neurath, *Empiricism and Sociology,* trans. Paul Foulkes and Marie Neurath, ed. Marie Neurath and Robert S. Cohen (Dordrecht: Reidel, 1973), 317.

59. See the articles collected in *Proceedings of the American Academy of Arts and Sciences* 80 (1951–1954) and *The Scientific Monthly* for September, October, November 1954 and January and February 1955. The latter were republished as Frank, *The Validation of Scientific Theories.*

60. See esp. the section entitled The Present State of Operationism, and Henry Guerlac, "Science During the French Revolution," in Frank, *The Validation of Scientific Theories*, 37–94, 171–191.

61. Reisch, *How the Cold War Transformed Philosophy of Science*, 308–310, 318–330.

62. Office of Scientific Research and Development, *Science, the Endless Frontier: A Report to the President by Vannevar Bush* (Washington, DC: United States Government Printing Office, 1945).

63. Ibid., 18–20.

64. James Bryant Conant, *On Understanding Science: An Historical Approach* (Oxford: Oxford University Press, 1947), 23.

65. Paul H. Buck et al., *General Education in a Free Society: Report of the Harvard Committee* (Cambridge, MA: Harvard University Press, 1945).

66. Conant, *My Several Lives,* 363–371; Conant, "President's Report," *Official Register of Harvard University* 41 (September 26, 1944), 9–14.

67. James Bryant Conant, "Oration at the Solemn Observance of the Tercentenary of Harvard College" (1936), reprinted in Conant, *My Several Lives,* 654–656.

68. Quoted in Hershberg, *James B. Conant,* 409.

69. James Bryant Conant, "President's Report," *Official Register of Harvard University* 45 (December 1, 1948), 13.

70. For a particularly clear statement of Conant's Hendersonian position, which runs through all of his books on general education in science, see James Bryant

Conant, "Introduction," in *Harvard Case Histories in Experimental Science,* vol. 1, ed. Conant (Cambridge, MA: Harvard University Press, 1957), x–xiv.

71. Conant, *My Several Lives,* 87–88.

72. James Bryant Conant, "Introduction," in Conant, *Harvard Case Histories,* 1:viii.

73. Conant, *On Understanding Science,* xiii.

74. Ibid., 11–12, 18.

75. Ibid., 18–19. For explicit, if largely noncommittal remarks on the Bernal-Polanyi split in the social study of science, see 129–130, n. 12.

76. Ibid., 18–28.

77. Ibid., 24. See also 101–103.

78. Ibid., 14. See also Conant, *Science and Common Sense,* 43. For a similar complaint against Mach, see 114–115, n. 5. In his attack on Mach and Pearson's conception of science, Conant was siding with the position the German physicist Max Planck had taken again Mach's instrumental view of science. See John T. Blackmore, *Ernst Mach: His Work, Life, and Influence* (Berkeley: University of California Press, 1972), 217–227; Fuller, *Thomas Kuhn,* 96–124.

79. For a general history of the teaching of science in postwar general education programs, see Peter S. Buck and Barbara Gutmann Rosenkrantz, "The Worm in the Core: Science and General Education," in *Transformation and Tradition in the Sciences: Essays in Honor of I. Bernard Cohen,* ed. Everett Mendelsohn (Cambridge: Cambridge University Press, 1984), 371–394.

80. *Official Register of Harvard University* 45 (September 15, 1948), 19. See also *Official Register of Harvard University* 46, No. 24 (September 1949), 20. In regard to Natural Sciences 11a, the number indicates that the course was intended for upperclassmen, who had already concentrated, not freshmen who had yet to choose their major, as was the case with Natural Sciences 4. See *Official Register of Harvard University* 44 (September 9, 1947), 23

81. James Bryant Conant, "Introduction," in Conant, *Harvard Case Histories,* 1:viii.

82. Hershberg, *James B. Conant,* 409; Conant, *My Several Lives,* 373.

83. James Bryant Conant, "Robert Boyle's Experiments in Pneumatics" and "The Overthrow of the Phlogiston Theory: The Chemical Revolution of 1775–1789," in Conant, *Harvard Case Histories,* 1:1–115; Conant, ed., "Pasteur's Study of Fermentation" and "Pasteur's and Tyndall's Study of Spontaneous Generation," in *Harvard Case Histories in Experimental Science,* vol. 2, ed. James Bryant Conant (Cambridge, MA: Harvard University Press, 1957), 437–539.

84. Conant, "Overthrow of the Phlogiston Theory," 67.

85. Conant, *On Understanding Science,* 132, n. 20; Leonard K. Nash, *The Nature of the Natural Sciences* (Boston: Little, Brown, 1963); Struan Jacobs, "J. B. Conant's Other Assistant: Science as Depicted by Leonard K. Nash, Including Reference to Thomas Kuhn," *Perspectives on Science* 18 (2010): 328–351.

86. Thomas S. Kuhn, "Subjective View: Thomas S. Kuhn, on Behalf of the Recent Student, Reflects on the Undergraduate Attitude," *Harvard Alumni Bulletin* 48, Special Issue on "Education in a Free Society" (September 22, 1945): 29–30.

87. Aristides Baltas et al., "A Discussion with Thomas Kuhn" (1997), reprinted in Thomas Kuhn, *The Road Since Structure: Philosophical Essays, 1970–1993, with an Autobiographical Interview,* ed. James Conant and John Haugeland (Chicago: University of Chicago Press, 2000), 275.

88. Kuhn, "What are Scientific Revolutions?" 16.

89. Ibid.; Kuhn, "Preface," xi.

90. Kuhn, "What are Scientific Revolutions?" 16.

91. Ibid., 19.

92. Kuhn, "Preface," xiii.

93. On "dogmatic initiation" in science pedagogy, see Thomas Kuhn, "The Essential Tension: Tradition and Innovation in Scientific Research" (1959), reprinted in Kuhn, *The Essential Tension: Selected Studies in Scientific Tradition and Change* (Chicago: University of Chicago Press, 1977), 229; Kuhn, "The Function of Dogma in Scientific Research," in *Scientific Change: Historical Studies in the Intellectual, Social, and Technical Conditions for Scientific Discovery and Technical Invention, from Antiquity to the Present,* ed. A. C. Crombie (New York: Basic Books, 1963), 351.

94. Baltas et al., "A Discussion," 256–257.

95. Ibid., 267–269.

96. Thomas Kuhn, "The Metaphysical Possibilities of Physics," n.d., box 1, TSKP.

97. Baltas et al., "A Discussion," 264.

98. Thomas Kuhn to Mrs. Ivan Fischer, July 27, 1943, box 12, TSKP. Kuhn attached to this letter a note dated May 31, 1991, which reads: "This is a letter recently returned to me by the addressee, my favorite aunt, Emma K. Fischer. I had written it to her shortly after my twenty-first birthday (which occurred on 18 July 1943). At that time I had recently graduated from Harvard and was beginning War work on radar countermeasures at the Radio Research Laboratory, located at Harvard."

99. Baltas et al., "A Discussion," 272–273.

100. See Thomas Kuhn, An Analysis of Causal Connexity," 1945, box 1, TSKP; and "A Comparison of the Logic of Propositions with That of Ascriptives," n.d., box 1, TSKP.

101. Thomas Kuhn, "Objectives of a General Education Course in the Physical Sciences," May 1947, box 1, TSKP.

102. Ibid.

103. Thomas Kuhn, "Natural Sciences 11 (a)," autumn 1947, box 1, TSKP.

104. Baltas et al., "A Discussion," 278.

105. Thomas Kuhn, "Notes & Ideas," 1949, box 1, TSKP.

106. Kuhn indicated that he had "read in toto" the books by Tarski, Woodger, Ayer, Wiener, Langer, Merton, Piaget, Schrödinger, and Werner.

107. Kuhn, "Notes & Ideas," TSKP.

108. Ibid. Kuhn had read a selection of Weber's epistemological and methodological writings, published in English translation as Max Weber, *The Methodology of the Social Sciences,* trans. and ed. Edward Shils and Henry A. Finch (Glencoe, IL: Free Press, 1949).

109. Thomas Kuhn to George F. Lombard, July 26, 1949, box 1, TSKP.

110. Thomas Kuhn, "Introduction: Textbook Science and Creative Science," 1951, Lowell Institute Lectures, box 3, TSKP. Cf. Kuhn, *Structure,* 1.

111. See, esp., Thomas Kuhn to David Owen, January 6, 1951, box 3, TSKP.

112. Kuhn, "Preface," xvi. On Kuhn's Lowell Lectures, see Jensine Andresen, "Crisis and Kuhn," *Isis* 90, Supplement (1999): S61–S63.

113. Kuhn, "Preface," xvi.

114. Of special importance for Kuhn were "The Function of Measurement in Modern Physical Science" (1961), reprinted in *Essential Tension,* 178–224, which marked the debut of the notion that would become "normal science," and "The Essential Tension" (1959), reprinted in *Essential Tension,* 225–239, which contained the first public use of the concept of the paradigm. See also Kuhn, "Function of Dogma."

115. Kuhn, "Preface," x.

116. Thomas Kuhn, *The Copernican Revolution: Planetary Astronomy in the Development of Western Thought* (Cambridge, MA: Harvard University Press, 1957).

117. Kuhn, "Preface," xviii–xix; Kuhn, *Structure,* ix–x.

118. Kuhn, "Preface," xviii–xix.

119. Kuhn to Owen, TSKP.

120. Kuhn, "Introduction," Lowell Lectures, TSKP.

121. Thomas Kuhn to Philipp Frank, n.d., unsent, box 25, TSKP. See also Philipp Frank to Kuhn, December 2, 1952, box 25, TSKP.

122. Baltas et al., "A Discussion," 291–292.

123. Thomas Kuhn to Charles Morris, July 31, 1953, box 25, TSKP.

124. Thomas Kuhn to Lawrence S. Kubie, March 17, 1955, box 25, TSKP.

125. Baltas et al., "A Discussion," 291. On the claims of Kuhn's book, see Robert S. Westman, "Two Cultures or One? A Second Look at Kuhn's *The Copernican Revolution,*" *Isis* 85 (1994): 79–115; N. M. Swerdlow, "An Essay on Thomas Kuhn's First Scientific Revolution, *The Copernican Revolution,*" *Proceedings of the American Philosophical Society* 148 (2004): 64–120.

126. Kuhn, *Copernican Revolution,* 1–2, 90, 190–200.

127. Kuhn, *Structure,* 43–46.

128. Thomas Kuhn, "Introduction," 1951, Lowell Lectures, box 3, TSKP.

129. Thomas Kuhn to Lombard, July 26, 1949, box 1, TSKP; Kuhn to Kubie, March 17, 1955, box 25, TSKP.

130. Kuhn, *Structure,* 111–135. On Kuhn's "X-rated" reflections in this part of the book (Chapter 10), see Peter Godfrey-Smith, *Theory and Reality: An Introduction to the Philosophy of Science* (Chicago: University of Chicago Press, 2003), 96–98.

131. Many of these documents are undated and even untitled, but they are gathered together among Kuhn's materials for the drafting of *Structure* after 1958. Both some scattered dates in the papers and the terminology of the notes indicate that they must have been composed in the late 1950s,

132. Thomas Kuhn, "New Outline—Chapter 1," n.d., box 4, TSKP.

133. See Thomas Kuhn, "SSR—New Outline," March 12, 1959, box 4, TSKP.

134. Thomas Kuhn, "I. Significance of Consensus," n.d., box 4, TSKP.

135. Thomas Kuhn, "Chapter II—The Normal Practice of Mature Science," n.d., box 4, TSKP.

136. Thomas Kuhn, untitled typescript, n.d., box 4, TSKP.

137. Thomas Kuhn, "Penultimate Draft of Structures before June 1960," n.d., box 4, TSKP.

138. See Kuhn, "Essential Tension," 229.

139. Kuhn, "Penultimate Draft," 39, 41, 43.

140. Kuhn, *Structure,* 43–47.

141. Ibid., 44–45. On the Berkeley context of the reception of Wittgenstein, see Stanley Cavell, *The Claim of Reason: Wittgenstein, Skepticism, Morality, and Tragedy,* new ed. (New York: Oxford University Press, 1979), xxiii–xxiv.

142. This feature of Kuhn's text is examined at greater length in Joel Isaac, "Kuhn's Education: Wittgenstein, Pedagogy, and the Road to *Structure,*" *Modern Intellectual History* (forthcoming 2012).

143. This work has only recently been undertaken. See David Kaiser, ed., *Pedagogy and the Practice of Science: Historical and Contemporary Perspectives* (Cambridge, MA: MIT Press, 2005); Kathryn M. Olesko, *Physics as a Calling: Discipline and Practice in the Konigsberg Seminar for Physics* (Ithaca, NY: Cornell University Press, 1991); William Clark, *Academic Charisma and the Origins of the Research University* (Chicago: University of Chicago Press, 2006).

144. *Official Register of Harvard University* 48 (September 10 1951), 22. It is worth noting that Kuhn and Nash changed the name of the course to "Research Patterns in Physical Science," a title that reflected Kuhn's growing preoccupation with the structure of scientific investigation.

145. See Thomas Kuhn, "Washington University Conference," May 15, 1949, box 12, TSKP; Kuhn, untitled talk prepared for Faculty Conference on General Education, State University of New York, June 17–23, 1951, box 12, TSKP.

146. See esp. Thomas Kuhn, "Can the Layman Know Science," December 13, 1955, box 12, TSKP.

Epilogue

1. James T. Kloppenberg, *The Virtues of Liberalism* (New York: Oxford University Press, 1998), 155–156.

2. Richard J. Bernstein, *The Restructuring of Social and Political Theory* (London: Methuen, 1976), 88; Clifford Geertz, *Available Light: Anthropological Reflections on Philosophical Topics* (Princeton, NJ: Princeton University Press, 2000), 165.

3. See, e.g., Gary Gutting, *What Philosophers Know: Case Studies in Recent Analytic Philosophy* (Cambridge: Cambridge University Press, 2009), 151; Harold Kincaid, *Philosophical Foundations of the Social Sciences: Analyzing Controversies in Social Research* (Cambridge: Cambridge University Press, 1996), 30–37.

4. Ludwig Wittgenstein, *Philosophical Investigations,* trans. G. E. M. Anscombe, 3rd ed. (Oxford: Blackwell, 2001), §80.

5. Ibid., §77.

6. Allan Janik, "Impure Reason Vindicated," in *Wittgenstein: The Philosopher and His Works,* ed. Alois Pichler and Simo Säätelä (Bergen: WAB, 2005), 275–276.

7. Thomas Kuhn, *The Structure of Scientific Revolution,* vol. 2, no. 2, *International Encyclopedia of Unified Science* (Chicago: University of Chicago Press, 1962), 23.

8. I say "somewhat" unwittingly, because Kuhn's appeal to the *Investigations* was not, of course, random. It is likely it came up in his discussions with Cavell and other colleagues in the philosophy Department at Berkeley.

9. Geertz, "Preface," in Geertz, *Available Light,* xi–xii.

10. For a helpful review of some of the early volumes in this series, see Walter Cerf, "Studies in Philosophical Psychology," *Philosophy and Phenomenological Research* 22 (1962): 537–558.

11. Peter Winch, *The Idea of a Social Science, and Its Relation to Philosophy* (London: Routledge and Kegan Paul, 1958).

12. For a volume that includes essays outlining an "American Wittgenstein," see Alice Crary and Rupert Read, eds., *The New Wittgenstein* (London: Routledge, 2000).

13. John H. Zammito, *A Nice Derangement of Epistemes: Post-Positivism in the Study of Science from Quine to Latour* (Chicago: University of Chicago Press, 2004), 66.

14. Dudley Shapere, "The Structure of Scientific Revolutions," *Philosophical Review* 73 (1964): 383–394; Shapere, "Meaning and Scientific Change," in *Mind and Cosmos: Essays in Contemporary Science and Philosophy,* ed. Robert Colodny (Pittsburgh, PA: University of Pittsburgh Press, 1966), 41–85; Israel Scheffler, *Science and Subjectivity* (Indianapolis, IN: Bobbs-Merrill, 1967).

15. Zammito, *Nice Derangement of Epistemes,* 72–77.

16. See esp. Thomas Kuhn, "Commensurability, Comparability, Communicability," (1983), reprinted in Kuhn, *The Road Since Structure: Philosophical Essays, 1970–1993, with an Autobiographical Interview* (Chicago: University of Chicago Press, 2000), 33–57.

17. Thomas Kuhn, "Dubbing and Redubbing: The Vulnerability of Rigid Designation" in *Scientific Theories,* ed. C. Wade Savage (Minneapolis: University of Minnesota Press, 1990), 298–318.

18. Aristides Baltas et al., "A Discussion with Thomas S. Kuhn," in Kuhn, *Road since Structure,* 276.

19. Kuhn, *Structure,* 98, 125, 136.

20. See Alan Richardson, "'That Sort of Everyday Image of Logical Positivism': Thomas Kuhn and the Decline of Logical Empiricist Philosophy of Science," in *The Cambridge Companion to Logical Empiricism,* ed. Alan Richardson and Thomas Uebel (Cambridge: Cambridge University Press, 2007), 346–369.

21. For a helpful outline of this "anti-positivist" moment in the study of science, see Peter Galison, "History, Philosophy, and the Central Metaphor," *Science in Context* 2 (1988): 197–212.

22. See esp. Carl Hempel, *Aspects of Scientific Explanation, and other Essays in the Philosophy of Science* (New York: Free Press, 1965). For a survey of the literature on scientific explanation from Hempel on, see Wesley C. Salmon, *Four Decades of Scientific Explanation* (1989; Pittsburgh: University of Pittsburgh Press, 2006).

23. Carl Hempel, "The Function of General Laws in History," *Journal of Philosophy* 39 (1942): 35–48.

24. See Samuel James, "Louis Mink, 'Postmodernism,' and the Vocation of Historiography," *Modern Intellectual History* 7 (2010): 151–184.

25. Kincaid, *Philosophical Foundations,* 191–221.

26. See the Prologue and Chapter 4 for further references.

27. For a flavor of Geertz's early interest in the program of *Toward a General Theory of Action,* see Clifford Geertz, "Religion as a Cultural System" (1996) and "Ritual and Social Change: A Javanese Example" (1959), reprinted in Geertz, *The Interpretation of Cultures* (New York: Basic Books, 1973), 87–125, 142–169.

28. Joy Harvey, "History of Science, History and Science, and Natural Sciences: Undergraduate Teaching of the History of Science at Harvard, 1938–1970," *Isis* 90, Supplement (1999): S289–S290.

29. On the failure of Frank's attempt to revive the social program of scientific philosophy and the subsequent rise of professional philosophy of science, see George Reisch, *How the Cold War Transformed Philosophy of Science: To the Icy Slopes of Logic* (Cambridge: Cambridge University Press, 2005), 294–310; Anita Burdman Feferman and Soloman Feferman, *Alfred Tarski: Life and Logic* (Cambridge: Cambridge University Press, 2004), 246–253; Peter Galison, "Constructing Modernism: The Cultural Location of the *Aufbau*," in *Origins of Logical Empiricism,* ed. Ronald N. Giere and Alan W. Richardson (Minneapolis: University of Minnesota Press 1996), 40.

30. The classic attacks on Parsonian sociology are C. Wright Mills, *The Sociological Imagination* (New York: Oxford University Press, 1959), 25–49; and Alvin Gouldner, *The Coming Crisis of Western Sociology* (London: Heinemann, 1971). On the Leary controversy, see Rebecca Lemov, *World as Laboratory: Experiments with Mice, Mazes, and Men* (New York: Hill and Wang, 2005), 228–234.

31. Morton Keller and Phyllis Keller, *Making Harvard Modern: The Rise of America's University* (New York: Oxford University Press, 2001), 298–299.

32. Ralph Tyler, "Study Center for Behavioral Scientists," *Science* 123 (1956): 406–407. For a personal account of the Center in its early years, see George Mandler, *Interesting Times: An Encounter with the Twentieth Century 1924—* (Mahwah, NJ: Erlbaum, 2002), 170–171. A more sociological description of the Center is to be found in Robert K. Merton, "The Sociology of Science: An Episodic Memoir," in *The Sociology of Science in Europe,* ed. Robert K. Merton and Jerry Gaston (Carbondale: Southern Illinois University Press, 1977), 102–105.

33. Clifford Geertz, "School Building: A Retrospective Preface," in *Schools of Thought: Twenty-Five Years of Interpretive Social Science,* ed. Joan W. Scott and Debra Keates (Princeton, NJ: Princeton University Press, 2001), 1–11.

34. The historical and sociological literature on institutes for advanced study and humanities centers is extremely thin. Many recent collections on the history of the postwar academy ignore them altogether. I have drawn on Björn Wittrock, "Institutes for Advance Study: Ideas, Histories, Rationales," unpublished lecture, 2002; Philip C. Converse, "Centres for Advanced Study: International/Interdisciplinary," in *International Encyclopedia of the Social and Behavioral Sciences,* ed. Neil Smelser and Paul Baltes (Amsterdam: Elsevier, 2001), 1613–1615; Robert W. Conner, "Do Centers for Advanced Study Deserve a History?" *Ideas* 9, no. 1 (2002), 26–32; Geoffrey Galt Harpham, "The National Humanities Center as an Institute for Advanced Study," unpublished paper, 2004.

35. See Roger L. Geiger, "Organized Research Units—Their Role in the Development of University Research," *Journal of Higher Education* 61 (1990): 1–19.

36. See Joy Rohde, "Gray Matters: Social Scientists, Military Patronage, and Democracy in the Cold War," *Journal of American History* 96 (2009): 99–122.

37. See the Prologue in this volume, n. 41 and 42.

38. See Prologue, n. 75 and 76.

39. Lorraine Daston and Peter Galison, *Objectivity* (New York: Zone Books, 2007).

40. I have adapted this point from a parallel remark by John Dunn about the theory of political obligation. See John Dunn, "Introduction," in Dunn, *Political Obligation in Its Historical Context: Essays in Political Theory* (Cambridge: Cambridge University Press, 1980), 2. For a broader account of "realistic" thinking about philosophical concepts, see Cora Diamond, "Realism and the Realistic Spirit," in Diamond, *The Realistic Spirit: Wittgenstein, Philosophy, and the Mind* (Cambridge, MA: MIT Press, 1991), 39–72.

Acknowledgments

Material support for this project was provided by several funding bodies and employers. In the early stages of research, I received crucial assistance from the Arts and Humanities Research Council; Trinity College, Cambridge; the Sara Norton Fund of the Faculty of History, University of Cambridge; the Royal Historical Society; and the Gilder Lehrman Institute of American History. In more recent years, my work has been sponsored by the British Academy and by the research network on the history of interdisciplinary social science since 1945, which is run out of the École normale supérieure de Cachan and funded by the Agence nationale de recherche.

Three academic homes have sustained my research over the past seven years. Selwyn College, Cambridge, appointed me to the Keasbey Research Fellowship in American Studies and gave me a term of unadulterated research at Brown University. My thanks to Charlie Craig and to my colleagues at Selwyn for the time and encouragement I was given to undertake the archival work that forms the basis of the present study. I am equally grateful to the School of History at Queen Mary, University of London, where I held a lectureship between 2007 and 2011. In addition to providing a rich intellectual environment, Queen Mary also gave me a semester of sabbatical leave during which the bulk of the manuscript was written. I owe a great deal to all of my friends and colleagues at Queen Mary. Finally, the Faculty of History at the University of Cambridge and Christ's College have given me a stimulating context in which to bring the book to completion.

This study draws on a range of archival materials, and I am grateful to a number of archives and their custodians for access to collections pertaining to the recent history of the human sciences. I am especially indebted to the staff of three archives at Harvard University: the Harvard University Archives in the Pusey Library, the Houghton Library, and the Baker Library at the Harvard Business School. My thanks also to the staff of the Institute Archives at the Massachusetts Institute of Technology. This book draws, here and there, on aspects of my earlier research. Parts of Chapters 4 and 5 were presented in earlier versions as, respectively, "W. V. Quine and the Origins of Analytic Philosophy in the United States,"

Modern Intellectual History 2 (2005): 205–234; and "Theorist at Work: Talcott Parsons and the Carnegie Project on Theory, 1949–1951," *Journal of the History of Ideas* 71 (2010): 287–311.

I could not have written this book without an extraordinary network of family and friends. Its members have given me untold encouragement, personal and intellectual, since I began my training as a historian. My parents, Paul and Nicola Isaac, have been instrumental in everything I have done in my career; indeed, my mother was my history teacher in secondary school. Her lessons inspired me to become a historian. My sister, Melanie Isaac-Åkerman, and her husband, Johan Åkerman, have provided much needed reassurance, and a place to get away from it all, during the long march from graduate studies to the completion of this book. For friendship and encouragement over many years, I am grateful to my brother-in-law, Leigh Aitken, and his wife, Melanie Green. Helen and Paul Aitken gave me love and support when I started in the road that led to this book. Their help is beyond thanks.

It would be very easy to wax lyrical about my friends and the sacrifices they have made on my behalf. While it is almost absurd to list them alphabetically, this is the only alternative I can find to a multivolume tribute. My heartfelt thanks, then, to Duncan Bell, Stephen and Sarah Brooks, Sean Campbell, Jamie Cohen-Cole, Sarah Fine, Daniel Geary, Helen Graham, Sam James, Duncan Kelly, Richard King, Leah Loughnane, Peter Mandler, Daniel Matlin, Antonio Maturo, David Milne, Josh Newman, Michael and Patricia O'Brien, Jennifer Ratner-Rosenhagen, Amber Steele, Michael Stevenson, Dan Stone, Robin Vandome, and Martin Woessner.

For comments, suggestions, and provocations on various aspects of the book project, I am grateful to Roger Backhouse, Charles Capper, John Carson, David Engerman, Philippe Fontaine, Daniel Geary, Peter Gordon, David Hollinger, Sam James, Richard King, Bruce Kulick, Philip Mirowski, Jennifer Ratner-Rosenhagen, Daniel Rodgers, Mark Solovey, and Martin Woessner. A handful of hardy souls read complete drafts of the manuscript and made many pertinent criticisms. Although I doubt that I have been able fully to meet the high standards they have set, I can at least give my thanks to Duncan Bell, Jamie Cohen-Cole, Henry Cowles, Duncan Kelly, Peter Mandler, David Milne, and Michael O'Brien. James Kloppenberg and Samuel Moyn read the manuscript for Harvard University Press and made several valuable suggestions for improvements to my argument. My editor, Kathleen McDermott, has aided me in the writing of this book in innumerable ways, large and small.

Tony Badger and Michael O'Brien were my PhD supervisors at the University of Cambridge. They have seen me labor on this research project for almost a decade. Tony has been a champion of my work and a much-valued advisor. In Michael I have found a mentor whose friendship and indefatigable support have been critical to my entry into the historical profession. I have no chance of repaying the colossal intellectual debt I owe him, but it's nice to be able to show him a book at last.

Without the love, advice, and encouragement of my wife, Cara, I would certainly not have been able to write this book—or, indeed, to have pursued what began as distant professional goals and speculative intellectual impulses. The Latin motto *sine qua non*—which roughly translates into English as "without which not"—expresses just how vital she has been to everything I have tried to achieve as a historian. Whatever merits it may be thought to possess, this book cannot represent more than a tiny fraction of the care and support she has shown me. My love for her, and for our daughter, Aletheia, is inexpressible.

Index